Neuropsychological
Function and
Brain Imaging

Critical Issues in Neuropsychology

ASSESSMENT ISSUES IN CHILD NEUROPSYCHOLOGY
Edited by Michael G. Tramontana and Stephen R. Hooper

BRAIN ORGANIZATION OF LANGUAGE AND COGNITIVE PROCESSES
Edited by Alfredo Ardila and Feggy Ostrosky-Solis

HANDBOOK OF CLINICAL CHILD NEUROPSYCHOLOGY
Edited by Cecil R. Reynolds and Elaine Fletcher-Janzen

MEDICAL NEUROPSYCHOLOGY: The Impact of Disease on Behavior
Edited by Ralph E. Tarter, David H. Van Thiel, and Kathleen L. Edwards

NEUROPSYCHOLOGICAL FUNCTION AND BRAIN IMAGING
Edited by Erin D. Bigler, Ronald A. Yeo, and Eric Turkheimer

NEUROPSYCHOLOGY, NEUROPSYCHIATRY, AND BEHAVIORAL NEUROLOGY
R. Joseph

RELIABILITY AND VALIDITY IN NEUROPSYCHOLOGICAL ASSESSMENT
Michael D. Franzen

A Continuation Order Plan is available for this series. A continuation order will bring delivery of each new volume immediately upon publication. Volumes are billed only upon actual shipment. For further information please contact the publisher.

Neuropsychological Function and Brain Imaging

Edited by

Erin D. Bigler

Austin Neurological Clinic
and University of Texas
Austin, Texas

Ronald A. Yeo

University of New Mexico
Albuquerque, New Mexico

and

Eric Turkheimer

University of Virginia
Charlottesville, Virginia

Plenum Press • New York and London

Library of Congress Cataloging in Publication Data

Neuropsychological function and brain imaging.

(Critical issues in neuropsychology)
Includes bibliographies and index.
1. Neuropsychology — Research — Methodology. 2. Brain — Imaging. I. Bigler, Erin
D., 1949–   . II. Yeo, Ronald A. III. Turkheimer, Eric. IV. Series.
[DNLM: 1. Brain — radiography. 2. Magnetic Resonance Imaging. 3. Neuropsychol-
ogy. 4. Tomography, Emission Computed. 5. Tomography, X-Ray Computed. WL 141
N4847]
QP360.N4936   1989   152   88-32470
ISBN 0-306-43045-2

© 1989 Plenum Press, New York
A Division of Plenum Publishing Corporation
233 Spring Street, New York, N.Y. 10013

Printed in the United States of America

Jan, Kathy, and Carol

Contributors

ERIN D. BIGLER, PH.D. · Austin Neurological Clinic and Department of Psychology, University of Texas at Austin, Austin, Texas 78705

C. MUNRO CULLUM, PH.D. · Department of Psychiatry, University of California at San Diego, La Jolla, California 92103; *present address:* Department of Psychiatry, University of Colorado Health Sciences Center, Denver, Colorado 80262

K. Y. HAALAND, PH.D. · Psychology Service, Veterans Administration Medical Center, and University of New Mexico, Albuquerque, New Mexico 87108

W.-D. HEISS, M.D. · Department of Neurology, Max Planck Institute of Neurological Research, Cologne University Clinics, D-5000 Cologne 41, Federal Republic of Germany

DAVID S. KNOPMAN, M.D. · Department of Neurology, University of Minnesota, Minneapolis, Minnesota 55455

J. R. LOWE, PH.D. · Department of Pediatrics, University of New Mexico, Albuquerque, New Mexico 87131

THOMAS G. LUERSSEN, M.D. · Department of Neurosurgery, University of California at San Diego, La Jolla, California 92161

RICHARD I. NAUGLE, PH.D. · Section of Neuropsychology, Department of Psychiatry, Cleveland Clinic Foundation, Cleveland, Ohio 44106

G. PAWLIK, M.D. · Department of Neurology, Max Planck Institute of Neurological Research, Cologne University Clinics, D-5000 Cologne 41, Federal Republic of Germany

SARAH RAZ, PH.D. · Department of Psychology, University of Texas at Austin, Austin, Texas 78712

ALAN B. RUBENS, M.D. · Department of Neurology, University of Arizona, Tucson, Arizona 85721

RONALD M. RUFF, PH.D. · Departments of Psychiatry and Neurosurgery, University of California at San Diego, La Jolla, California 92103

J. N. RUTLEDGE, M.D. · Capital Radiology of Austin and University of Texas at Austin, Austin, Texas 78705

OLA A. SELNES, PH.D. · Department of Neurology, Johns Hopkins University School of Medicine, Baltimore, Maryland 21205

viii

ERIC TURKHEIMER, PH.D. · Department of Psychology, University of Virginia, Charlottesville, Virginia 22903

RONALD A. YEO, PH.D. · Department of Psychology, University of New Mexico, Albuquerque, New Mexico 87131

Preface

Over the past two decades researchers and clinicians in the neurosciences have witnessed a literal information explosion in the area of brain imaging and neuropsychological functioning. Until recently we could not view the nervous system except through the use of invasive procedures. Today, a variety of imaging techniques are available, but this technology has advanced so rapidly that it has been difficult for new information to be consolidated into a single source. The goal of this volume is to present information on technological advances along with current standards and techniques in the area of brain imaging and neuropsychological functioning.

The quality of brain imaging techniques has improved dramatically. In 1975 one had to be content with a brain image that only offered a gross distinction between ventricular cavities, brain, and bone tissue. Current imaging techniques offer considerable precision and approximate gross neuroanatomy to such an extent that differentiation between brain nuclei, pathways, and white–gray matter is possible.

These technological advances have progressed so rapidly that basic and clinical research have lagged behind. It is not uncommon, particularly in longitudinal research, for the technical methodology of a study to become obsolete while that study is still in progress. This has hampered certain aspects of systematic research and has also produced the need for a textbook that could address contemporary issues in brain imaging and neuropsychology.

Given these issues we feel that this text will meet an important need in the area of neuropsychological function and imaging of the brain. To that end this volume focuses on the current status of computerized tomography, magnetic resonance imaging, and positron emission tomography in providing a neuroanatomical and pathological basis for scientific study of human neuropsychological function. We realize that while this text is being prepared for publication there will be advances that will alter the findings reviewed in this book. We have attempted to make all material as current as possible, but in a rapidly changing field such an attempt will always be a losing proposition.

There are several people that we would like to acknowledge for their help and assistance. Drs. David R. Steinman and Nancy L. Nussbaum reviewed various stages of this project and their feedback is greatly acknowledged. Several of the photographs were done by the expert hand of David Matson.

Erin D. Bigler
Austin, Texas
Ronald A. Yeo
Albuquerque, New Mexico
Eric Turkheimer
Charlottesville, Virginia

Contents

1. Neuropsychological Function and Brain Imaging: Introduction and
   Overview 1
   ERIN D. BIGLER, RONALD A. YEO, and ERIC TURKHEIMER

2. Neuroanatomy and Neuropathology: Computed Tomography
   and Magnetic Resonance Imaging Correlates 13
   J. N. RUTLEDGE

   Introduction 13
   Embryology 13
   Neuroimaging 17
   Normal Anatomy 23
      Hindbrain, Rhombencephalon 23
      Midbrain 25
      Diencephalon 26
      Telencephalon 27
   Neuropathology 30
      Congenital Disease 31
      Vasculopathies 32
      Trauma 36
      Neoplasms 36
      Infectious Diseases 40
      White-Matter Disease 44
   Conclusion 45
   Appendix 45
   References 45

3. Techniques of Quantitative Measurement of Morphological
   Structures of the Central Nervous System 47
   ERIC TURKHEIMER

   Introduction 47
   Measuring the Volume of Brain Structures 48
   Ventricular Atrophy Using Internal Representation 49
   The Ventricle–Brain Ratio 50

Comparisons among the Methods 50
Hemispheric Volume and Asymmetry 51
Mean Brain Density Numbers 52
Measurement of Cortical Atrophy 53
Measurement of Lesion Location 54
    Qualitative Approaches 54
    Quantitative Approaches 55
Conceptualizing Lesion Location 56
    A Quantitative Model of Lesion Extent 57
    Localization Tomography 58
    The Quantitative Basis of Double Dissociation 58
    A Polynomial Solution 59
    An Example with Simulated Data 61
Conclusion 62
References 62

4. Positron Emission Tomography and Neuropsychological Function .. 65

G. PAWLIK and W.-D. HEISS

Introduction 65
Methods 66
    Principles of PET 66
    Brain Energy Metabolism 66
    Brain Blood Flow 67
    PET Image Analysis 68
Results 68
    Normal Functional States 68
    Clinical Syndromes 84
Conclusions 129
References 131

5. Computed Tomographic Scanning and New Perspectives in Aphasia 139

DAVID S. KNOPMAN, OLA A. SELNES, and ALAN B. RUBENS

Introduction 139
Methods 139
    Computed Tomographic Methods 140
    Language Methods 141
    Analytic Methods 142
Results 144
    Single-Word Comprehension 144
    Sentence Comprehension 146
    Confrontation Naming 151
    Sentence Repetition 152
    Nonfluency 155
Discussion 157
References 158

6. Brain Imaging and Neuropsychological Outcome in Traumatic
   Brain Injury 161

RONALD M. RUFF, C. MUNRO CULLUM, and THOMAS G. LUERSSEN

Introduction 161
Neuroimaging in the Acute and Chronic Stages of TBI 162
   Acute Treatment Stages 162
   Chronic Treatment Stages 163
Patterns of Neuropathology 167
Neuroimaging Correlates of Neuropsychological Functioning 171
   Ventricular Enlargement 171
   Atrophy 173
   Hematoma 176
   Magnetic Resonance Imaging Studies 177
Direction of Future Studies 180
References 181

7. Brain Imaging and Neuropsychological Identification of Dementia
   of the Alzheimer's Type 185

RICHARD I. NAUGLE and ERIN D. BIGLER

Introduction 185
Clinical Characteristics of DAT 185
Incidence and Prevalence 188
Neuropathology and Morphological Brain Changes of DAT 189
Quantifying Cerebral Measurements 195
Neuropsychological Findings 198
Neuropsychological and CT Scan Interrelationships 199
Presenile versus Senile Dementia 204
Subgroups of DAT Patients 206
Prediction of Deterioration 212
Conclusion 213
References 213

8. Neuropsychological and Neuroanatomic Aspects of Complex
   Motor Control 219

K. Y. HAALAND and RONALD A. YEO

Introduction 219
Limb Apraxia: Behavioral Characterization and Lesion Location 220
Experimental Hand Posture Tasks: Behavioral Characterization and Lesion Location . 229
Motor Skills: Behavioral Characterization and Lesion Location 236
Summary and Conclusions 241
References 242

9. Structural Brain Abnormalities in the Major Psychoses 245

SARAH RAZ

Introduction 245
Clinicopathological and Biochemical Characteristics and CT
        Abnormalities in the Psychoses. 247
    Psychotic Symptomatology and CT Abnormalities 248
    Genetic and Environmental Factors and CT Abnormalities in
        the Psychoses 251
    Cognitive Factors and CT Abnormalities in the Psychoses. 255
    Drug Response and CT Abnormalities in the Psychoses. 256
    Neurochemical Dysregulation and CT Abnormalities in
        the Psychoses 258
Integration of Findings 260
References 262

10. Cerebral Imaging and Emotional Correlates 269

C. MUNRO CULLUM

Introduction 269
Research Considerations 271
Focal and Asymmetric Emotional Processing Evidence 272
    Epilepsy Studies 272
    Psychiatric Disorders 273
    Affective Asymmetry and Localized Cerebral Damage 273
Personality Assessment with Neurological Patients 276
    The MMPI and Lateralized Cerebral Lesions 278
    Recent MMPI Findings in Patients with Lateralized Cerebral Lesions
        Identified by CT 281
    Lateralized Cerebral Lesions and the Effects of Etiology 283
    Lesion Size Effects 285
Conclusions 288
Appendix: MMPI Scales 288
References 289

11. Individual Differences 295

RONALD A. YEO

Introduction 295
Sex Differences 301
Age Differences 308
Anatomic Asymmetries 311
Summary 312
References 312

12. Structural Anomalies and Neuropsychological Function 317

ERIN D. BIGLER, J. R. LOWE, and RONALD A. YEO

Introduction 317
Congenital Disorders 317
Congenital Cerebrovascular Abnormalities and Subsequent Stroke 324
Periventricular Intraventricular Hemorrhage 328
Learning Disorder 336
Summary 337
References 337

13. Neuropsychological Functioning and Brain Imaging: Concluding
Remarks and Synthesis 341

ERIN D. BIGLER, RONALD A. YEO, and ERIC TURKHEIMER

Index 349

Neuropsychological Function and Brain Imaging
Introduction and Overview

ERIN D. BIGLER, RONALD A. YEO, and ERIC TURKHEIMER

The study of the effects of known brain lesions on behavior is crucial to progress in understanding the relationship between cognition and the brain. Before the advent of modern brain-imaging techniques, the precise effects of brain lesions could be studied only in animals. This period (circa 1930 to 1975) was extremely productive for animal model research (Kolb & Whishaw, 1985), but progress in human neuropsychology was plagued by the inability to quantify the location, size, or extent of a brain lesion in the living individual.

These particular problems are best exemplified by reviewing some of the methodologies of this past research era in human neuropsychology. For example, in the now classic symposium held in 1962 on the "Frontal Granular Cortex and Behavior" (see Warren & Akert, 1964), Teuber (1964) presented a paper on the behavioral effects of frontal lesions in man. Aspects of his research were based on patients who had penetrating head injuries, and skull x rays were used to document the location and position of entry (see Figure 1). Based on these skull films, the extent of underlying cerebral damage was inferred. Obviously, no details of the cerebral structures can be obtained from skull films, and thus, no precision could be obtained in such research in terms of the extent and location of actual damage. In 1962 this was as close as one could come in giving a noninvasive *in vivo* appraisal of locus of brain damage from cerebral trauma. In contrast, similar studies recently have been done by Salazar, Grafman, Vance, *et al.* (1986) using strict computerized tomography (CT) quantifications methods which carefully outline underlying structural pathology (see Figure 2). This research (see also Grafman, Salazar, Weingartner, Vance, & Amin, 1986; Salazar *et al.*, 1985; Salazar, Grafman, Schlesselman, *et al.*, 1986), because it permits direct quantification of areas and structures affected, has resulted in a much clearer understanding of the behavioral effects of basal forebrain lesions.

At the same symposium at which Teuber discussed the frontal cortex, Reitan (1964) presented his now oft-cited work on the effects of frontal lesions on a collection of tests that would eventually become the basis of the Halstead–Reitan Neuropsychological Test Battery (Reitan & Wolfson, 1985). In establishing the criteria for inclusion of patients with frontal lobe damage,

ERIN D. BIGLER ● Austin Neurological Clinic and Department of Psychology, University of Texas at Austin, Austin, Texas 78705.   RONALD A. YEO ● Department of Psychology, University of New Mexico, Albuquerque, New Mexico 87131.   ERIC TURKHEIMER ● Department of Psychology, University of Virginia, Charlottesville, Virginia 22903.

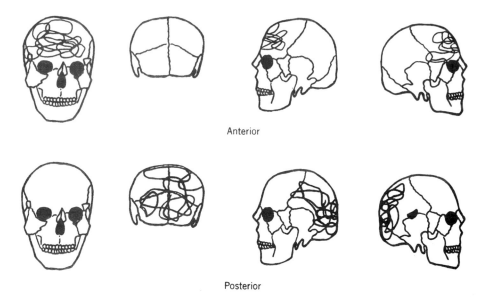

Anterior

Posterior

FIGURE 1. Composite diagrams showing outlines of skull defects in 20 men with anterior (frontal) and 20 with posterior (parieto-occipital) missile wounds of brain (from Teuber, 1964). This illustration depicts the methodology in the pre-brain-imaging era, when one had to infer the locus of damage by the tract or trajectory of the missile by noting entrance and exit points from x-ray skull films. Although this technique provides some index of whether "anterior" or "posterior" brain regions were involved, no precision as to the actual locus or extent of the lesion can be given.

Reitan had to rely on the opinion of "fully qualified neurologists and neurological surgeons," who, in turn, based their clinical judgment on the neurosurgical findings at the time of operation or on neurological examination (i.e., physical findings such as paralysis), direct ventriculography, cerebral arteriography, or pneumoencephalography. Although such standards permitted gross distinctions, it was virtually impossible to study specific effects of discrete lesions on neuropsychological performance.

Prior to the advent of CT techniques to image the brain, one had to rely on inferential methods based on imaging techniques of the day, namely, pneumoencephalography (PEG), cerebral arteriography, and radioactive isotope brain scanning. Pneumoencephalography, introduced by Dandy in 1918 (Oldendorf, 1980), can outline ventricular structures but not image brain tissue directly (see Figure 3). Thus, in interpreting PEG results, one has to rely on inferential methods based on shifts and distortions that could be detected in the cerebral ventricles (see Figure 4). For example, a large tumor might shift the ventricular system laterally and collapse the part of the ventricle closest to the tumor. Similarly, cerebral arteriography can directly image the cerebral vasculature but not brain tissue. In interpreting cerebral arteriography results, one again relies on some distortion or change in the position of the cerebral blood vessels.

Radioisotope scanning techniques (see Oldendorf, 1980) were being developed as early as 1947, but even by the 1960s, these techniques gave only the grossest image of the brain (see Figure 5) and had only limited clinical utility. During this period, attempts were made (see Figure 6) to image the brain via ultrasonic waves. The technique, called echoencephalography (see Hovind, Galicich, & Matson, 1967), could only detect large distortions or aberrations in brain structure and yielded no information about anatomic detail. Thus, the inferential methods of angiography/arteriography or PEG were the primary imaging methods of that period, and the popularity of these techniques continued right up until the early 1970s. But, as Oldendorf (1980) states, both

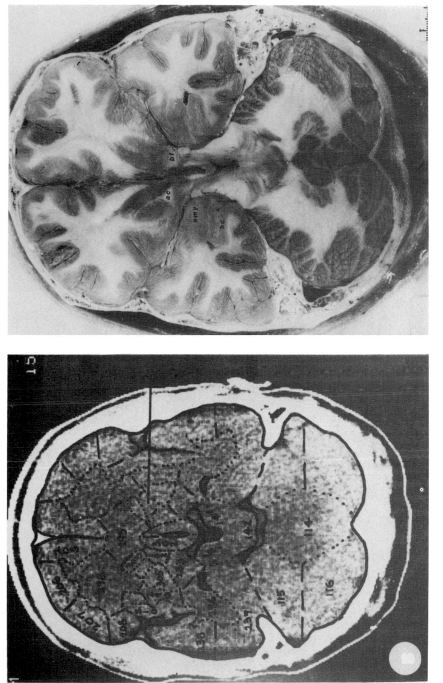

FIGURE 2. (Left) Horizontal CT section at the level of the basal forebrain (ac, anterior commissure; amy, amygdala; bf, basal forebrain; hc, hippocampus). (Right) Anatomic section from the same brain scanned on the left. The anatomic section was utilized to establish a template (the lines superimposed over the CT image on the left) that in turn permitted structure and lesion location. By using such techniques, precise lesion-location studies can be accomplished. (From the work of Salazar et al., 1985, 1986, and used by permission of Dr. Salazar.)

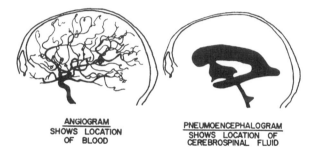

ANGIOGRAM
SHOWS LOCATION
OF BLOOD

PNEUMOENCEPHALOGRAM
SHOWS LOCATION OF
CEREBROSPINAL FLUID

FIGURE 3. Schematic depictions of cerebral angiogram and pneumoencephalogram (reproduced by permission from Oldendorf, 1980).

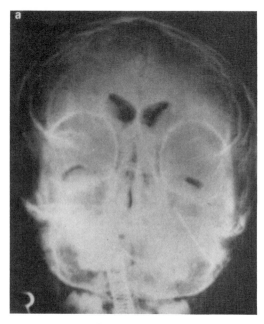

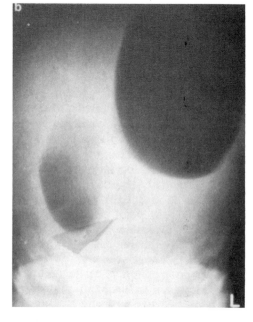

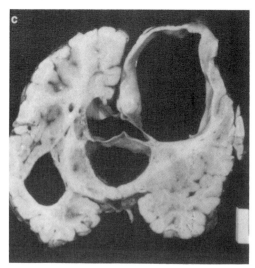

FIGURE 4. Early brain-imaging methods. (a) Pneumoencephalogram, anteroposterior views in brow-up position, shows that right temporal horn is slightly narrower than the left. This suggests partial collapse of the right temporal horn, which was a result of compression from a tumor. (From Falconer, 1970; reprinted by permission from the *Journal of Neurosurgery*). Note that with PEG the ventricular space in a single plane is relatively well defined, but no detailed image of the brain can be visualized. (b) Ventriculogram (similar to PEG except that air is injected directly into the ventricle) demonstrates a large porencephalic cyst. At postmortem (c), the coronal brain section demonstrates the cyst visualized in the ventriculogram, but as with PEG, no anatomic details can be ascertained from ventriculogram studies. (From Salmon, 1970; reprinted by permission from the *Journal of Neurosurgery*.)

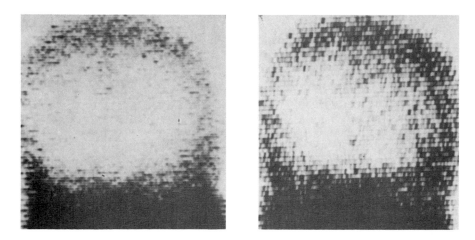

FIGURE 5. Radioisotope brain scan in an adult who sustained a closed-head injury. The image on the left was 1 week after injury and appears normal (i.e., even distribution and absorption of the isotope). The image on the right was taken 3 weeks after injury and demonstrates an abnormal distribution on the left side, which proved to be a hematoma. (From Cowan, Maynard, & Lassiter, 1970; by permission of the *Journal of Neurosurgery*.)

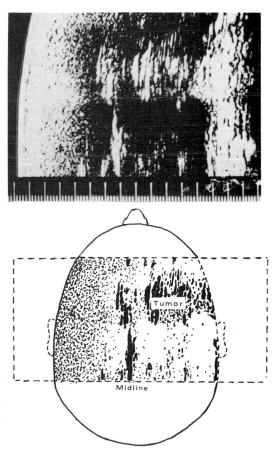

FIGURE 6. Ultrasonic-beam-generated echoencephalogram is presented in the top view. Bottom view is a diagramatic representation of the echoencephalogram findings in relation to the head. Note that the echoencephalogram gave only the crudest image of brain structures. (From Hovind *et al.*, 1967; reprinted by permission from *Neurology*.)

these techniques shared two major limitations: ''Both were traumatic (and as a result, not readily repeated), and both showed only tissue compartments (blood and cerebrospinal fluid, respectively) that were seldom of clinical interest. Structural information about the brain tissue itself had to be inferred (p. 89).'' Because of these limitations they were not very useful in doing clinical lesion-location neuropsychological research.

During the same period, numerous psychological and psychodynamic theories concerning the basis of such disorders as schizophrenia, autism, major affective disorders (the so-called manic–depressive illness spectrum), and learning disorders, to name a few, continued to flourish. For example, prior to 1970, psychodynamic and psychoanalytic theories predominated the literature concerning schizophrenia, autism, and manic–depressive illness (Strauss & Carpenter, 1981). However, with the advances in the psychopharmacological treatment of these disorders in the 1950s and 1960s (Valzelli, 1973), the research focus began to switch to more biological explanations (see Flor-Henry, 1983). Rather unexpectedly the first CT studies in schizophrenic patients (Johnstone, Crow, Frith, Husband, & Krel, 1976) demonstrated ventricular enlargement, a finding that had been suggested by PEG and postmortem studies, but as already pointed out there were severe limitations in interpreting such studies. Thus, the work by Johnstone *et al.* (1976) was the first *in vivo* documentation of possible structural brain abnormalities in psychiatric patients who were thought to have no ''organic'' dysfunction or deficit. Based on our current knowledge and understanding, a wide variety of diagnostic groups may have subsets of patients who demonstrate structural abnormalities or irregularities in CT or magnetic resonance imaging (MRI) scan results (see Figure 7).

In a similar vein, known neurological disorders such as Alzheimer's and closed-head injury were difficult to study prior to CT scanning because of the unavailability of *in vivo* studies. Again, research had to await postmortem investigation before structural changes in these disorders could be assessed. Even this was quite unsatisfactory because the testing would have been done prior, sometimes years prior, to the death of the patient, requiring further inference to establish what the brain structure might have been like at the time of evaluation. Current technology is very different. We now have CT, MRI, positron emission tomography (PET), and related technologies

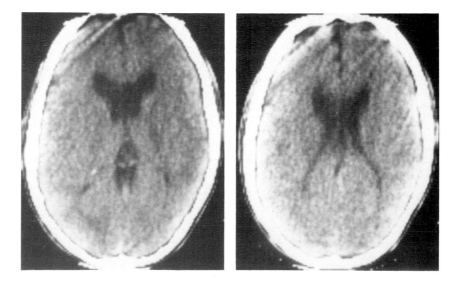

FIGURE 7. Abnormal CT scan in an 21-year-old-high-school-educated male schizophrenic. Note the ventricular enlargement, particularly of the anterior horns, and the frontal pole atrophy.

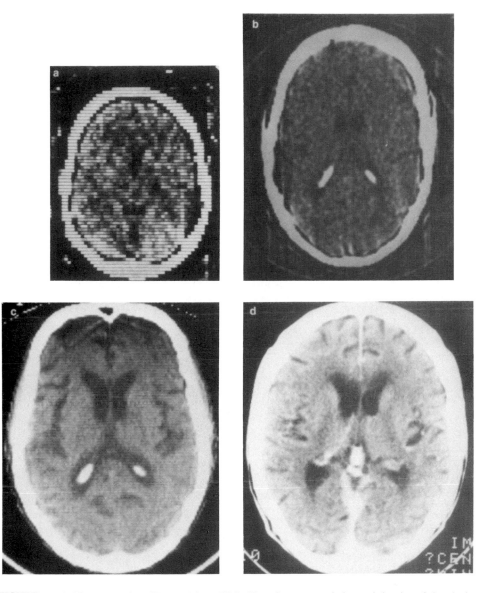

FIGURE 8. (a) First-generation CT scan (circa 1974). Note the poor resolution and the size of the pixels. (b) Second-generation CT scan, which demonstrates better definition of ventricular space but still poor resolution of brain substance; however, even this poor resolution was dramatically better than that of PEG. (c) Improved resolution on the same scanner as in B and on the same patient 5 years later. (This patient had Alzheimer's disease, and the second scan demonstrates atrophy.) Note the better differentiation of cerebral structures in this CT scan. These improvements relate to increased computer sophistication in signal processing and resolution. (d) Current-generation CT scan, which demonstrates some degree of white and gray matter differentiation and better sulcal identification along with ventricular and cistern spaces.

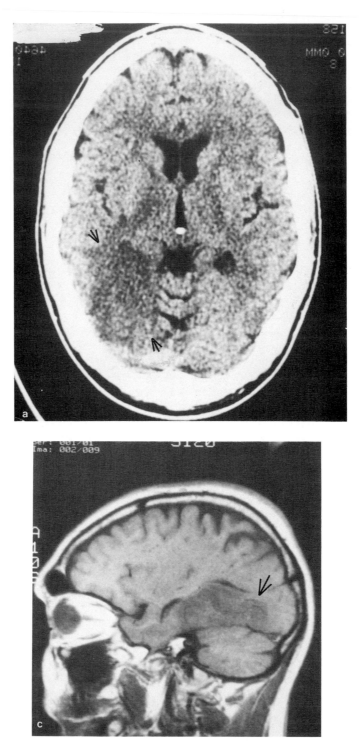

FIGURE 9. Current-generation CT scan (a), which demonstrates area of infarction (arrow). (b) Magnetic resonance imaging scan of the same patient; note the detailed resolution of the MRI results in terms of normal and abnormal anatomy and the extent of the cerebral infarction (arrow). The MRI also permits imaging in the sagittal (c) and coronal (d) planes, which allows three-dimensional viewing of the lesion (arrows). The MRI image is very close to an actual gross anatomic cross section.

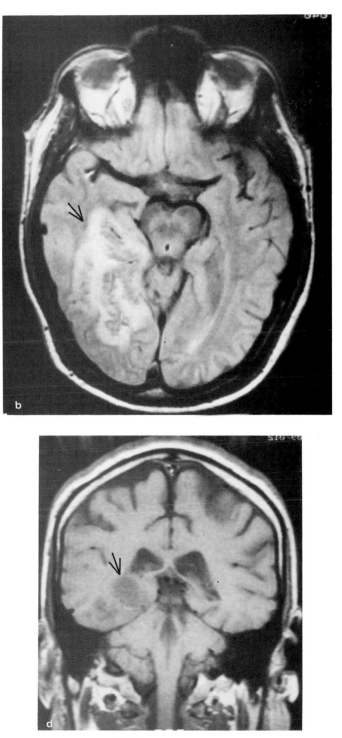

FIGURE 9. (*continued*)

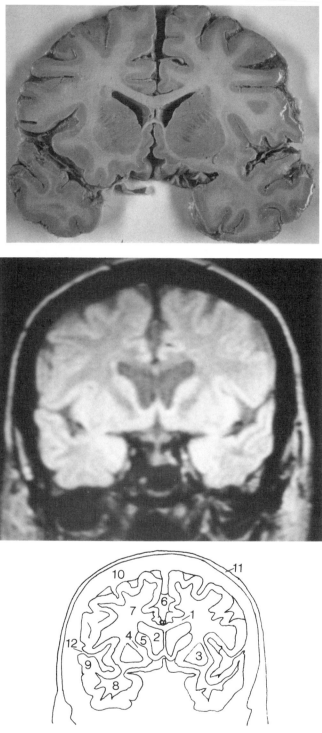

FIGURE 10. (Top) Coronal anatomic section. (Middle) Similar coronal view from MRI scan. (Bottom) Schematic drawing of major anatomic sites as depicted by MRI: (1) corpus callosum, (2) anterior horn of lateral ventricle, (3) putamen/globus pallidus complex, (4) internal capsule, (5) caudate nucleus, (6) interhemispheric fissure, (7) central white matter, (8) temporal lobe, (9) gray matter, cortical mantle, (10) skull, (11) scalp, and (12) Sylvian fissure. Note the anatomic precision that can be obtained by the use of the MRI.

along with methods to quantify various brain parameters (Bottomley, 1984; Naeser, 1985; Yeo, Turkheimer, Raz, & Bigler, 1987).

The past two decades have witnessed previously unimaginable advances in nontraumatic *in vivo* brain imaging. We can now accomplish with humans what heretofore could only be accomplished using animal models, leading to a more precise understanding of human brain function by studying naturally occurring lesions and neurological states. An era of clinical neuroscience research is emerging that promises to provide a clearer definition of human brain–behavior relationships. In fact, in clinical neuropsychology, we have had to dismiss a considerable amount of research conducted prior to 1975. As Swiercinsky and Leigh (1979) have pointed out, CT scan results provide a much more thorough and reliable indicator of actual underlying "organic" damage than electroencephalography or the neurological examination. Since the EEG and neurological examination typically provided the only criteria used in much of the clinical neuropsychological research of this period, it leaves many of the pre-1975 conclusions suspect. Also, there has been considerable refinement in the quality of the CT image (see Figure 8), and with the advent of MRI scanning even finer precision can be achieved (Figure 9 and 10), which has further advanced the use of this technique in studying structural and anatomic effects.

The goals of this text are to review the major neuropsychobiological disorders utilizing current brain-imaging techniques. We hope this endeavor will reduce some of the ambiguity in the cognitive neurosciences by providing more objective criteria for examining the relationship between *in vivo* abnormal brain structure and function.

REFERENCES

Bottomley, P. A. (1984). NMR in medicine. *Computerized Radiology, 8,* 57–77.

Cowan, R. J., Maynard, C. D., & Lassiter, K. R. (1970). Technetium-⁹⁹m pertechnetate brain scans in the detection of subdural hematomas: A study of the age of the lesion as related to the development of a positive scan. *Journal of Neurosurgery, 32,* 30–34.

Falconer, M. A. (1970). Significance of surgery for temporal lobe epilepsy in childhood and adolescence. *Journal of Neurosurgery, 33,* 233–248.

Flor-Henry, P. (1983). *Cerebral basis of psychopathology.* Bristol, England: John Wright & Sons.

Grafman, J., Salazar, A., Weingartner, H., Vance, S., & Amin, D. (1986). The relationship of brain-tissue loss volume and lesion location to cognitive deficit. *The Journal of Neuroscience, 6,* 301–307.

Hovind, K. H., Galicich, J. H., & Matson, D. D. (1967). Normal and pathologic intracranial anatomy revealed by two-dimensional echoencephalography. *Neurology, 17,* 253–262.

Johnstone, E. C., Crow, T. J., Frith, D. C., Husband, J., & Krel, L. (1976). Cerebral ventricular size and cognitive impairment in chronic schizophrenia. *Lancet, 2,* 924–926.

Kolb, B., & Whishaw, I. Q. (1985). *Fundamentals of human neuropsychology.* New York: W. H. Freeman.

Naeser, M. A. (1985). Quantitative approaches to computerized tomography in behavioral neurology. In M.-M. Mesvlam (Ed.), *Principles of behavioral neurology,* pp. 363–383. Philadelphia: F. A. Davis.

Olendorf, W. H. (1980). *The quest for an image of brain.* New York: Raven Press.

Reitan, R. (1964). Psychological deficits resulting from cerebral lesions in man. In J. M. Warren & K. Akert (Eds.), *The frontal granular cortex and behavior,* pp. 259–312. New York: McGraw-Hill.

Reitan, R. M., & Wolfson, D. (1985). *The Halstead–Reitan Neuropsychological Test Battery.* Tucson: Neuropsychology Press.

Salazar, A. M., Jabbari, B., Vance, S. C., Grafman, J., Amin, D., & Dillon, J. D. (1985). Epilepsy after penetrating head injury. I. Clinical correlates: A report of the Vietnam head injury study. *Neurology, 35,* 1406–1414.

Salazar, A. M., Grafman, J., Schlesselman, S., Vance, S. C., Mohr, J. P., Carpenter, M., Pevsner, P., Ludlow, C., & Weingartner, H. (1986). Penetrating war injuries of the basal forebrain: Neurology and cognition. *Neurology, 36,* 459–465.

Salazar, A. M., Grafman, J. H., Vance, S. C., Weingartner, H., Dillon, J. D., & Ludlow, C. (1986). Consciousness and amnesia after penetrating head injury: Neurology and anatomy. *Neurology, 36,* 178–187.

Salmon, J. H. (1970). Isolated unilateral hydrocephalus. *Journal of Neurosurgery, 32,* 219–224.

Strauss, J. S., & Carpenter, W. T. (1981). *Schizophrenia.* New York: Plenum Press.

Swiercinsky, D. P., & Leigh, G. (1979). Comparison of neuropsychological data in the diagnosis of brain impairment with computerized tomography and other neurological procedures. *Journal of Clinical Psychology, 35,* 242–246.

Teuber, H. L. (1964). The riddle of frontal lobe function in man. In J. M. Warren & K. Akert (Eds.), *The frontal granular cortex and behavior,* pp. 410–444. New York: McGraw-Hill.

Valzelli, L. (1973). *Psychopharmacology.* Flushing, NY: Spectrum Publications.

Warren, J. M., & Akert, K. (1964). *The frontal granular cortex and behavior.* New York: McGraw-Hill.

Yeo, R., Turkheimer, E., Raz, N., & Bigler, E. D. (1987). Volumetric asymmetries of the human brain: Intellectual correlates. *Brain and Cognition, 6,* 15–23.

2

Neuroanatomy and Neuropathology
Computed Tomography and Magnetic Resonance Imaging Correlates

J. N. RUTLEDGE

INTRODUCTION

The science of neuroanatomy and the basis of description of the brain presuppose that behavior is a reflection of that structure (Adams, 1984; Adams & Sidman, 1968; Carpenter, 1976; Goss, 1973; Nieuwenhuys, Vougd, & Van Huijzen, 1979; Williams, 1975). Although the complex function of the brain has not allowed a point-by-point assignment of specific characteristics as once hoped, much of gross function is understood and can be localized. The revolution in imaging in the past few years has allowed observation of the living brain and furthered this understanding. To know the brain's basic anatomy, nomenclature, and techniques used to obtain images of that anatomy is necessary for its prudent exploration.

EMBRYOLOGY

The formation of the brain and neuroaxis begins in the third to fourth week of embryonic development (Carpenter, 1985; Cowan, 1979; Jacobson, 1978). A portion of the embryonic disk called the neural plate is activated, and opposing folds from the neural tube, which develops into the spinal cord and brain vesicles (Figure 1). The terms rostral (front), caudal (end), ventral (belly), and dorsal (back) come from the description of the neural tube, as do many of the anatomic descriptions and divisions of the brain.

The formation of the three primary vesicles or divisions (forebrain, prosencephalon; midbrain, mesencephalon; and hindbrain, rhombencephalon) of the brain occurs at the rostral end of the tube by the fourth week of development.

By the fifth week additional vesicles form, dividing the forebrain into the telencephalon (*telos* means end) and diencephalon (*di* means two); and the hindbrain into the metencephalon (*met* means after) and myelenencephalon (*myelos* means related to spinal cord) (see Figure 2). During this time the major components also undergo a bending (flexures) caused by rapid unequal growth and further demarcate the divisions. The first occurs at the junction of the hindbrain and the spinal

J. N. RUTLEDGE ● Capital Radiology of Austin and University of Texas at Austin, Austin, Texas 78705.

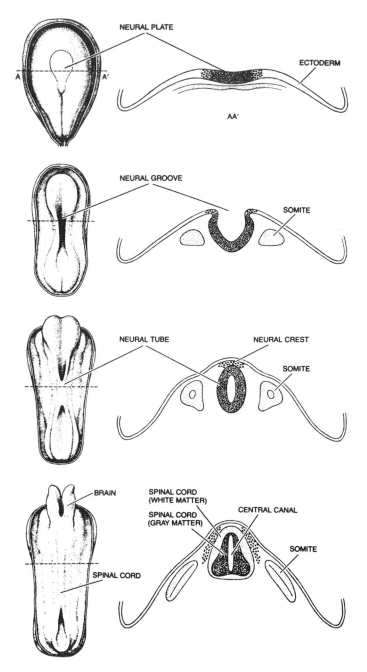

FIGURE 1. Genesis of the nervous system from the ectoderm, or outer cell layer, of a human embryo during the third and fourth weeks after conception is represented in these four pairs of drawings, which show both an external view of the developing embryo (left) and a corresponding cross-sectional view at about the middle of the future spinal cord (right). The central nervous system begins as the neural plate, a flat sheet of ectodermal cells on the dorsal surface of the embryo. The plate subsequently folds into a hollow structure called the neural tube. The head end of the central canal widens to form the ventricles, or cavities, of the brain. The peripheral nervous system is derived largely from the cells of the neural crest and from motor-nerve fibers that leave the lower part of the brain at each segment of the future spinal cord. (From Cowan, 1979; reproduced by permission from *Scientific American*.)

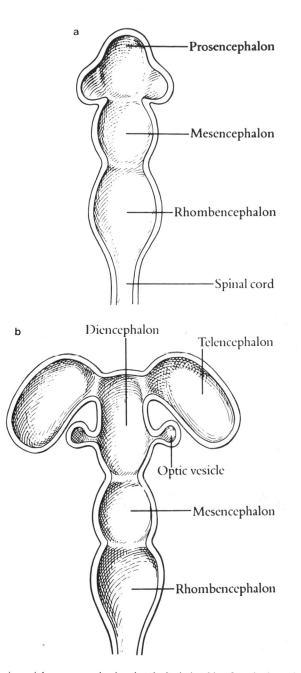

FIGURE 2. Primary brain vesicles are an early sign that the brain is taking form in the embryo of a vertebrate. They are a set of three bulbous swellings at the forward end of the neural tube, the precursor of the brain and the spinal cord. The swellings, from back to front, are the prospective rhombencephalon, or hindbrain; the prospective mesenencephalon, or midbrain; and the prospective prosencephalon, or forebrain. In a human embryo the primary brain vesicles are apparent before the fifth week of gestation (a). Then, in the fifth week, the diencephalon, or central chamber of the forebrain, begins to develop side chambers: the telencephalon, or cerebral hemispheres (b). (From Nauta & Feirtag, 1986; reproduced by permission from W. H. Freeman.)

cord (cervical flexure), the second occurs in the midbrain (midbrain flexure), and the third occurs later, producing the pontine flexure.

This rapid growth and thickening dorsally in the metencephalon becomes the cerebellum and gives the flattened appearance to the fourth ventricle, the residual vesicle. The diencephalon's rapid growth produces the epithalamus, thalamus, and hypothalamus. The enlargement of the thalamus narrows the vesicle at this level, forming a cleft, the third ventricle.

The telencephalon undergoes extensive change and growth with a cellular migration that forms the gray matter of the cortex. The midbrain undergoes the least change, and in the adult the constricted portion of the vesicle becomes the aqueduct of Sylvius. The endothelial lining of the neural tube forms the choroid plexus, which secretes cerebrospinal fluid.

In the telencephalon's growth, it divides into two halves called hemispheres, each containing a diverticulum of the primitive vesicles, the lateral ventricles. As the hemisphere expands rostrally to form the frontal lobes, dorsally to produce the parietal lobes, and posterior and inferior to form the occipital and temporal lobes, a C-shaped configuration is produced (see Figure 3). As the

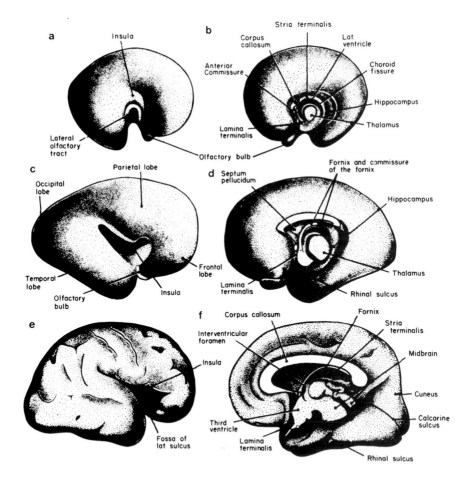

FIGURE 3. Development of human cerebral hemisphere.   (a, b) Lateral and medial surfaces of the hemisphere in a fetus of 3 months.   (c, d) Lateral and medial surfaces of the hemisphere in a fetus at the beginning of the fifth month.   (e, f) Lateral and medial surfaces of the hemisphere at the end of the seventh month. Note the C-shaped appearance of the developing cerebral hemispheres. (From Carpenter & Sutin, 1983; reproduced by permission from Williams & Wilkins.)

surface of the hemisphere develops, it loses its smooth appearance and forms multiple gyri and sulci. This allows an increase in cortex without an increase in cranial volume. As a result of the dorsal expansion of the cerebral hemispheres and lateral ventricles, the underlying structures, the corpus striatum and the limbic system (parahippocampus and cingulate gyrus), also have a C-shaped appearance. As the hemisphere grows, the cortex of the corpus striatum is overgrown. This forms a hidden portion within the lateral sulcus called the insula.

A stratified cellular anatomy of the brain also results from the folding of the neural plate. The three layers—the endoderm, the innermost layer; the mesoderm, the middle layer; and the ectoderm, the outermost layer—form a basis of pathological diagnosis from imaging. Nerve and glial cells come from the outermost layer, and the choroid plexus and the endothelial lining of the ventricular system come from the innermost layer. The anatomic location and resulting cellular composition for any area, age, and history provide the clues used in differential diagnosis.

NEUROIMAGING

The history of imaging of the living brain began serendipitously. An accidental head trauma with resulting pneumocephalus (air around the brain) was seen and reported on a plain-film x ray by Luckett (1913). This was subsequently developed by Dandy (1918, 1919) into the procedure pneumoencephalography. By injecting air through trephine openings cut into the skull or lumbar subarachnoid puncture, the ventricular system and cisterns surrounding the brain could be examined.

With the development of nontoxic vascular dye by Moniz (1927), arteriography superseded pneumoencephalography, yielding more information (e.g., mass effect, vascularity) with less morbidity (Oldendorf, 1980). Like computed tomography and magnetic resonance imaging that would follow, advancing techniques directly visualizing the tissue or characteristic of interest supplanted the more indirect or invasive techniques.

The development of computed tomography by Hounsfield in 1973 revolutionized neuroimaging by allowing direct evaluation of the brain parenchyma. In conventional plain film x ray, images are formed by the attenuation of x rays as they pass through various tissues and are expressed in a single plane by the degree of blackening on x-ray film (Christensen, Curry, & Dowdey, 1978).

Computed tomography in a simplified manner uses the same principle: a thinly calibrated beam of x rays is passed through a subject and attenuated depending on the type of tissue (Christensen et al., 1978; Lee & Rao, 1983; Newton & Potts, 1983; Oldendorf, 1980; Ramsey, 1987). As the x rays emerge from the patient/subject, they arrive at an image receptor (scintillation crystal) and are recorded. As the x rays are rotated around the subject, a computer algorithm (calculation schema) can reconstruct the attenuation changes into a two-dimensional image.

There are two major configurations used today for the x-ray tube and detectors: a "third-generation" gantry system in which a fan of x rays passes through the subject onto an array of detectors and both the x-ray tube and the detector are rotated around the subject, and a "fourth-generation" gantry system in which detectors are fixed in a complete circle around the subject and only the x-ray tube rotates. Both third- and fourth-generation configurations can produce relatively equivalent images today (see Figure 8, Chapter 1). The CT images are displayed using a gray scale in which the greatest attenuation of x rays is represented as white and the least as black (bone is white, water is gray, air is black) (see Figures 4–7). Magnetic resonance imaging uses similar mathematical algorithms to produce and display images but is based on a completely different form of data.

Magnetic resonance (MR) imaging (or MRI), previously called nuclear magnetic resonance (NMR) imaging, uses different principles to image the brain (Brant-Zawadzki & Norman, 1987; Pyckett, 1982, 1983). Because atoms with an odd number of nucleons have a magnetic dipole, they tend to align or "spin" with a polar axis pointed north like a compass needle when placed

FIGURES 4–7. Normal gross anatomy on computed tomographic images. 1, Frontal horn of the lateral ventricle; 2, temporal horn of the lateral ventricle; 3, pons; 4, fourth ventricle; 5, suprasellar cistern; 6, frontal lobe; 7, temporal lobe; 8, cerebellum; 9, vermis of the cerebellum; 10, midbrain tectum; 11, quadrigeminal cistern; 12, thalamus; 13, third ventricle; 14, lenticular nucleus (putamen and globus pallidus); 15, caudate nucleus; 16, insula; 17, Sylvian fissure; 18, body of lateral ventricle; 19, choroid plexus; 20, supracerebellar cistern; 21, parietal lobe; 22, occipital lobe; 23, corona radiata; 24, corpus callosum; 25, centrum semiovale; 26, interhemispheric fissure; 27, falx cerebrum.

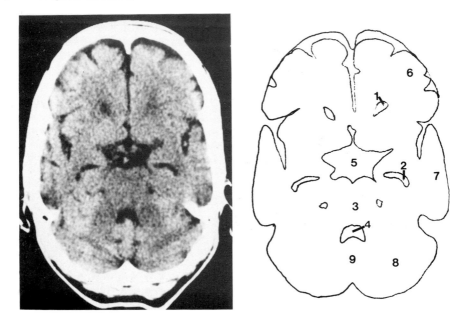

FIGURE 4. Section through cerebellum, fourth ventricle, and basal cisterns.

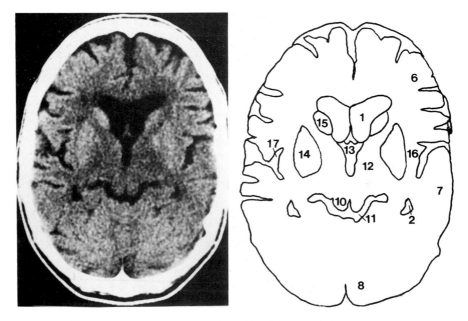

FIGURE 5. Section through basal ganglia and low ventricular system.

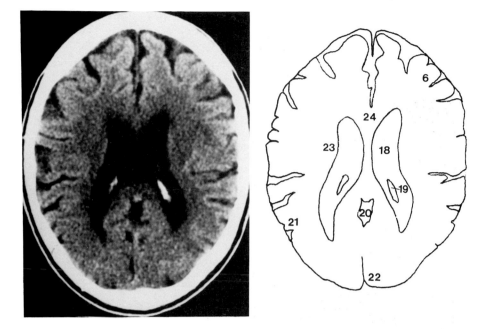

FIGURE 6. Section through high ventricular system.

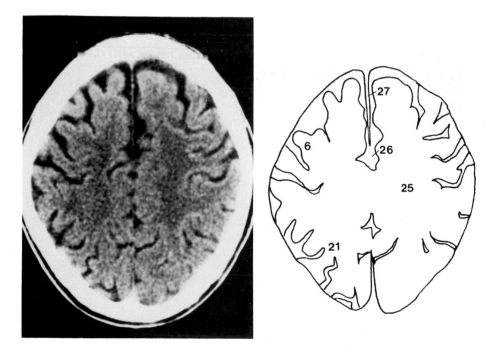

FIGURE 7. Section through high cortex.

in a magnetic field. When selected sections are stimulated with radiofrequency (RF) waves, the resulting energy absorption is manifested by a change in the orientation of the original alignment. This process is called "resonance." When the RF source is turned off, the energy is released in a process known as "relaxation." This release of energy can be detected as RF signals, measured, and processed into images (see Figure 8).

The actual change in alignment is really relatively complex with precession (spinning like a top) and certain vibrations (see Figure 8). To aid in understanding, this motion's description has been simplified by freezing it in time and measuring the angle of alignment relative to the magnetic field in a cartesian coordinate system (x,y). The speeds at which the x and y axes realign with the magnetic field or relax are called the T_1 and T_2 characteristic, respectively (see Figure 5).

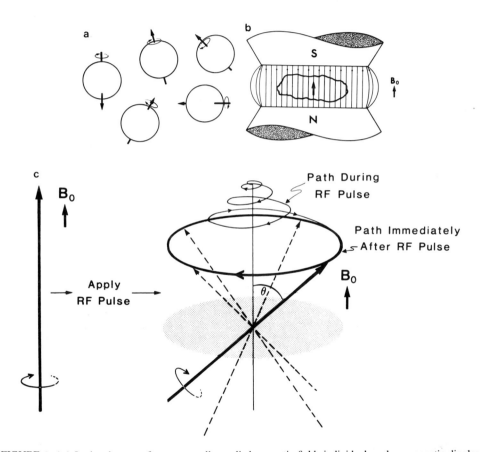

FIGURE 8. (a) In the absence of any externally applied magnetic field, individual nuclear magnetic dipoles within a sample are randomly oriented. (b) When a uniform, static magnetic field (B_0) is applied, a preponderance of NMR-sensitive nuclei will align themselves with the lines of magnetic induction in such a way that a net magnetic moment is generated within a sample (N, north; S, south). (c) On application of a radiofrequency (RF) pulse, the net magnetic moment is perturbed from its equilibrium position and, because of its properties of spin, begins to precess about the static field direction. The angle between the z axis and the magnetization vector continues to increase as long as the pulse remains on. When it is turned off, the vector precesses freely at the final angle, θ, and its rotation describes the wall of a cone. The component of magnetization that rotates in the x,y plane (shaded area) generates the nuclear signal. (From Pykett, Buonanno, Brady, & Kistler, 1983; reproduced by permission from the Radiologic Society of North America.)

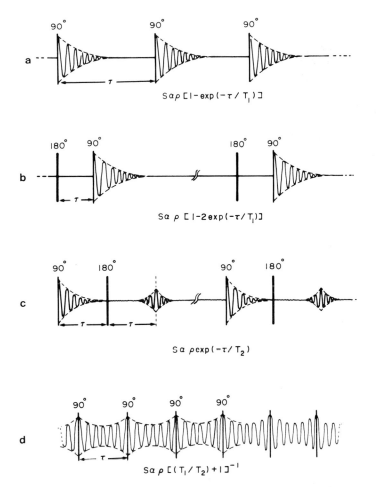

FIGURE 9. Many different types of pulse sequences can be used to elicit an NMR signal. In the saturation-recovery (a) and inversion-recovery (b) sequences, the image intensity S generally contains T_1 information. The spin-echo sequence (c) will yield a T_2-weighted image. The steady-state free-precession pulse sequence (d) yields an image for which the intensity is a more complicated function involving the ratio T_1/T_2. (From Pykett et al., 1983; reproduced by permission from Pergamon Press.)

The importance of this is that, as in CT images where the gray scale represents different tissues, mapping of T_1 and T_2 characteristics displayed on a gray scale may also be associated with tissue types.

In its most basic form, T_1 is theorized to represent the interaction between macromolecules and a single bound hydration layer, and T_2 represents the exchange diffusion between that bound layer and a free water phase (see Figure 9). Fortunately, the tissue type bears the most influence in both these interactions.

T_1-weighted images show as bright or white on the gray scale fat or subacute hemorrhage, and black represents fluid such as CSF. On the T_2-weighted images, fat appears gray, iron black, and fluid white. Substances that are always without signal or black include cortical bone, blood and CSF flow, and air (see Figures 10 and 11).

Advantages of magnetic resonance imaging lie in the nature of life itself. The MRI looks directly at water and so is exquisitely sensitive to pathological changes in water distribution such as vasogenic or cytotoxic edema. The more traditional radiological CT imaging is derived from x-ray studies, which are most sensitive to calcium and less sensitive to attenuation changes caused by fat, water, and air.

FIGURES 10–12. Normal gross neuroanatomy on sagittal magnetic resonance images. 1, Corpus callosum; 2, lateral ventricle; 3, third ventricle; 4, pituitary; 5, optic tract; 6, mamillary body; 7, midbrain; 8, pons; 9, medulla; 10, fourth ventricle; 11, cerebellum; 12, supracerebellar cistern; 13, pineal; 14, quadrigeminal plate; 15, cingulate gyrus; 16, calcarine sulcus; 17, parietal–occipital sulcus; 18, frontal lobe; 19, parietal lobe; 20, occipital lobe; 21, temporal lobe; 22, central sulcus; 23, Sylvian fissure; 24, internal cerebral vein; 25, straight sinus; 27, horizontal fissure; 28, primary fissure; 29, Broca's area; 30, Wernicke's area.

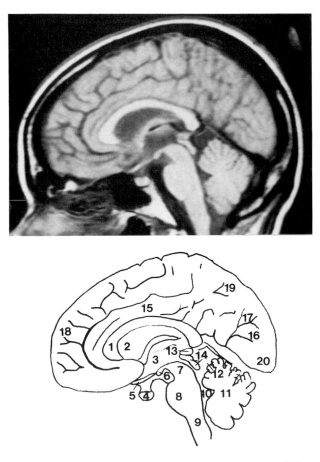

FIGURE 10. T_1-weighted midline image showing fat as increased signal and CSF as decreased signal.

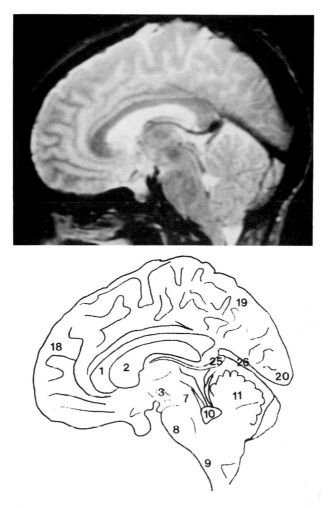

FIGURE 11. T_2-weighted midline image showing fat as decreased signal and CSF as increased signal. Note that bone, air, and flow always have decreased signal.

NORMAL ANATOMY

Because of the accuracy of modern imaging in depicting the living brain, we can use these images to review the normal adult anatomy. From the review of embryology, we know that the brain can be divided into five major components, and each is discussed separately.

Hindbrain, Rhombencephalon

In the adult the hindbrain can be separated into three major components: the medulla, the pons, and the cerebellum (Figures 6, 10, 13, 22). The medulla oblongata is continuous caudally with the spinal cord, which it joins at the foramen magnum (the opening in the bottom of the skull), and rostrally is continuous with the pons, ending at the pontomedullary sulcus. It is approximately 2.5 cm in length.

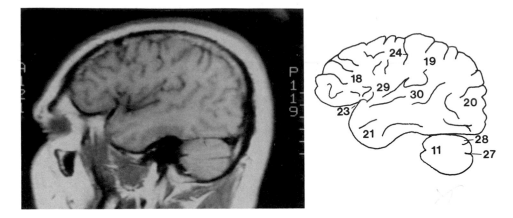

FIGURE 12. Off-midline sagittal image showing sulcal pattern.

Surface features of the medulla include the midline ventral and dorsal grooves called the anterior median fissure and the dorsal median sulcus. Two lateral sulci are also present ventrally, and the pyramid is the structure formed between these two grooves. Posterior and lateral to the pyramid and separated by the multiple roots of the hypoglossal nerve is a swelling called the olive. Dorsally, a lateral sulcus also exits and separates the cuneate and gracile tubercles. Posterior and lateral to the olive is the inferior cerebellar peduncle. In the groove separating the two is the origin for the ninth, 10th, and 11th cranial nerves.

The pons is the most rostral part of the hindbrain and is characterized by a ventral swelling of its surface. It is separated superiorly or rostrally by the superior pontine sulcus from the midbrain. Its dorsal surface is flattened and forms the floor of the fourth ventricle. Originating laterally is the trigeminal nerve, which runs anteriorly. Dorsal lateral stands form the middle cerebellar peduncle on which the cerebellum rests. Structures such as the medial lemniscus and spinal thalamic tract run in the dorsal portion of the pons beneath the floor of the fourth ventricle. In the basis pointis, the belly of the pons, multiple bundles and longitudinal fibers are present.

The seventh and eighth (vestibular–cochlear) cranial nerves originate between the floor and the lateral wall of the fourth ventricle, and the sixth (abducens) cranial nerve originates just anterior on the floor of the fourth ventricle (see Figure 13).

The cerebellum consists of several parts. The midline narrow portion is the vermis, two large lateral parts are the cerebellar hemispheres, and two smaller inferior lateral parts are called the flocculonodulus. The cerebellum is wrapped around the pons and forms the roof of the fourth

FIGURES 13–18. Normal axial neuroanatomic T_2-weighted magnetic resonance images. 1, Fourth ventricle; 2, pons; 3, cerebellar hemisphere; 4, vermis; 5, internal auditory canal; 6, temporal lobe; 7, petrous portion of the temporal bone; 8, cerebellar pontine angle cistern; 9, aqueduct of Sylvius; 10, red nucleus; 11, substantia nigra; 12, frontal lobe; 13, thalamus; 14, globus pallidus; 15, putamen; 16, caudate; 17, third ventricle; 18, occipital lobe; 19, insula; 20, internal capsule; 21, frontal horn of lateral ventricle; 22, occipital horn of lateral ventricle; 23, pineal; 24, parietal lobe; 25, optic radiations; 26, body of lateral ventricle; 27, corpus callosum; 28, corona radiata; 29, occipital lobe; 30, centrum semiovale; 31, interhemispheric fissure; 32, Sylvian fissure; 33, hypothalamus; 34, brachium pontis (middle cerebellar peduncle); 35, atrium of lateral ventricle; 36, splenium of corpus callosum; 37, medulla.

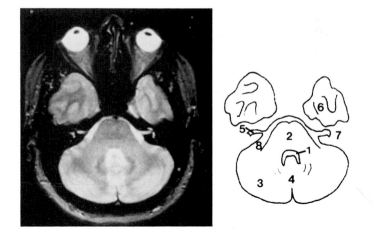

FIGURE 13. Section through fourth ventricle, cerebellum, and temporal lobes.

ventricle. In size it occupies the majority of the posterior fossa, which is a space or compartment defined superiorly by the tentorium cerebelli and laterally by the petrous pyramids of the temporal bone and the occipital bones. The cerebellum is attached to the midbrain ventrally by the brachium pontis and superiorly by the cerebellar peduncles or brachium conjunctivium. Because of the complexity of the cerebellum a number of names are relevant. Important fissures are the primary fissure, which overlies the superior portion of the cerebellum and separates the neocerebellum from the middle cerebellum, and the horizontal fissure, which lies close to the middle of the cerebellum. The cerebellum is highly folded and fimbriated, which increases surface area without increasing volume. The fourth ventricle is formed by the pons ventrally and the cerebellum dorsally; it is a diamond-shaped cavity whose superior extent is the aqueduct of Sylvius in the midbrain and inferior extent is the foramen of Magendie at the level of the medulla. A tentlike extension into the cerebellum with the apex called the fastigium and the walls called the superior and inferior medullary velum is also present. The roof of the fourth ventricle contains choroid plexus, which produces CSF. Laterally the fourth ventricle opens into the cerebellar pontine angle formed by the petrous pyramids, the pons, and the cerebellum via the formina of Lushka.

Midbrain

The midbrain is a short cylindrical segment that undergoes little embryological change. It connects the pons inferiorly and the forebrain rostrally by dividing into the cerebral peduncles (Figures 5, 12, 14, 21). The ventral surface is formed by the cerebral peduncles and embraces the interpeduncular cistern. The midbrain also can be divided ventrally into the tegmentum and dorsally into the tectum by a line through the aqueduct of Sylvius joining the third and fourth ventricles.

Dorsally the midbrain has four bumps or colliculi—the corpora quadrigemina or the collicular plate. The superior colliculi are connected via a brachium with the lateral geniculate body and are a part of the visual system. The inferior colliculi are connected to the medial geniculate bodies and form part of the auditory system. The third cranial nerves originate from the interpeduncular cistern. The fourth nerves wrap around the midbrain after originating dorsally, adjacent to the superior cerebellar peduncle. Within the midbrain, the red nucleus and substanita nigra, both structures of the extrapyramidal motor system, can be visualized.

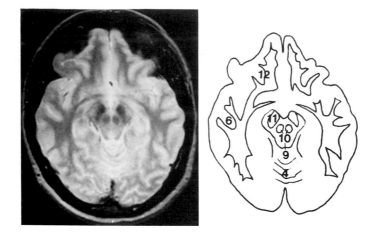

FIGURE 14. Section through midbrain and temporal lobes.

Diencephalon

The diencephalon consists of structures around the third ventricle: the thalamus, epithalamus, and hypothalamus (Figures 5, 10, 15, 20). The thalami are large egg-shaped structures adjacent to and forming the lateral walls of the third ventricle. A small gray-matter tract, the massa intermedia, connects the two thalami and extends through the third ventricle. In the roof of the third ventricle lie the formix and choroid plexus, which produces CSF. The floor of the third ventricle is the hypothalamus with its multiple nuclei that control endogenous secretion. Anteriorly and superiorly two openings form the third ventricle with the foramina of Monro connecting the lateral ventricles in the cerebrum to the third ventricle. Posteriorly and superiorly lie the habenular commissure, posterior commissure, and pineal gland. Recesses are formed by the third ventricle adjacent to the pineal, habenula, optic tracts, and infundibulum, and, in the past, these were impor-

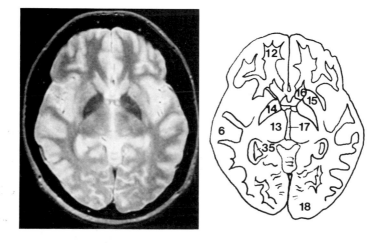

FIGURE 15. Section through low basal ganglia level.

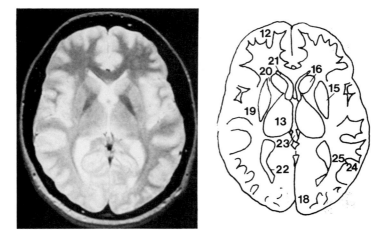

FIGURE 16. Section through high basal ganglia level.

tant pneumoencephalographic features. The hypothalamus also includes several small parts: the mamillary bodies, optic chasm, infundibulum, and hypophysis.

Telencephalon

This forms the largest portion of the brain and is divided into two parts, the right and left cerebral hemispheres (Figures 6, 7, 10–12, 15–18, 19–23). As previously discussed the surface is convoluted, which allows an increased surface area without a volume increase. The surface has been divided into lobes: frontal, parietal, occipital, insular, and temporal. The C-shaped embryologic growth (see Figure 3) causes a large fold or sulcus to separate the temporal lobe from the frontal and is called the Sylvian or lateral sulcus. The next largest deep or major sulcus is the central sulcus, which is a vertical separation of the anterior (frontal) and posterior (parietal) brain.

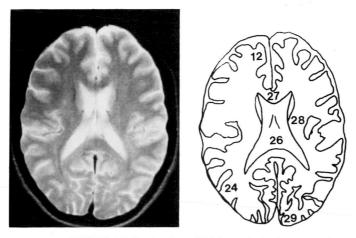

FIGURE 17. Section through high ventricular level.

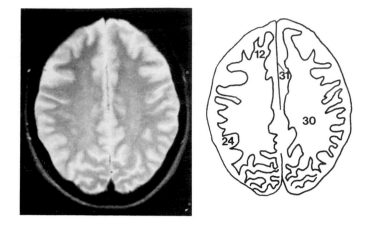

FIGURE 18. Section through high cortex.

The frontal lobe is the most rostral or anterior portion and is limited posteriorly by the central sulcus and inferiorly by the lateral sulcus. Medially the lobe extends to the midline, the interhemispheric fissure, separating the hemispheres. Each hemisphere has three layers, an outer gray, an inner white (centrum semiovale), and a central gray (the basal ganglia). The two hemispheres are connected by a large C-shaped white-matter tract, the corpus callosum (see Figures 10, 12, and 18).

The temporal lobe is separated from the rest of the hemisphere by the lateral sulcus medially and superiorly and lies within a fossa (cavity) of the skull, the middle fossa. The posterior superior extent of the lobe is poorly defined and blends with the parietal and occipital lobes.

The parietal lobe lies posterior to the frontal and superior to the temporal. The posterior boundary is the occipital lobe. The medial surface, the parietal–occipital sulcus, separates the two lobes. The anterior medial margin is an imaginary line continuous with the central sulcus.

The insula lies in the lateral or Sylvian sulcus. The margin of the sulci are called operculum (lips) superior and inferior, which are made up of the frontal and temporal lobes (see Figures 7, 15, 29).

Other important sulci that demarcate the brain are the calcarine sulcus dividing the medial portion of the occipital lobes and the cingulate sulcus separating the cingulate gyrus medial from the rest of the frontal lobe (see Figure 10).

Deep to the gray-matter gyrus is the centrum semiovale, the white matter plane. These association fibers connect portions of the gyri to one another. In the midline, fibers coalesce to connect the two hemispheres and are called the corpus callosum. Different white-matter fibers such as the corona radiata, the superior longitudinal bundle, and other short association fibers are not differentiated by present imaging techniques. One exception to this is the optic radiations, which can be seen on MRI (see Figure 15).

FIGURES 19–22. Normal coronal neuroanatomic T_2-weighted magnetic resonance images. 1, Fourth ventricle; 2, pons; 3, cerebellar hemisphere; 4, vermis; 5, internal auditory canal; 6, temporal lobe; 7, petrous portion of the temporal bone; 8, cerebellar pontine angle cistern; 9, aqueduct of Sylvius; 10, red nucleus; 11, substantia nigra; 12, frontal lobe; 13, thalamus; 14, globus pallidus; 15, putamen; 16, caudate; 17, third ventricle; 18, occipital lobe; 19, insula; 20, internal capsule; 21, frontal horn of lateral ventricle; 22, occipital horn of lateral ventricle; 23, pineal; 24, parietal lobe; 25, optic radiations; 26, body of lateral ventricle; 27, corpus callosum; 28, corona radiata; 29, occipital lobe; 30, centrum semiovale; 31, interhemispheric fissure; 32, Sylvian fissure; 33, hypothalamus; 34, brachium pontis (middle cerebellar peduncle); 35, atrium of lateral ventricle; 36, splenium of corpus callosum; 37, medulla.

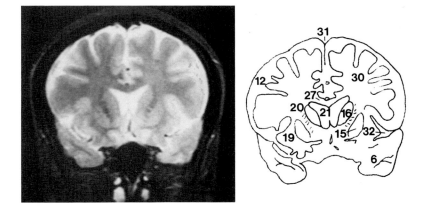

FIGURE 19. Section through frontal horns of lateral ventricles.

Embedded deep in the white matter are the basal ganglia, which also represent the gray matter of the telencephalon. The basal ganglia are comprised of identifiable structures on both CT and MR. The foremost is a white-matter tract, the internal capsule, which in the axial or horizontal plane can be seen as a V-shaped structure with the bend pointing medially. This embraces the neostriatum and paleostriatum—the caudate, putamen, and globus pallidus. On the MR images the globus pallidus has a decreased signal. The caudate and putamen seen in the sagittal plan have a comma- or C-shape appearance that corresponds to the floor of the lateral ventricle (see Figures 5, 15, 16, 19, 20).

Another set of structures associated with the C shape and lateral ventricle includes the hippocampus and parahippocampal gyrus, which form the medial border of the temporal lobe. The lateral ventricles are also present within the cerebral hemispheres and occupy a relatively central position. The frontal horns and bodies are in the frontal lobe. The antrum connects the extensions into the occipital lobe, the occipital horn, the temporal lobe, and the temporal horn. Images in the sagittal plane show a C-shaped configuration formed by the frontal and temporal ventricles.

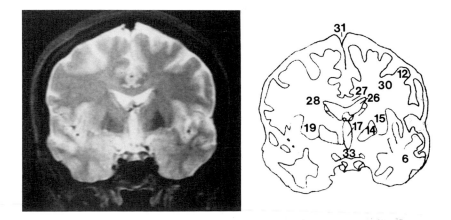

FIGURE 20. Section through third ventricle.

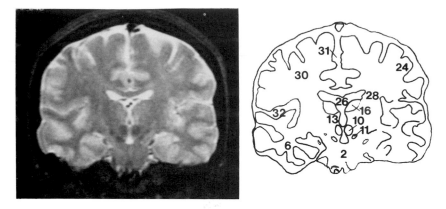

FIGURE 21. Section through brainstem.

NEUROPATHOLOGY

An overview of neuroimaging of brain pathology requires an understanding of the imaging techniques as previously discussed. In computed tomography, recognition of disease is dependent on attenuation changes of tissue, calcification, and enhancement by contrast agents demarking areas of blood–brain barrier breakdown. In magnetic resonance imaging, changes depict pathology by changes in water (cytotoxic or interstitial edema), tissue density, paramagnetic deposition, and fat concentration. On both CT and MR, knowledge of gross anatomy to identify any change such as mass effect is pivotal in pathological differentiation. The most important diagnostic information is then the location of the pathological change.

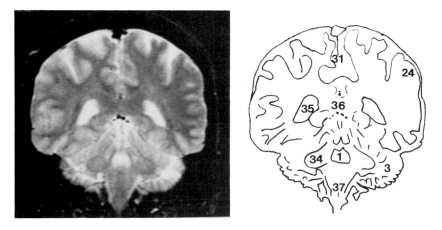

FIGURE 22. Section through fourth ventricle and cerebellum.

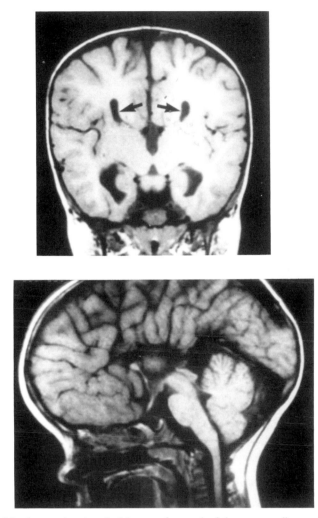

FIGURE 23. T_1-weighted coronal and sagittal images of agenesis of the corpus callosum. Note lateral displacement of bodies of lateral ventricles (arrows), the lack of a cingulate gyrus, and elevation of the third ventricle.

After any imaging changes are localized and characterized, a differential diagnosis can be considered in one of the following groups: congenital lesions, vascular lesions, traumatic lesions, neoplastic lesions, and infectious/inflammatory lesions. The length restrictions on a single chapter prevent detailed discussion of each group. Representative cases and diseases are provided as models of pathological change.

Congenital Disease

Midline defects are some of the most common congenital malformations and can be represented by agenesis of the corpus callosum (Adams & Sidman, 1968; Jacobson, 1978) (see Figure

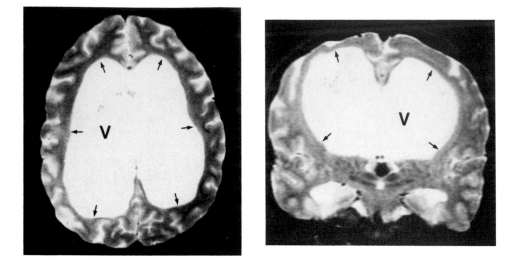

FIGURE 24. T_2-weighted axial and coronal images of communicating hydrocephalus. The lateral ventricles (V) are massively dilated and compress the normal brain parenchyma.

23). This defect results in a lateral displacement of the lateral ventricles, giving them a parallel configuration. Absence of the cingulate gyrus is frequently associated with this abnormality.

A similar midline abnormality is found in Dandy–Walker syndrome. This abnormality results from agenesis of the vermi of the cerebellum, giving a cystlike appearance to the fourth ventricle in the posterior fossa and splitting the cerebellar hemispheres. Other midline abnormalities include encephaloceles (see Figure 1 in Chapter 12) and meningoceles, where incomplete closure of the neural plate leaves bleblike extensions, which may not contain neural material. Closely associated with these changes are Chiari malformations. Herniation of the cerebellar tonsils through the foramen magnum causes symptoms by compression of the medulla. These symptoms are sometimes confused with multiple sclerosis.

Congenital arachnoid cysts (see Figure 7 in Chapter 12) are probably related to birth trauma but will be described here. They are usually found around the base of the brain and are clearly defined without mass effect or displacement of adjacent structures.

Another midline abnormality is a lipoma, a benign fatty tumor, usually with a high T_1 signal intensity and causing no mass effect or symptoms.

Hydrocephalus or dilation of the ventricular system can be "communicating," involving all the ventricles, or "obstructive," with a pathological lesion causing a limited enlargement of the ventricular chambers by trapping the CSF production of the choroid plexus (see Figure 24). Aqueductal stenosis is a form of obstructive hydrocephalus and can be distinguished by lack of fourth ventricle dilatation in an otherwise dilated system.

Vasculopathies

Vascular insults to the brain can be ischemic or anoxic (Davis, 1985). Although classically called strokes, they can result from multiple etiologies (embolic, vascular occlusion, vasculitis). Identification on CT or MR is helped by typical appearance of decreased attenuation or increased T_2 signal in the distribution of blood vessels (see Figure 25). Mass effect from edema can accompany these lesions to a limited extent. The presence of associated hemorrhage is also important clinically and can be seen by increased attenuation (white) on CT and increased signal (white) on

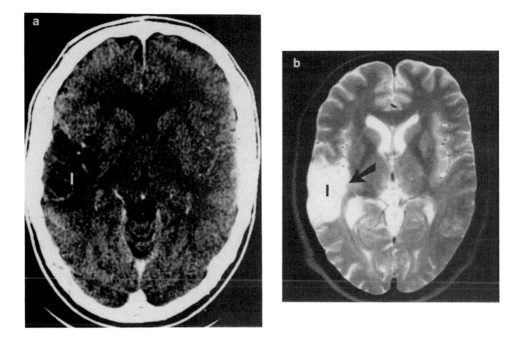

FIGURE 25. (a) CT and (b) T_2-weighted MRI images of a right middle cerebral artery infarct. A decrease in attenuation (black) is seen on the CT, and an increase signal is seen on the MR (white).

T_1 MR images (see Figures 25 and 26). Resolution of the hemorrhage over time shows a gradual decrease in attenuation on CT and variable changes on MR (T_1 isointense signal acute, increase signal subacute 1–3 days; T_2 increase signal subacute with ring of decrease signal representing hemosiderin deposition gradually replacing the increased signal in chronic lesions) (see Figures 27 and 28).

Other etiologies for strokes include primary intracranial bleeds, usually localized to the basal ganglia with dramatic mass effect (see Figure 2 in Chapter 13), or are the result of vascular malformations. Arterial–venous malformations (AVM) are a congenital spaghettilike tangle of arteries and veins with rapid flow that enlarge with age. These may spontaneously bleed or result in a steal of blood, producing local ischemia. On CT the vessels are outlined by contrast, whereas MR shows a decreased T_1 and a T_2 signal flow void. Aneurysms (blebs of blood vessels, usually occurring at bifurcations of vessels around the base of the brain) may also spontaneously bleed (see Figure 29). Focal collections of blood help localize these lesions, which also may be associated with vascular spasm and infarct.

The most common causes of infarct are stenosis (narrowing of vessels) and emboli (small arteriosclerotic plaques or blood clots occluding vessels). Generalized hypoxia may also result in infarcts in "watershed" distribution areas of most distal perfusion and are usually subcortical.

The increased sensitivity of MRI also has shown a new type of vascular change not seen on CT. This consists of subcortical and periventricular focal patchy areas of increased signal on T_2 images. It has been speculated that these represent focal demyelination, ischemic changes, and/or spaces around the vessels (*état criblé*). Whatever the result, the underlying etiology is small-vessel disease, either hyalinization from hypertension or arteriosclerosis. It is interesting to speculate how these findings relate to dementia with a threshold effect reached after gradual loss of

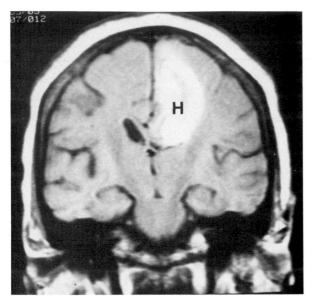

FIGURE 26. T_1-weighted coronal MR image of intraparenchymal hematoma resulting from hypertensive hemorrhage. At 3 days of age an increased signal is present.

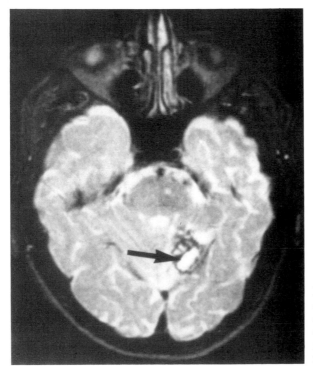

FIGURE 27. T_2-weighted axial MR image of subacute intraparenchymal hemorrhage showing peripheral decreased signal caused by hemosiderin and the increased signal caused by lysis and reabsorption of the clot.

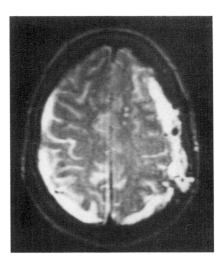

FIGURE 28. T_2-weighted MR image of chronic hemorrhage showing residual hemosiderin causing decreased signal.

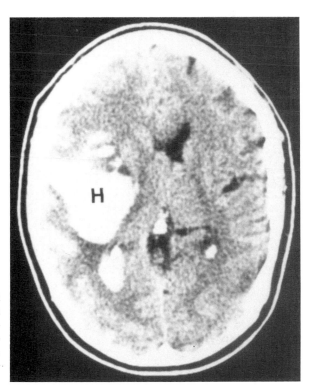

FIGURE 29. Axial CT image of a right intrasylvian hematoma resulting from rupture of a middle cerebral artery aneuryism. The blood has an increased attenuation (white) and causes a mass effect, compressing the basal ganglia and shifting the midline structures.

the subcortical association pathways (see Figure 30). Prominence of cerebral sulci and the ventricular system can be seen in focal and diffuse patterns, and both are probably the results of vascular change and neuronal loss. A generalized atrophy can sometimes give the false impression of hydrocephalus and is called hydrocephalus *ex vacuo*.

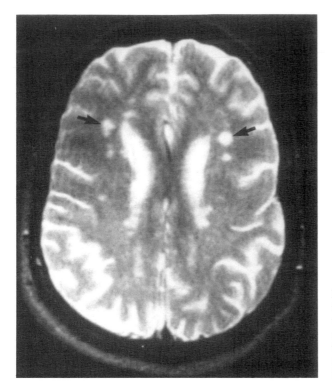

FIGURE 30. Axial T_2-weighted MR image showing multiple focal periventricular lesions felt to represent a microangiopathic leukoencephalopathy caused by small-vessel atherosclerosis or hyalinization resulting from hypertension.

Trauma

A broad range of abnormalities can be observed in the traumatized patient both acutely and chronically. Trauma may be manifest by fractures, hemorrhage, or focal atrophy.

In ascending order of severity, cerebral contusion can be recognized on CT by poorly defined areas of decreased attenuation representing edema. Also, petechial hemorrhage may be present and seen as increased attenuation. Magnetic resonance shows the edema as an increased T_2 signal, and the petechial hemorrhage as increased T_1 signal in subacute injuries. Traumatic intracerebral hematomas do not appear different from other causes, and history and location help in its assessment.

Hemorrhage may also occur extraaxially, outside the brain, with blood present in the subdural or epidural spaces. Subdurals present with a concave appearance with CT and MR characteristics of blood and associated mass effect (see Figure 31). Epidural hematomas present as sharply marginated biconvex collections of blood with similar CT and MR characteristics. A bilateral occurrence is not infrequent and is caused by coup/contracoup injury. A CT of patients in a subacute or an anemic state may be confusing because of an isointensity of blood and brain parenchyma, making distinction difficult. Mass effect and effacement of cerebral sulci are always present.

Unique to MR, in severe traumatic cases, subtle white-matter fractures may be seen, and these may explain severe dementia in the face of otherwise normal examinations.

Neoplasms

Tumors may present with multiple diagnostic characteristics on MR and CT. One key finding is mass effect or displacement of surrounding structures. Other changes seen on CT are areas of

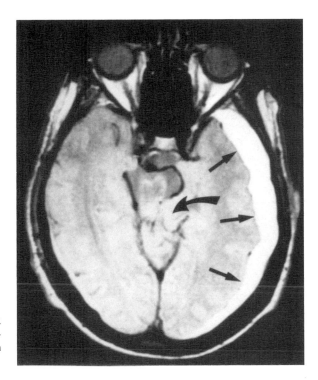

FIGURE 31. Mixed-weighted axial MR image of subdural hematoma (small arrows) causing transtentorial herniation (large arrow).

increased attenuation, which may represent hypercellularity, and areas of decreased attenuation representing areas of necrosis or fluid collection. Dystrophic calcification is not infrequently present and documents a longstanding process. Hemorrhage seen as an increased attenuation may be present and help in differentiation. The MR demonstrates these changes with additional sensitivity to edema and subacute hemorrhage and without the artifacts that limit CT in the posterior and middle fossas. In tumor differentiation, the location and the underlying tissue cell type of the region provide the most help.

One of the largest groups of brain tumors are the gliomas. Several grades or degrees of aggressiveness exist. A grade I glioma can be seen to be an ill-defined, usually homogeneous lesion with little associated edema or mass effect. With increasing malignancy, increased mass effect, edema, and more inhomogeneous imaging characteristics are found. The highest-grade tumor is called glioblastoma multiformi (see Figures 32–34).

Other gliomas include oligodendroglioma, which frequently has chunky diffuse calcification and a frontal lobe location, and medulloblastoma and ependymoma, which may show uniform contrast enhancement and a posterior fossa location.

Meningiomas are usually slow-growing homogeneous tumors associated with diffuse punctate calcification and basilar locations. The tumors are extraaxial and tend to have homogeneous contrast enhancement (see Figure 35).

Metastatic tumors are usually aggressive and demonstrate inhomogeneous imaging characteristics and multiplicity, which helps in differentiation (see Figure 36). Contrast enhancement and the presence of hemorrhage are also common. Highly vascular metastases such as nephroma, melanoma, and choriocarcinoma are associated with hemorrhage. Differential considerations should include multiple abscesses, which may also present with similar imaging patterns.

Suprasellar tumors include chromophobic adenomas of the pituitary, usually homogeneous in character and centered in the pituitary. Another common tumor, the craniopharyngioma, contains a large amount of cholesterol and other fatty materials, giving an increased T_1 signal and

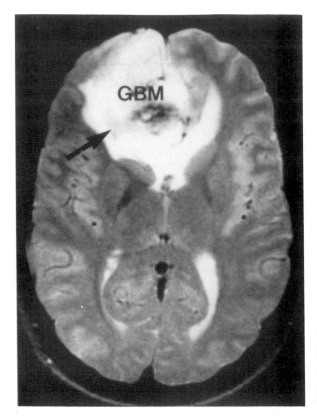

FIGURE 32. T_2-weighted MR image of glioblastoma multiforme (GBM) extending across corpus callosum and compressing the basal ganglia. Inhomogeneity and extensive edema (white) suggest aggressiveness of tumor.

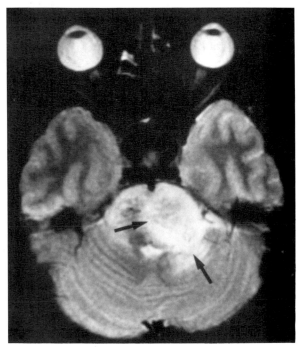

FIGURE 33. T_2-weighted MR image of brainstem glioma (arrows) showing a diffuse infiltrating lesion with mild mass effect.

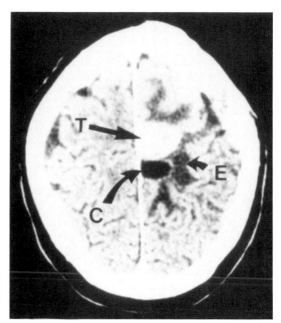

FIGURE 34. Axial CT contrast-enhanced image of grade II glioma in the left vertex showing enhancement (T), associated cyst (C), and edema (E).

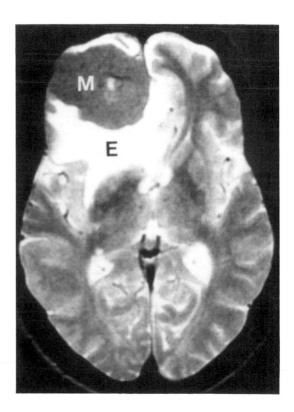

FIGURE 35. T_2-weighted axial MR image of a right frontal meningioma (M) showing relatively homogeneous slightly decreased signal characteristics, mass effect, and edema (E).

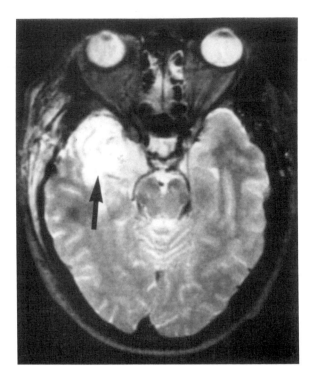

FIGURE 36. Axial T_2-weighted MR image of right temporal metastasis showing similar characteristics to other tumors: increased signal and mass effect (arrow).

decreased attenuation on CT. Calcification is also common in craniopharyngiomas and should be differentiated from calcification associated with aneurysms adjacent to the sella.

Another extraaxial tumor is the acoustic neuroma. This tumor may present with hearing or balance difficulty, as it arises from the seventh or eighth cranial nerve, usually in the internal auditory canal. The tumor is frequently symptomatic when very small and is seen by loss of fluid in the internal auditory canal. Lymphoma has been a relatively rare primary intracranial tumor but is now increasing in frequency as an associated process of immunity disorders. Its presentation is variable with no uniform pattern and so is difficult to describe. There is, however, a predilection for involvement of the basal ganglia and periventricular white matter with a mass producing a usually homogeneous lesion surrounded by edema. Differentiation from other primary brain tumors and infectious processes can be difficult.

Infectious Diseases

Brain abscesses usually present as a multifocal cystic pattern of decreased tissue (necrosis) density surrounded by enhancement. On CT this appears as a decrease in attenuation surrounded by enhancement; on MR the necrotic portion has a decreased signal on T_1 and an increased signal on T_2. As with tumors, a mass effect may be present, but usually to a mild degree (see Figure 37). Also frequently noted are pointing and thinning of the wall of the abscess closest to the ventricle surface. Multiple etiologies for abscesses exist and are usually interrelated by hematogenous seeding of infectious particles or by a form of direct extension.

AIDS may present as multiple abscesses. However, no assurance of a single infectious process can be made. It may also present as lymphoma or a generalized viremia with atrophy (see

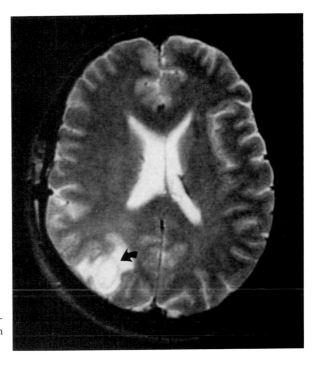

FIGURE 37. T_2-weighted axial MR image of right parietal abscess (arrow) with ring lesion and associated edema.

Figure 38). Other infectious agents such as parasites, fungi, and tuberculosis produce a granulomatous response with time. The indolence of these processes causes little mass effect and frequently results in dystrophic calcification as well as the ringlike enhancement from disruption of the blood–brain barrier as seen in other abscesses.

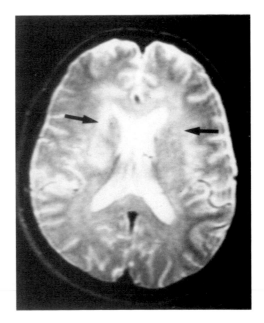

FIGURE 38. T_2-weighted axial MR image showing diffuse patchy periventricular viral leukoencephalopathy in AIDS patients.

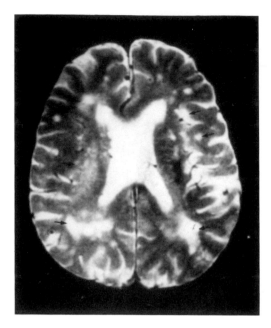

FIGURE 39. T_2-weighted axial MR image of multiple sclerosis showing multiple periventricular focal plaques.

Cerebritis, diffuse infection, is usually seen as a poorly defined area of decreased attentuation on CT and increased signal on MR. Slight mass effect, vascular congestion, and edema are responsible for these changes. In contrast to the diffuse involvement seen in most bacterial and viral infections, herpes simplex has a predilection for early involvement of the subfrontal and medial temporal lobes, and this characteristic may provide differentiation.

Infectious processes may also involve extraaxial spaces and cause epidural or subdural emphyema or meningitis. These processes are usually distinguishable by enhancement with contrast agents. Meningitis is the most difficult to image because of the resolution limitations on seeing meningeal thickening. Similar to epidural and subdural hematomas, empyemas have a concave or biconvex appearance and so are easily distinguished.

FIGURE 40. Templates of computed tomographic cuts with Brodmann's areas marked (in numbers) on left (see/compare with Fig. 41) and vascular territories on right. Branches of anterior cerebral artery: Mesial frontal and parietal regions indicated by open circles, triangles and squares, Ts, Xs, and plus signs, cross-hatched areas of the corpus callosum (CC) and shading of the Heubner's artery (HA) region. Branches of the middle cerebral artery: Central and lateral regions of the frontal, temporal, and parietal regions indicated by closed triangles, crossed circles and squares, stars, verticle scribble, slanted lines, vertical lines, and cross hatched lines in the region of the insular branch (IS) and shaded area of the lenticulostriate (LS). Branches of the posterior cerebral artery: posterior (splenium) of the CC in X, mesial occipital and temporal lobe regions in dotted circles and Xs and the shaded area designated THP (thalamoperforating, thalamogeniculate, and posterior choroidal). Al indicates anterior internal frontal; Ml, middle internal frontal; Pl, posterior internal frontal; PAC, paracentral; IP, superior and interior internal parietal; HA, Heubner's artery; CC, arteries of corpus callosum (small branches of either pericallosal artery or medial branch of posterior cerebral artery); PF, prefrontal; PC, precentral; CS, central sulcus; PP, anterior and posterior parietal; AN, angular; AT, anterior temporal; MT, middle temporal; PT, posterior temporal; TO, temporo-occipital; LS, lenticulostriate; IB, insular branches (small branches of major branches of middle cerebral artery); LB, lateral branch (includes temporal branches); MB, medial branch (includes calcarine and parieto-occipital arteries); THP, thalamoperforating, thalamogeniculate, and posterior choroidal; and ACH, anterior choroidal and additional small branches of internal carotid artery. (With permission from Damasio, 1983.)

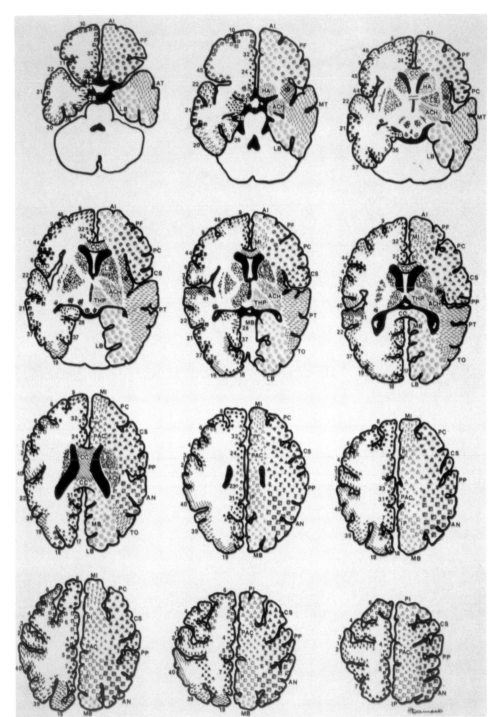

White-Matter Disease

Most white-matter diseases have inflammatory characteristics similar to infectious diseases. Two types of processes exist, myelinoclastic and dysmyelination. The most prevalent disease is multiple sclerosis, of unknown etiology, which presents as multiple focal white-matter lesions in a periventricular distribution (see Figure 39). Acute lesions show enhancement. Both CT and MR demonstrate characteristics of decreased tissue density in chronic lesions. Multiplicity and asym-

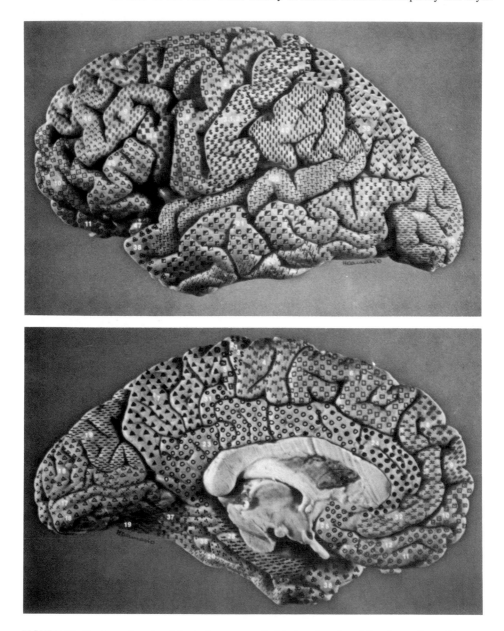

FIGURE 41. Brodmann's cytoarchitectonic areas, marked in lateral and mesial surfaces of actual human brain. Coded areas correspond to those in Fig 40. (With permission from Damasio, 1983.)

metry help differentiate multiple sclerosis from other white-matter diseases such as progressive multifocal leukoencephalopathy, which occurs in immunosuppressed patients, and congenital diseases such as metachromatic leukodystrophy, which results from an enzyme difficiency.

CONCLUSION

Neuroimaging's impact in the few years since its inception has been considerable. Its primary use has been the detection of disease through its ability to delineate normal anatomy. Its true potential, however, lies in its use to understand the function of the brain and its metabolism. To that end, it is first necessary to have an intimate understanding of that anatomy.

APPENDIX

Figures 40 and 41 (from Damasio, 1983) depict the major cerebral vascular territories present on the horizontal views in CT and MRI. The nomenclature for branches of the major cerebral arteries is based on the comparative works of Lazorthes, Gouaze, and Salamon (1976), Salamon and Huang (1976), Waddington (1974), and Wilson (1972). Gross anatomic structures were identified in accordance with the atlases of Matsui and Hirano (1978) and Palacios, Fine, and Henghton (1980).

REFERENCES

Adams, J. H. (Ed.) (1984). *Greenfield's neuropathology* (4th ed.). New York: Wiley Medical Publications.

Adams, R. D., & Sidman, F. L. (1968). *Introduction to neuropathology*. New York: McGraw-Hill.

Brant-Zawadzki, M., & Norman, D. (Eds.) (1987). *Magnetic resonance imaging of the central nervous system*. New York: Raven Press.

Carpenter, M. B. (1976). *Human neuroanatomy* (7th ed.). Baltimore: Williams & Wilkins.

Carpenter, M. B. (1985). *Core text of neuroanatomy* (3rd ed.). Baltimore: Williams & Wilkins.

Carpenter, M. B., & Sutin, S. (1983). *Human neuroanatomy* (8th ed.) Baltimore: Williams & Wilkins.

Christensen, E. E., Curry, T. S., & Dowdey, J. E. (1978). *An introduction to the physics of diagnostic radiology*. Philadelphia: Lea & Febiger.

Cowan, W. M. (1979). The development of the brain. *Scientific American, 3*, 112–133.

Damasio, H. (1983). A computed tomographic guide to the identification of cerebral vascular territories. *Archives of Neurology, 40*, 138–142.

Dandy, W. E. (1918). Ventriculography following the injection of air into the cerebral ventricles. *Annals of Surgery, 18*, 5.

Dandy, W. E. (1919). Roentgenography of the brain after injection of air into the spinal canal. *Annals of Surgery, 70*, 347.

Goss, C. M. (Ed.) (1973). *Gray's anatomy* (29th American ed.). Philadelphia: Lea & Febiger.

Jacobson, M. (1978). *Developmental neurobiology* (2nd ed.). New York: Plenum Press.

Lazorthes, G., Gouaze, A., & Salamon, G. (1976). *Vascularisation et circulation de l'encephale*. Paris: Masson.

Lee, S. H., & Rao, K. C. V. G. (1983). *Cranial computed tomography*. New York: McGraw-Hill.

Luckett, W. H. (1913). Air in the ventricles of the brain, following a fracture of the skull. *Surgery, Gynecology & Obstetrics, 17*, 237–240.

Matsui & Hirano, A. (1978). *An atlas of the human brain for computerized tomography*. Tokyo: Igaku-Shoin Ltd.

Moniz, E. (1927). L'encephalographie arterille. Son importance dans la localisation des tumeurs cérébrales. *Review of Neurology (Paris)*, 72–90.

Nauta, W. J. H., & Feirtag, M. (1986). *Fundamental Neuroanatomy* New York: W. H. Freeman & Co.

Newton, T. H., & Potts, D. G. (1983). *Advanced imaging techniques*. San Anselmo: Clavadel Press.

Nieuwenhuys, Vougd, & Van Huijzen (1979). *The human central nervous sytem.* New York: Springer-Verlag.

Oldendorf, W. H. (1980). *The quest for an image of the brain.* New York: Raven Press.

Palacios, E., Fine, M., & Henghton, V. (1980). *Multiplanar anatomy of the head and neck for computed tomography.* New York: John Wiley & Sons.

Pykett, I. L., Newhouse, J. H., Buonanno, F. S., Brady, T. J. Goldman, M. R., Kistler, J. P., & Pohost, G. M. (1982). Principles of nuclear magnetic resonance imaging. *Radiology, 143,* 157–168.

Pykett, I. L., Buonanno, F. S., Brady, T. J., & Kistler, J. P. (1983). Techniques and approaches to proton NMR imaging of the head. *Computerized Radiology, 7,* 1–17.

Ramsey, R. G. (1987). *Neuroradiology.* Philadelphia: W. B. Saunders.

Salamon, G., & Huang, Y. P. (1976). *Radiologic anatomy of the brain.* Berlin: Springer-Verlag.

Waddington, M. W. (1974). *Atlas of cerebral angiography with anatomic correlation.* Boston: Little, Brown, & Co.

Williams, (1975). *Functional neuroanatomy.* Philadelphia: W. B. Saunders.

Wilson, M. (1972). *The anatomic foundation of neuroradiology of the brain.* Boston: Little, Brown, & Co.

3

Techniques of Quantitative Measurement of Morphological Structures of the Central Nervous System

ERIC TURKHEIMER

INTRODUCTION

The title of this chapter is borrowed from a well-known chapter written by Samuil M. Blinkov and Il'ya Glezer in 1968. Perusal of Blinkov and Glezer's work is an excellent means of appreciating the changes in human neuropsychology that have been introduced by the wide availability of high-quality *in vivo* images of the brain. The techniques Blinkov and Glezer described involved calculation of measurements of brain morphology from serial brain sections or casts of brain structures. Measures based on radiographic techniques receive scant attention; CT and MRI, of course, are not mentioned at all.

Today, *in vivo* imaging procedures are the most frequently used means of studying the relationship between brain morphometry and behavior. Many types of brain images are used in neuropsychological research. This chapter discusses those, primarily CT but recently MRI as well, that produce literal images of brain structures. Other techniques, such as positron emission tomography and regional cerebral blood flow, produce a map of brain function but not straightforward images of structures.

Paradoxically, many of the same qualities that make CT and MRI so immediately useful for clinical purposes introduce complications to their use as research data. Whether the internal digital representation of the image or the resulting film image is employed, the most immediate information provided by an image is analogue: lesions can be visualized in space, and the outlines of brain structures can be detected. This information is of great importance to the clinician. The researcher requires knowledge of the exact size, extent, and location of structures, and this more precise information is sometimes difficult to extract from the image.

An example may help to clarify the point. Suppose a neuroradiologist is shown a CT image and is asked to rate it for "cortical atrophy" on a scale from 1 to 5. Although the rating may be wanting for reliability and validity, it is at least numerical and may be entered for better or worse into statistical calculations. On the other hand, suppose a researcher endeavors to measure "cor-

---

ERIC TURKHEIMER • Department of Psychology, University of Virginia, Charlottesville, Virginia 22903.

tical atrophy'' directly. How does one go about doing this? The clinician's impression of the atrophied brain is visual, loosely based on the degree of widening of the cortical sulci and fissures. Should the researcher, therefore, measure the width of the widest sulci, the collective volume of all of them, the surface area of the cortex, or the average brain density in a region including the cortex? All have advantages, and only empirical investigation can reveal which is most useful.

We should not imagine, therefore, that the wealth of data provided by brain images frees the neuropsychologist from considerations of measurement, research design, and statistical analysis that are so common elsewhere is psychology. Indeed, the subtlety of brain image data makes such considerations all the more crucial.

MEASURING THE VOLUME OF BRAIN STRUCTURES

The first quantitative techniques to be developed for CT involved computation of the size of various brain structures. Many of the earliest algorithms for computing size were developed for the measurement of other bodily organs such as the liver and eyes (Cooper, 1985). Psychiatry, neurology, and psychology, however, first turned their attention to the cerebral ventricles, the most immediately apparent landmark in CT of the head. Indeed, an entire research area concerning the relationship of ventricle size to psychopathology was made possible by the development of these methods.

The earliest empirical studies of ventricular size used clinical ratings of "ventricular atrophy" (e.g., Roberts, Caird, Grossart, & Steven, 1976). These are not described in detail here. Other early techniques involved application of traditional linear measures such as the Evans ratio (Evans, 1942), the distance between the tips of the frontal horns divided by the width of the skull measured between its inner tables. These measurements were taken manually, using a ruler, from film images.

Obviously, linear measures of ventricle size on a single slice are far from perfectly valid indicators of the overall volume of a structure so irregularly shaped. Soon, therefore, investigators began to measure the area of slices of ventricles using a mechanical device known as a planimeter, which computes the area of an irregular closed curve (e.g., Synek & Reuben, 1976). A great many studies of ventricular size employed mechanical planimetric measures.

An alternate means of computing the area of the ventricles on a single slice is called digital planimetry (Turkheimer, Yeo, & Bigler, 1983). This technique involves converting the perimeter of the ventricles to a series of (x,y) coordinates using a common computer peripheral called a digitizing tablet. Given a set of k points (x_i, y_i) defining the perimeter of a closed curve (i.e., the set begins and ends with the same point), the area of the curves is given by

$$\frac{1}{2} \sum_{i=1}^{k} (x_c - x_{c+1})(y_c + y_{c+1}),$$

which may easily be calculated on a microcomputer.

An area of the ventricles at a single slice contains more information than a linear measure but is nonetheless an incomplete estimate of total ventricle volume. Given areas S_n of consecutive slices of the ventricles, the volume between any two may be calculated as $\frac{1}{2}d(S_1 + S_2)$, where d is the distance between the slices. If the structure is assumed to come to a point (i.e., with zero area) at the slices immediately above and below the highest and lowest slices in which it is visible, the volume of the structure equals

$$d \sum_{i=1}^{n} S_i$$

It should be noted that in most cases, differences in the axial dimension (inferior–superior or across horizontal slices) do not contribute to variance in ventricle volume between subjects. In most research applications, the number of slices and their thickness are fixed across subjects: subjects with larger ventricles do not have more or thicker slices. The relationship between planimetric and volumetric measures, therefore, is not the ratio of a quadratic function to a cubic, as would be expected on the basis of geometry if there were variation in all three dimensions; instead, the relationship is more like that between a one-item and a several-item test. Each slice, considered as an imperfect measure of ventricle volume, is like an item on a "volume scale." Summing these reduces the ratio of error variance to total variance, thus increasing the reliability (and potentially the validity) of the scale.

VENTRICULAR ATROPHY USING INTERNAL REPRESENTATION

All of the techniques discussed so far employ film images as their basis. These images, the most familiar products of brain imaging, are not the most immediate products of imaging processes. These processes are not light-based, and they bear little relation to photographic processes, which their products superficially resemble. In fact, brain images are a matrix of beam attenuations that index the density of the brain. These attenuation values are transformed to a gray scale in order to produce an image. The images are thus essentially digital.

Film-based techniques usually involve extensive manipulation of the image. Typically, structures of interest are first traced from a film image placed on a light box. The tracings are then measured using mechanical or digital planimetry as described above, and the digitized images are employed as the basis for calculations of area and volume. The techniques therefore involve converting the image from digital to analogue (the tracing) and back to digital again. This necessarily results in a loss of information and is, in addition, a time-consuming and laborious task.

Another class of quantification techniques involves direct manipulation of the digital representation of the image, offering advantages in speed and accuracy. Computational complications, however, are increased, mostly because of the necessity of developing computational means for outlining structures. Although their reliability may be questioned, human judges are quite facile at outlining a region of interest such as that produced by the ventricles; computers, in contrast, are eminently reliable but difficult to program to recognize shapes.

The most straightforward automated technique involves establishing a cutoff absorption value to divide an image into brain tissue and CSF. Because of differences between scanners and "machine drift" resulting in differences across time in a single scanner, the cutoff score cannot be established absolutely but must be estimated within each sample of scans (Williams, Bydder, & Kreel, 1980). The ambiguity involved in establishing an appropriate cutoff may be appreciated by considering the optimal values arrived at by different researchers: Brassow and Bauman (1978), Penn, Belanger, and Yasnoff (1978), Hacker and Artman (1978), Ito, Hatazawa, Yamaura, and Matsuzawa (1981), and Reveley (1985) used cutoffs of 12, 20, 10, 18, and 15, respectively. Jernigan, Zatz, and Naeser (1979) allowed for differences in average attenuation values between subjects and between slices within subjects by computing an average brain tissue attenuation value for each slice and considering all samples of four pixels with attenuation values significantly lower by t test to represent CSF. Pfefferbaum, Zatz, and Jernigan (1986) describe an interactive computer technique that allows an operator to find the optimal cutoff between brain tissue and CSF values on each slice for each subject.

A shortcoming of manual, film-based techniques and automated techniques using a single cutoff score is that they fail to account for partial voluming at the perimeter of structures. Partial voluming results from pixels at the perimeters of structures that comprise both brain tissue and CSF. The attenuation values for these pixels are between those of pure tissue and CSF, depending on the quantity of each in the pixel. This results in a blurring of the boundaries of structures.

The principal technique for accounting for partial voluming, called partial volume analysis (PVA), was first reported by Walser and Ackerman (1976). For purposes of this technique, it is assumed that a region of a brain image contains only two substances: brain tissue and CSF. If P_c denotes the proportion of the sample that is CSF, $1 - P_c$ represents the proportion that is brain tissue, and D_c and D_b represent the average densities of CSF and brain tissue, respectively, then the average density of the sample is given by $D_a = P_c D_c + (1 - P_c) D_b$, which may then be solved for P_c, yielding $(D_a - D_b)/(D_c - D_b)$. This provides an estimate of the proportion of the sample representing CSF that takes partial voluming into account.

Pentlow, Rottenberg, and Deck (1978) and Rottenberg, Pentlow, Deck, and Allen (1978) describe several methods of obtaining estimates of D_c and D_b using region-of-interest algorithms. Jernigan et al. (1979) assume a constant difference of 13 units between D_c and D_b, so only D_b need be estimated from the scan.

The most recent techniques combine automated measurement algorithms with programs that conduct the analyses without any intervention by an operator. Baldy et al. (1986) report a program that uses partial volume analysis and expert knowledge of where CSF filled spaces are likely to occur on each slice. Both Pfefferbaum et al. (1986) and Baldy et al. (1986) include several methods of image enhancement in the measurement program.

THE VENTRICLE–BRAIN RATIO

The most common outcome of planimetric or volumetric measurements of ventricle size is a ventricle–brain ratio (VBR), which is a ratio of ventricle size to total cranial size, using either volumes or areas. Use of VBR was first reported by Synek and Reuben (1976). It is intended to correct ventricle sizes for differences resulting from overall head size. Zatz and Jernigan (1983) reported that 44% of the variance in ventricle volume was explained by the linear and quadratic effects of cranial volume. Zatz, Jernigan, and Ahumada (1982a) recommend removing the effects of cranial volume by partial correlation instead of dividing it out in a ratio.

COMPARISONS AMONG THE METHODS

Progress from linear measures taken from film images with a ruler to automated volumetric measures taken from magnetic tape entails increases in both accuracy and complexity. Linear measures may be taken by anyone with access to film images and a light box; mechanical and digital planimetric methods require film images and fairly simple equipment; methods based on internal digital representations require access to the imaging equipment itself and sophisticated programs to analyze the data. It is important to investigate the degree of sophistication of measurement that is required to obtain a suitable degree of accuracy.

It should be noted at the outset that not all means of measuring volume are actually measuring the same thing. Manual planimetric techniques measure the area of the perimeter of structures as they are visible on an image. Automated techniques, particularly those using partial volume analysis, measure the total amount of CSF within a specified region of interest. This may include a particular structure of interest such as the ventricles in addition to small cisterns, low-density areas of brain tissue, and scan artifacts that would be omitted by a clinician (Jernigan et al., 1979).

Comparisons of measurement methods can be addressed in several ways. With human subjects, the "true" volume of structures is unknown, so the validity of measures must be estimated from their reliabilities or from their validities relative to previously established techniques. Reliabilities of techniques are typically quite high. Even clinical ratings of ventricular atrophy have been shown to have interrater reliabilities of greater than 0.90 (Pfefferbaum et al., 1986). Automated techniques have even greater reliabilities, often approaching unity (Yeo, Turkheimer, & Bigler, 1983; Pfefferbaum et al., 1986; MacInnes, Franzen, Wray, Mahoney, & McGill, 1987).

Another approach is to use casts of known volume as a means of establishing validity. Rottenberg et al. (1978) showed that techniques using partial volume analysis produced only 3% error, whereas analyses using absorption thresholds between brain tissue and CSF produced 14% error. These errors were systematic: all measures were lower than actual volume and were correlated with each other at $r>0.99$. Techniques using partial volume analysis are closer to true values because they include partially volumed pixels at the perimeter of structures that are omitted by threshold techniques. Systematic differences of this type, however, would probably be unimportant in most research applications.

Still another approach is to examine correlations among different methods of measuring size. Correlations among different methods are usually between 0.80 and 0.90, consistent with the reliabilities of about 0.90 discussed above. Several studies, including Penn et al. (1978), Reveley (1985), Turkheimer, Yeo, and Bigler (1983), and MacInnes et al. (1987), have shown linear measures to have lower correlations with other measures, but Penn et al. (1978) show that these relationships are substantially increased if rank-order correlations are used. Inspection of Penn et al.'s scatterplots suggests that linear measures are systematically, but not linearly, related to planimetric and volumetric measures. They should therefore be squared before they are used in research applications.

A final means of assessing the validity of measurement techniques is to compare the measures' sensitivity to brain–behavior relationships. Reveley (1985), for example, compared the sensitivity of volumetric, planimetric, and linear measures of ventricle volume to differences in ventricle volume between schizophrenics and controls; all measures detected differences. Reveley also examined the intraclass correlations of the measures across a sample of monozygotic twin pairs. In this analysis, automated volumetric and planimetric techniques performed better ($r>0.75$) than mechanical planimetry or Evans' ratio ($r=0.73$ and 0.58, respectively). As discussed above, Evans' ratio might have done better had it been squared. Raz et al. (1988) showed that a volumetric reanalysis of scans previously analyzed planimetrically by Boronow et al. revealed differences between schizophrenics and controls that were not detected in the original analysis.

In summary, the various methods of measuring size of structures from brain images show considerable consistency. Although automated volumetric measures that take partial volume effects into account are no doubt to be preferred on the basis of speed and elegance, manual planimetric measures appear to provide excellent approximations of volumes for most purposes. Even linear measures (squared) and clinical ratings correlate quite highly with more sophisticated measurement techniques.

HEMISPHERIC VOLUME AND ASYMMETRY

Considering the widespread dissatisfaction with linear measurements of the cerebral ventricles, it is surprising that most studies of cerebral asymmetry continue to use linear measures. An additional problem of linear measures is made apparent in their application to cerebral asymmetry: researchers do not agree about where linear measurements are best applied. The width of an irregularly shaped structure like the frontal lobe can be taken at many different places, and different researchers have arrived at different techniques, introducing needless variance into the research literature.

Linear measures of cerebral asymmetry were first reported by Lemay (Lemay, 1976, 1977; Lemay & Kido, 1978), who established the convention of assessing "frontal" and "occipital" asymmetry on the basis of the widths and lengths of anterior and posterior sections of the brain. Lemay measured length by centering an overlay of concentric circles on the midpoint of the interhemispheric fissure: if one hemisphere had greater extension anteriorly or posteriorly, asymmetry (or "petalia") was considered to be present. Width was measured by the length of lines extending from the interhemispheric fissure at a point approximately 5 mm from the ends of the hemispheres (LeMay, 1976).

LeMay's basic technique was adapted by a number of researchers studying cerebral asymmetry in various kinds of psychopathology. The wide variance in methodology in these studies has been well summarized by Tsai, Nasrallah, and Jacoby (1983). Studies varied in how the midline was defined when it was not clearly visible; where on the midline width measurements were taken; and whether measurements were taken directly from the CT films or from magnified projections of them. In addition, Tsai *et al.* (1983) point out that each study selected only one location for the measurement, which is necessarily a poor estimate of total asymmetry. They recommend using measurements at two points along the hemispheric fissure on each of two consecutive CT slices. When they applied their technique and three variations (Luchins, Weinberger, & Wyatt, 1979; Andreason, Dennert, & Olsen, 1982; Jernigan, Zatz, Moses, & Cardellino, 1982) of the LeMay method to a sample of schizophrenic and manic patients, they found considerable variability in the assessment of asymmetry across techniques, despite reliabilities of greater than 0.90 for almost all of them.

All of these studies also continued to use a statistical convention introduced by LeMay in which asymmetry differences between groups are assessed by differences in frequencies of patients showing left and right asymmetries. Since the variable of asymmetry is essentially continuous, considerable information is lost by reducing varying degrees of asymmetry to a trichotomy among left, right, and no asymmetry. Furthermore, the trichotomization leads to unresolvable debate about the criterion for presence of asymmetry: Lemay established a criterion of at least 1 mm difference, but other researchers have advocated only considering statistically significant hemispheric differences (Tsai, Nasrallah, and Jacoby, 1984). Using continuous scores avoids the controversy. The most attractive alternative is to use a ratio of the difference between the hemispheres (whatever the measure used) to their sum. This appears to have been suggested first by Chui and Damasio (1980), although they did not use it in their statistical analyses.

Planimetric or volumetric measures of cerebral asymmetry completely solve the problems with single linear measurements that were partly ameliorated by the multiple linear measures of Tsai and colleagues. Three such studies have been reported. Luchins and Meltzer (1983) computed planimetric equivalents of linear methods described by Andreason *et al.* (1982). Yeo, Turkheimer, Raz, and Bigler (1987) computed separate volumes for left and right frontal and occipital regions using digital planimetry. One problem that remains with these techniques is the definition of ''frontal'' and ''occipital.'' Yeo *et al.* simply divided the brain into quadrants at the hemispheric fissure and a line roughly bisecting it; most other researchers have defined frontal and occipital regions to include only tissue close to the anterior and posterior extensions of the brain. The problem is not strictly resolvable, as the anatomic boundaries between the lobes are not easily visualized on CT.

MEAN BRAIN DENSITY NUMBERS

Another quantification technique that has been widely employed involves comparing measured brain density values either between or within (e.g., between hemispheres) patients. This technique is problematic from the outset because CT brain densities are measured on a somewhat arbitrary scale: they are a transformation of log attenuation values, the transformation varying with the window level and width. An advantage of mean density measurements, however, is that they do not involve definition of perimeters of structures, with all the problems that have already been discussed. In addition, they may be sensitive to comprehensive changes in brain composition that are not reflected in volumes of structures.

Walser and Ackerman (1976) published the first report of a system for analyzing attenuation values. Their system superimposed a circular region of interest on a scan image and reported the number of pixels in the region, the mean and standard deviation of the attenuation values, and a measure of ''roughness'' computed as the ''average sum of the absolute values of the differences of all adjacent points'' in the region.

Investigations using mean density values have produced diverse and somewhat perplexing results. Decreased brain densities are usually associated with normal aging (Zatz, Jernigan, & Ahumada, 1982b) and most forms of psychopathology (Golden et al., 1981), although equal (Coffman & Nasrallah, 1985) or greater (Dewan et al., 1983) densities are sometimes reported in psychopathological groups.

Much of this variability is probably methodological in origin. First of all, there is wide variability in how regions of interest for mean density measures are located. They are frequently based on anatomic markers not clearly visualizable on CT, and interrater reliabilities are not reported routinely. Equally important are sources of error based on the physical properties of the scanner itself (Williams et al., 1980; Jacobson, Turner, Baldy, & Lishman, 1985). As listed by Jacobson et al., these include the familiar partial volume effect, visually detectable artifacts such as streaking, and, most importantly, scanner drift, which is change in the mean level of attenuation values in a machine over time. Jacobson et al. analyzed mean densities of over 300 scans. Using careful procedures designed to maximize reliability, they obtained an interrater reliability of 0.9. They found substantial variability resulting from skull thickness (females, with thinner skulls, appeared to have denser brains), scanner drift, changes in head position, and the size of the region of interest used to determine the mean. They concluded that each of these artifacts was substantial enough to account for most of the substantive findings reported in the literature.

MEASUREMENT OF CORTICAL ATROPHY

The term "cortical atrophy" refers to widening of the sulci at the perimeter of the cortex. On CT, it is visualizable as an increase in the convoluted appearance of the perimeter of the brain. As discussed at the beginning of this chapter, measurement of cortical atrophy has been particularly problematic because it does not involve a unitary brain structure such as the ventricles.

Early methods of measuring cortical atrophy from CT involved ratings by neuroradiologists (Roberts et al., 1976; DeLeon, George, Ferris, Reisberg, & Kircheff, 1981; Huckman, Fox, & Topel, 1975). The first quantitative method that was developed involved computing the sum of the widths of the four largest sulci on a particular slice (Gonzalez, Lantieri, & Nathan, 1978; Brinkman, Sarwar, Levin, & Morris, 1981). This may be considered as a very rough estimate of the total volume of widened sulci across slices. It suffers from many of the same problems of arbitrariness and low reliability that beset other measures (Jacoby, Levy, & Dawson, 1980).

Planimetric and volumetric estimates of the total volume of sulci in the cortex have been calculated from both film images (Bigler, Hubler, Cullum, & Turkheimer, 1985) and internal digital representations (Jernigan et al., 1979). Calculations based on film images offer the advantage of flexibility in how the sulcal borders are determined and also permit calculation of different parameters, such as cortical surface area, that have been shown to correlate more highly with clinical ratings of atrophy than do volumetric measurements (Turkheimer et al., 1984). The increase in flexibility comes with a cost in reliability, however: reliabilities of sulcal measurements are around 0.80 (Turkheimer et al., 1984), compared to reliabilities of over 0.9 for volumes of larger structures such as ventricles. It also appears that although surface area measurements are better predictors of clinical ratings, volumetric measures are better predictors of behavioral change resulting from atrophy (Bigler et al., 1985).

Automated measurements based on digital representations, on the other hand, offer advantages in speed and reliability but provide estimates of the total CSF volume in the perimeter of the cortex rather than of sulcal volume per se. They also permit corrections for bone artifacts (Pfefferbaum et al., 1986; Baldy et al., 1986). Although specific comparisons among the various methods have not been undertaken, there are several reasons to believe that automated methods may be especially superior for measuring cortical atrophy. Sulci are small structures, so the proportion of their total volume affected by partial voluming is higher than for larger structures. The degree of sulcal widening that is visible is very dependent on scanner settings and is difficult to resolve into an area to be measured by planimetry.

MEASUREMENT OF LESION LOCATION

All of the studies discussed so far have used brain images in an attempt to find anatomic correlates of functionally defined syndromes such as schizophrenia and dementia. This is not the only image-based research design that is possible, however. An alternative approach is to select patients with demonstrable brain lesions and to study covariation between functional deficit and lesion size and location. In fact, Chapter 1 of this volume emphasizes the potential of brain images for recreating a classic design from animal neuroscience in which surgical brain lesions are used to investigate brain–behavior relationships.

Unfortunately, the analogy between the study of surgical lesions in animals and natural lesions in humans is not a simple one. Animal neuroscientists can control the location and size of lesions, but researchers using human subjects study lesions of uncontrollable size with irregular distribution across different areas of the brain.

Nevertheless, there is good justification to consider reversing the normal sequence of quantitative image-based research. Studies searching for anatomic correlates of known behavioral syndromes rely for success on the existence of fairly circumscribed structural defects that are visible on CT or MRI, such as enlarged ventricles in schizophrenia; the lack of consensus in much of this research suggests that anatomic correlates of psychopathology and dementia may be more subtle than had been hoped. Studies examining localization of function in patients with naturally occurring lesions, on the other hand, rely on functional differences among regions of the brain, which, though controversial in magnitude, are known to exist.

The major obstacle to progress in this area is the difficulty of quantifying lesion location. Naturally occurring lesions are not restricted to unitary brain structures, so some means of representing their location must be developed. Methods of measuring lesion location can be divided into qualitative and quantitative approaches. Qualitative approaches, which are the most common, represent location either by reference to visible brain structure or by simply outlining the shape of a lesion on a CT slice. Quantitative approaches involve numerical mesures of location that can be employed in statistical analyses.

Qualitative Approaches

The most straightforward means of representing lesion location is to classify lesions relative to large anatomic features of the brain. The most common approach of this type is to compare lesions in the left and right cerebral hemispheres. The major difficulty with this approach is that it ignores considerable variation of location within the larger structures.

A more exact alternative is to measure location in terms of fine anatomic features of the brain. Naeser and her colleagues, for example, have developed a detailed system for representing damage to several cortical and subcortical areas involved in language functions and aphasia (summarized in Naeser, 1983). The CT slices are coded on the basis of anatomic landmarks, so similar slices may be compared across subjects. Within the standardized slices, rules have been developed for the identification of brain structures, including the Broca and Wernicke cortical areas and a variety of small subcortical regions. Subjects are also classified on the basis of the presence or absence of several carefully defined aphasic syndromes.

Two strategies are then available for the analysis of brain–behavior relationships. Patients with similar behavioral manifestations can be compared for the presence or absence of damage to brain structures, or patients with similar patterns of damage can be compared for the presence of behavioral manifestations. The chief difficulty with either of these approaches is the absence of a quantitative method for making inferences. Naeser and colleagues, after making detailed observations of lesions and behavior, rely on case-by-case analysis to draw inferences about the importance of different structures. Case-by-case analysis limits the number of subjects that may be usefully employed and necessarily involves some degree of subjective interpretation of the data.

Another qualitative approach to lesion location is not keyed to specific structural features of the brain. Instead, each subject's lesion is outlined (Mazzocchi & Vignolo, 1979; Kertesz, 1983). Composite diagrams of the "overlap" of lesions from patients with similar behavioral deficits are then superimposed on a schematic CT slice. Comparison of lesion overlap in groups of patients with different behavioral deficits can provide an impression of how function is localized but is even more subjective than the Naeser *et al.* method of coding brain structures.

A disadvantage of all qualitative methods of measuring lesion location is that they preclude statistical analysis of brain–behavior relationships, which are complicated by several factors. Most naturally occurring lesions compromise several contiguous small brain structures and affect more than one dimension of neuropsychological function. This means that both the dependent variables (coding damage to anatomic areas) and the independent variables (measuring behavior) are likely to be substantially intercorrelated. These considerations make it likely that the data analysis will be complex; it may often not be possible to reach definitive conclusions, as we demonstrate below. Quantitative approaches to lesion localization have the advantage of permitting localization data to be analyzed with established statistical methods.

Quantitative Approaches

Basso, Luzzatti, and Spinnler (1979) divided the anteroposterior dimension of CT images into 60 equal parts and recorded the furthest anterior and posterior extension of each subjects' lesion. They then produced histograms of the number of patients with lesions at each point on the grid for groups of patients defined in terms of behavioral deficits. Comparison of these histograms gives an impression of the importance of different regions of the anterior–posterior dimension to different functional groups. Although this method provides a somewhat more precise measure of location, at least in one dimension, the analysis of data remains impressionistic.

In a series of studies of mood disorders in stroke patients, Robinson and his colleagues (most recently Starkstein, Robinson, & Price, 1987) have measured the average distance between the anterior border of the lesion and the frontal pole as a percentage of the total length of the anterior–posterior dimension of the brain. This measure is then correlated with behavioral measures of mood for both left and right lesions.

A difficulty of this measure is the complexity of its relationship with lesion size. Large lesions will tend to have greater anterior extension than small ones. In fact, a lesion involving an entire hemisphere would be measured as very anterior (since its greatest anterior extension would be far from the frontal pole) despite involving the entire frontal region of the hemisphere. Robinson *et al.* avoid this difficulty by discarding subjects with lesions that are anterior to 40% of the anterior–posterior dimension and posterior to 60% of the anterior–posterior dimension. This, however, comes at a cost in sample size and eliminates subjects with primarily central lesions who might otherwise contribute to the analysis.

An additional difficulty involves analyzing measures of this type with linear correlations. If a region close to the middle of the anterior–posterior dimension (e.g., the motor strip) is particularly important to a function, linear correlations between function and anterior extension will not reveal it. Sinyor *et al.* (1986), for example, found a curvilinear relationship between the Robinson *et al.* location measure and depression scores among right hemisphere patients. Sinyor *et al.* included the centrally distributed lesions that Robinson *et al.* discard, which may have enhanced their ability to detect the particular importance of the middle of the AP dimension.

Our own group has used several quantitative measures of lesion location. One approach we have experimented with is calculating the centroid of the lesion (i.e., the center of gravity) in terms of an *(x,y,z)* coordinate axis, with the *z* axis running in the inferior–superior dimension across scan slices (Yeo, Turkheimer, & Bigler, 1984). The *(x,y,z)* coordinates of lesion centroids can then be used as three independent variables in the prediction of behavioral deficits. This approach has several advantages over other methods. It quantifies location in three dimensions instead of simply along the anterior–posterior dimension. The centroid of a lesion is the best single measure of overall lesion location. Nevertheless, results using this method have been somewhat disappointing.

Another approach our group has employed involves dividing a CT slice into thirds along the anterior–posterior dimension and computing the percentage of each third that is compromised (Yeo, Melka, & Haaland, 1988). Correlations of these three percentages with function provide a quantitative measure of localization along the anterior–posterior dimension.

A difficulty with this method involves the correlations among the three regional percentages. These correlations will be high in groups with large lesions or lesions close to the borders between the regions, which will both tend to produce lesions extending into more than one region. In one study these correlations have been as high as 0.88 (Yeo, Melka, *et al.,* 1987).

CONCEPTUALIZING LESION LOCATION

Variation in the location of naturally occurring lesions can be conceptualized in three dimensions: locus, size, and shape. By locus, we refer to the qualitative sense of where a lesion is. Locus can be represented either in terms of anatomic markers (e.g., left versus right lesions, parietal lesions, lesions involving Broca's area) or by quantitative measures (e.g., centroids or greatest extent along a dimension).

Lesion size refers to the volume of a lesion or, on a single slice, its area. Most of the methods for measuring the volume of other brain structures that were reviewed above may be applied to lesion size.

Existing methods of measuring lesion location are limited to analysis of lesion location and size, but these two parameters do not provide a complete description of lesion location. The two hypothetical lesions in Figure 1, for example, have identical centroids and volumes but involve

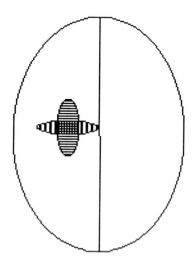

FIGURE 1. Two lesions with the same size and locus but different extents.

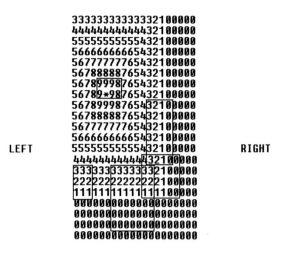

FIGURE 2. Five hypothetical rectangular lesions superimposed on an importance grid.

different tissue and could be expected to show very different behavioral consequences. This suggests that the third dimension required for a complete description of lesion location is shape. None of the lesion localization techniques that have been developed include any consideration of lesion shape. We refer to the combination of lesion locus, size, and shape, comprising a complete description of lesion location, as the extent of a lesion.

A Quantitative Model of Lesion Extent

Consider a CT slice divided into an arbitrarily fine array. Suppose each point in the array is assigned a number representing the importance of the point to some behavioral function, with higher numbers indicating greater importance. An example of an importance array is given in Figure 2. The function illustrated is localized in the left-frontal region.

Suppose that a lesion is represented as a contiguous region of the importance array. Some examples of lesion representations are shown as rectangles in Figure 2. According to the model, the predicted deficit produced by each lesion is equal to the sum of the importance values contained in the lesion. Table 1 shows the predicted deficits of the lesions in Figure 2. The lesion at the lower left, for example, contains importance values of 3, 3, 3, 2, 2, 2, 1, and 1. The predicted deficit for this patient would be 18. According to the model, therefore, larger lesions will produce greater deficits than smaller ones, and lesions in regions with high importance values will produce greater deficits than those in regions with small importance values.

TABLE 1. Observed Deficits for Five Hypothetical Lesions

Lesion	x_1	x_2	y_1	y_2	Observed deficit
1	5	8	7	8	71
2	12	16	9	14	36
3	1	3	15	17	18
4	12	17	14	17	30
5	6	13	15	20	42

Localization Tomography

In a study of brain–behavior relationships using naturally occurring lesions, we are given the extent of a lesion and a behavioral measure of deficit for each patient. The problem is to recover the importance values on the basis of the given data. Interestingly, this problem—estimating the individual values in an array on the basis of a finite number of sums of values in regions of the array—is an application of tomography, the same mathematical technique that is used in the creation of the CT images.

In CT the attenuation of each beam passing through the brain is a function of the sum of the densities of brain tissue along its path. The task is to recover the interior densities from a finite set of these linear sums. In the present problem, we are given sums of regions rather than of paths.

The Quantitative Basis of Double Dissociation

Figure 3 is an illustration of the classic paradigm of double dissociation. Two measures of behavioral deficit (function 1 and function 2) are taken from each of two lesioned patients (patient 1 and patient 2, with lesions 1 and 2 defining regions 1 and 2, respectively), with no overlap between their lesions. Let D_{ij} equal the behavioral deficit for patient i on function j, and I_{ij} equal to the importance of region i for function j. Patient 1 shows a deficit of D_{11} for function 1 and no deficit for function 2; patient 2 shows no deficit for function 1 and a deficit of D_{22} for function 2. The region defined by patient 1's lesion has an importance of I_{11} to function 1 and an importance of I_{12} to function 2; the region defined by patient 2's lesion, conversely, has importance values of I_{21} and I_{22} to the two behavioral measures. We can then state unambiguously that

$$I_{11}=D_{11}; \; I_{12}=0; \; I_{21}=0; \; I_{22}=D_{22}$$

This is simply a way of saying that the region in lesion 1 is uniquely associated with function 1, and the region in lesion 2 is uniquely associated with function 2: the possibility of such an unambiguous conclusion is why double dissociation is such a highly sought after result.

Figure 4, however, is a more realistic representation of studies of naturally occurring lesions. The two patients' lesions have been superimposed on one slice to emphasize that they overlap. In addition, both patients show some deficit on both behavioral measures. Three regions are now defined: one unique to each lesion (denoted 1 and 2, as above) and one defined by their overlap

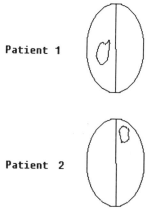

Patient 1

Patient 2

FIGURE 3. Double dissoiation. Patient 1 shows a deficit on function 1; patient 2 shows a deficit on function 2. Lesions do not overlap.

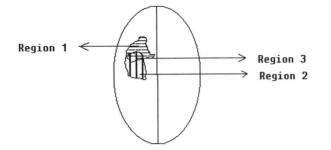

FIGURE 4. Superimposed lesions from two hypothetical patients. Lesions overlap, and patients show deficits on both functions.

(region 3). The deficit of patient 1 on function 1 is determined by the sum of the two regions contained in the lesion; that is,

$$D_{11} = I_{11} + I_{31}$$

Similarly, the deficit of patient 2 on function 1 is determined by the sum of the two regions contained in the lesion:

$$D_{21} = I_{21} + I_{31}$$

Thus, the importance of the three regions can be represented as two equations in three unknowns, which is to say that the solution to the problem is underdetermined: when lesions overlap, an infinite number of importances can be assigned that fit the observed data. This conclusion may explain much of the ambiguity in the results of studies of naturally occurring lesions.

A Polynomial Solution

It is generally true that solutions to tomographic problems based on a finite number of sums will be underdetermined; some additional means of constraining the solution must be adopted (Natterer, 1986). Below, we demonstrate one method of accomplishing this restriction. We simplify the problem in several ways to assist in the presentation of the method. First, we limit the problem to two-dimensional representations of lesions on a single CT slice. Second, we represent the shapes of lesions as rectangles. Considering lesions as complex shapes in three dimensions complicates the solution algebraically but does not alter the basic approach.

Let the importance array be represented as an importance function in two variables, x and y, with x representing the left–right dimension and y the anterior–posterior dimension. Suppose further that the importance function is a third-degree polynomial in x and y; that is,

$$I = a_1 x^3 + a_2 y^3 + a_3 x^2 + a_4 y^2 + a_5 xy + a_6 x + a_7 y + a_8$$

where I is the importance of any point (x,y).

According to the model, the predicted deficit of each patient is equal to the sum of the (x,y) points inside the lesion. The sum of a continuous function within a region A is equal to the integral of the function over the region. Since each lesion is represented as a rectangle from x_1 to x_2 and y_1 to y_2, this predicted deficit for each patient equals

$$\int_{x_1}^{x_2} \int_{y_1}^{y_2} a_1 x^3 + a_2 y^3 + a_3 x^2 + a_4 y^2 + a_5 xy + a_6 x + a_7 y + a_8$$

Evaluating this integral, we get

$$\frac{1}{4} (x_2^4 - x_1^4) (y_2 - y_1) \, a_1 + \frac{1}{4}(y_2^4 - y_1^4) (x_2 -$$
$$x_1) \, a_2 + \frac{1}{3} (x_2^3 - x_1^3) (y_2 - y_1) \, a_3 + \frac{1}{3} (y_2^3$$
$$- y_1^3) (x_2 - x_1) a_4 + \frac{1}{4} (x_2^2 - x_1^2) (y_2^2 - y_1^2)$$
$$a_5 + \frac{1}{2} (x_2^2 - x_1^2) (y_2 - y_1) \, a_6 + \frac{1}{2}(y_2^2 - y_1^2)$$
$$(x_2 - x_1) + (x_2 - x_1) (y_2 - y_1)$$

We now have the predicted deficit for each patient in terms of the original importance function and an observed deficit for each patient from the behavioral measure. We need to find values for the parameters a, b, c, d, and e in the importance function that maximize the fit between the predicted and observed deficits.

A simple way to accomplish this is with ordinary least-squares regression. The coefficients for the parameters of the importance function may be calculated from the observed parameters of the lesion. (Representing lesions as complex polygons instead of as rectangles complicates these calculations, but they remain conceptually straightforward.) The observed deficit variable (the behavioral measure) may be regressed on these coefficients, resulting in values for the original importance function that maximize the fit between the predicted and observed deficits.

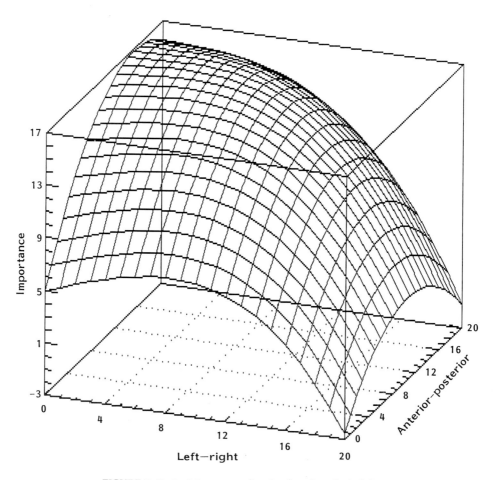

FIGURE 5. Derived importance function from hypothetical data.

An Example with Simulated Data

We had the computer simulate 100 random rectangular "lesions" on the hypothetical importance array illustrated in Figure 2. The extents of these lesions are listed in Table 1, and the first five have been superimposed on the importance array in Figure 2. The "observed deficits" have been calculated as the sum of the importance values within each lesion. The polynomial importance function we fit to these data is continuous; the difference between the continuous solution and the discrete given data will introduce a little error into the analysis.

The given data are the lesion parameters, x_1, x_2, y_1, y_2, and the observed deficits. The task is to recover the original importance array from these data.

The stepwise regression program from the STATGRAPHICS statistical package (STSC, 1986) was used to regress the observed deficits onto the calculated lesion variables. We have found it useful to begin with the single variable representing the size of the lesion and then add variables for the linear, quadratic, and cubic effects of x and y one at a time, stopping when no variables add significant fit to the model.

In the present example, the given importance array can be represented by a cubic polynomial in x and y, that is, as

$$0.002x^3 + 0.052y^2 - 0.421x - 1.569y + 0.023xy - 4.946$$

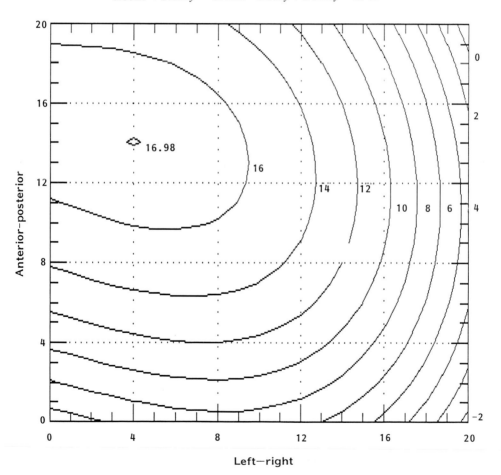

FIGURE 6. Derived importance function from hypothetical data.

The graph of this polynomial is shown in Figure 5. A contour plot of the same graph is given in Figure 6. In this plot, it can be seen that the procedure has correctly recovered the point of highest importance, in the left frontal region. The R^2 between the observed and predicted importance values is 0.98.

CONCLUSION

Methods of quantifying brain images have progressed along with the imaging methods themselves. Techniques permitting research into many aspects of brain–behavior relationships are within the reach of anyone with access to film images of scans. Methods for measuring volumes of large structures such as the cerebral ventricles are particularly accessible. The cerebral hemispheres, in particular, can be easily and accurately measured: techniques developed for measuring ventricles should be applied to studies of asymmetry of the hemispheres. Smaller structures such as sulci present greater methodological problems. Use of raw beam-attenuation values as the dependent variable in research presents problems that probably cannot be solved, at least without intimate knowledge of the imaging hardware and hands-on access to it. Results in research using mean density values have in any case not been as impressive as those in other areas.

Brain imaging technology and the techniques to analyze images have both advanced to the point where technological possibilities no longer limit the possibilities of research. Instead, the greatest need is for theoretically compelling theories of brain–behavior relationships that images can be used to test. Many of the theories that have been tested using brain images, such as ventricular enlargement in schizophrenia and cortical atrophy in dementia, have been developed in response to technological possibilities: theory has been driven by method. This may be putting the methodological cart before the theoretical horse.

REFERENCES

Andreason, N. C., Dennert, J. W., & Olsen, S. A. (1982). Hemispheric asymmetries and schizophrenia. *American Journal of Psychiatry, 139*, 427–430.

Baldy, R. E., Brindley, G. S., Ewusi-Menshah, I., Jacobson, R. R., Reveley, M. A., Turner, S. W., & Lishman, W. A. (1986). A fully-automated computer-assisted method of CT brain scan analysis for the measurement of cerebrospinal fluid spaces and brain absorption density. *Neuroradiology, 28*, 109–117.

Basso, A., Luzzatti, C., & Spinnler, H. (1980). Is ideomotor apraxia the outcome of damage to well defined regions of the left hemisphere? *Journal of Neurology, Neurosurgery, and Psychiatry, 43*, 118–126.

Bigler, E. D., Hubler, D. W., Cullum, C. M., & Turkheimer, E. (1985). Intellectual and memory impairment in dementia: CT volume correlations. *Journal of Nervous and Mental Disease, 173*, 347–352.

Blinkov, S. M., & Glezer, I. I. (1968). *The human brain in figures and tables: A quantitative handbook.* New York: Plenum Press.

Brassow, F., & Bauman, K. (1978). Volume of brain ventricles in man determined by computer tomography. *Neuroradiology, 16*, 187–189.

Brinkman, S. D., Sarwar, M., Levin, H., & Morris, H. H. (1981). Quantitative indexes of computed tomography in dementia and normal aging. *Neuroradiology, 138*, 89–92.

Chui, H. C., & Damasio, A. R. (1980). Human cerebral asymmetries evaluated by computed tomography. *Journal of Neurology, Neurosurgery, and Psychiatry, 43*, 873–878.

Coffman, J. A., & Nasrallah, H. A. (1985). Relationships between brain density, cortical atrophy and ventriculomegaly in schizophrenia and mania. *Acta Psychiatrica Scandinavica, 72*, 126–132.

Cooper, W. C. (1985). A method for volume determination of the orbit and its contents by high resolution axial tomography and quantitative image analysis. *Transactions of the American Ophthalmological Society, 83*, 546–609.

DeLeon, M. J., George, A. E., Ferris, S. H., Reisberg, B., & Kircheff, I. I. (1981). Computer tomographic evaluation of brain–behavior relationships in senile dementia. *International Journal of Neuroscience, 12*, 246–247.

Dewan, M. J., Pandurangi, A. K., Lee, S. H., Ramachandran, T., Levy, B. F., Boucher, M., Yozawitz, A., & Major, L. (1983). Cerebellar morphology in chronic schizophrenic patients: A controlled computed tomographic study. *Psychiatry Research, 10,* 97–103.

Evans, W. A. (1942). An encephalographic ratio for estimating ventricular enlargement and cerebral atrophy. *Archives of Neurology and Psychiatry, 47,* 931–937.

Golden, C. J., Moses, J. A., Zelazowski, M. A., Graber, B., Zatz, L. M. Horvath, T. B., & Berger, P. A. (1981). Cerebral ventricular size and neuropsychological impairment in young chronic schizophrenics. *Archives of General Psychiatry, 37,* 619–623.

Gonzalez, C. F., Lantieri, R. L., & Nathan, R. J. (1978). The CT scan appearance of the brain in a normal elderly population: A correlative study. *Neuroradiology, 16,* 120–122.

Hacker, H., & Artman, H. (1978). The calculation of CSF spaces in CT. *Neuroradiology, 16,* 190–192.

Huckman, M. S., Fox, J., & Topel, J. (1975). The validity of criteria for the evaluation of cerebral atrophy by computed tomography. *Neuroradiology, 116,* 85–92.

Ito, M., Hatazawa, J., Yamaura, H., & Matsuzawa, T. (1981). Age-related brain atrophy and mental deterioration—a study with computed tomography. *British Journal of Radiology, 54,* 384–390.

Jacobson, R. R., Turner, S. W., Baldy, R. E., & Lishman, W. A. (1985). Densitometric analysis of scans: Important sources of artefact. *Psychological Medicine, 15,* 879–889.

Jacoby, R. J., Levy, R., & Dawson, J. M. (1980). Computed tomography in the elderly: I. The normal population. *British Journal of Psychiatry, 136,* 249–255.

Jernigan, T. L., Zatz, L. M., & Naeser, M. A. (1979). Semiautomated methods for quantitating CSF volume on cranial computed tomography. *Radiology, 132,* 463–466.

Jernigan, T. L., Zatz, L. M., Moses, J. A., & Cardellino, J. P. (1982). Computed tomography in schizophrenics and normal volunteers: II. Cranial asymmetry. *Archives of General Psychiatry, 39,* 771–773.

Kertesz, A. (1983). Right-hemisphere lesions in constructional apraxia and visuospatial deficit. In A. Kertesz (Ed.), *Localization in neuropsychology,* pp. 455–470. New York: Academic Press.

Lemay, M. (1976). Morphological cerebral asymmetries of modern man, fossil man, and nonhuman primate. *Annals of the New York Academy of Sciences, 280,* 349–360.

Lemay, M. (1977). Asymmetries of the skull and handedness: Phrenology revisited. *Journal of the Neurological Sciences, 32,* 243–253.

Lemay, M., & Kido, D. K. (1978). Asymmetries of the cerebral hemispheres on computed tomograms. *Journal of Computer Assisted Tomography, 2,* 471–476.

Luchins, D. J., & Meltzer, H. Y. (1983). A blind, controlled study of occipital cerebral asymmetry in schizophrenia. *Psychiatry Research, 10,* 87–95.

Luchins, D. J., Weinberger, D. R., & Wyatt, R. J. (1979). Schizophrenia: Evidence of a subgroup with reversed cerebral asymmetry. *Archives of General Psychiatry, 36,* 1309.

MacInnes, W. D., Franzen, M. D., Wray, A. B., Mahoney, P. D., & McGill, J. E. (1987). Interrelationships between various methods of measuring ventricular–brain ratio. *The International Journal of Clinical Neuropsychology, 9,* 56–61.

Mazzocchi, F., & Vignolo, L. A. (1979). Localisation of lesions in aphasia: Clinical–CT scan correlations in stroke patients. *Cortex, 15,* 627–654.

Naeser, M. (1983). CT lesion size and lesion locus in cortical and subcortical aphasias. In A. Kertesz (Ed.), *Localization in neuropsychology,* pp. 63–119. New York: Academic Press.

Natterer, F. (1986). *The mathematics of computed tomography.* New York: John Wiley & Sons.

Penn, R. D., Belanger, M. G., & Yasnoff, W. A. (1978). Ventricular volume in man computed from CAT scans. *Annals of Neurology, 3,* 216–223.

Pentlow, K. S., Rottenberg, D. A., & Deck, M. D. F. (1978). Partial volume summation: A simple approach to ventricular volume determination from CT. *Neuroradiology, 16,* 130–132.

Pfefferbaum, A., Zatz, L. M., & Jernigan, T. L. (1986). Computer-interactive method for quantifying cerebrospinal fluid and tissue in brain CT scans: Effects of aging. *Journal of Computer Assisted Tomography, 10,* 571–578.

Raz, S., Raz, N., Weinberger, D. R., Boronow, J., Pickar, D., Bigler, E. D., & Turkheimer, E. (1987). Morphological brain abnormalities in schizophrenia determined by computerized tomography: A problem of measurement? *Psychiatry Research, 22,* 91–98.

Reveley, M. A. (1985). Ventricular enlargement in schizophrenia: The validity of computerized tomographic findings. *British Journal of Psychiatry, 147,* 233–240.

Roberts, M. A., Caird, F. I., Grossart, K. W., & Steven, J. L. (1976). Computerized tomography in the diagnosis of cerebral atrophy. *Journal of Neurology, Neurosurgery, and Psychiatry, 39,* 909–915.

Rottenberg, D. A., Pentlow, K. S., Deck, M. D. F., & Allen, J. C. (1978). Determination of ventricular volume following metrizamide CT ventriculography. *Neuroradiology, 16,* 136–139.

Sinyor, D., Jacques, P., Kaloupek, D. G., Becker, R., Goldenberg, M., & Coopersmith, H. (1986). Post-stroke depression and lesion location: An attempted replication. *Brain, 109,* 537–546.

Starkstein, S. E., Robinson, R. G., & Price, T. R. (1987). Comparison of cortical and subcortical lesions in the poststroke mood disorders. *Brain, 110,* 1045–1059.

STSC (1986). *Statgraphics.* Rockville, MD: STSC, Inc.

Synek, V., & Reuben, J. R. (1976). The ventricular–brain ratio using planimetric measurement of EMI scans. *British Journal of Psychiatry, 49,* 233–237.

Tsai, L. Y., Nasrallah, H. A., & Jacoby, C. G. (1983). Hemispheric asymmetries on computed tomographic scans in schizophrenia and mania: A controlled study and critical review. *Archives of General Psychiatry, 40,* 1286–1289.

Tsai, L. Y., Nasrallah, H. A., & Jacoby, C. G. (1984). Cerebral asymmetry in subtypes of schizophrenia. *Journal of Clinical Psychiatry, 45,* 423–425.

Turkheimer, E., Yeo, R. A., & Bigler, E. D. (1983). Digital planimetry in APLSF. *Behavior Research Methods and Instrumentation, 15,* 471–473.

Turkheimer, E., Cullum, C. M., Hubler, D. W., Paver, S. W., Yeo, R. A., & Bigler, E. D. (1984). Quantifying cortical atrophy. *Journal of Neurology, Neurosurgery, and Psychiatry, 47,* 1314–1318.

Walser, R. L., & Ackerman, L. V. (1976). Determination of volume from computerized tomograms: Finding the volume of fluid-filled brain cavities. *Journal of Computer Assisted Tomography 1,* 117–130.

Williams, G., Bydder, G. M., & Kreel, L. (1980). The validity and use of computed tomography attenuation values. *British Medical Bulletin, 36,* 279–287.

Yeo, R. A., Turkheimer, E., & Bigler, E. D. (1983). Computer analysis of lesion volume: Reliability, utility and neuropsychological applications. *Clinical Psychology, 5,* 45.

Yeo, R. A., Turkheimer, E., & Bigler, E. D. (1984). The influence of sex and age on unilateral cerebral lesion sequelae. *International Journal of Neuroscience, 23,* 299–301.

Yeo, R. A., Melka, B. E., & Haaland, K. Y. (1988). *Intra-hemispheric differences in motor control: A re-examination of the Semmes hypothesis.* Paper presented at the meeting of the International Neuropsychological Society, New Orleans.

Yeo, R. A., Turkheimer, E., Raz, N., & Bigler, E. D. (1987). Volumetric asymmetries of the human brain: Intellectual correlates. *Brain and Cognition, 6,* 15–23.

Zatz, L. M., & Jernigan, T. L. (1983). The ventricular–brain ratio on computed tomography scans: Validity and proper use. *Psychiatry Research, 8,* 207–214.

Zatz, L. M., Jernigan, T. L., & Ahumada, A. J. (1982a). White matter changes in cerebral computed tomography related to aging. *Journal of Computer Assisted Tomography, 6,* 19–23.

Zatz, L. M., Jernigan, T. L. & Ahumada, A. J. (1982b). Changes on computed cranial tomography with aging: Intracranial fluid volume. *American Journal of Neuroradiology, 3,* 1–11.

4

Positron Emission Tomography and Neuropsychological Function

G. PAWLIK and W.-D. HEISS

INTRODUCTION

The cerebral localization not only of elementary sensory or motor functions but also of complex mental activities and human behavior has been dominated traditionally by neuroanatomic concepts derived from correlative clinicopathological studies. However, these morphologically biased views culminating in Kleist's largely speculative brain charts (Kleist, 1937) lacked the physiological dimension. In contrast, physiological experiments with intraoperative electrical stimulation of the cerebral cortex, as performed most systematically by Penfield and Rasmussen (1950), provided only sketchy information on the highly complex regional interactions associated with neuropsychological functions. For a while, this dilemma favored by methodological one-sidedness caused a resurgence of the old universalistic notion that localization is an artificial observer-made attribute of the brain (Gooddy & McKissock, 1951). Recent interdisciplinary research, though, has shown much evidence to the contrary, but at a more sophisticated conceptual level of stochastic cybernetics.

Significant progress was made possible by the introduction of radiotracer methods for the measurement of regional cerebral blood flow (CBF) that, in general, is closely related to brain activity. Using a gamma camera or multidetector equipment, these methods, like electrophysiological brain mapping, yield projected views only at relatively poor spatial resolution and at an almost undefined depth of field. However, they provide physiological information on a regional basis, at least on major cerebral cortical areas, and they have contributed substantially to current neuropsychological knowledge (Ingvar & Lassen, 1975).

It was not before x-ray computed tomography (CT) became available that the living brain could be visualized in three dimensions and at acceptable spatial resolution. Although in principle capable of tomographic blood flow imaging (on at most three to five slices, and at poor signal-to-noise ratio) using the local kinetics of radiopaque stable xenon as recorded on sequential scans (Meyer, Haymann, Yamamoto, Sakai, & Nakajima, 1980), because of its physical characteristics CT remains suited primarily for structural imaging. Likewise, the more modern techniques of magnetic resonance imaging (MRI), featuring excellent spatial resolution, high contrast, and minimal biological risk, are being used predominantly for morphological proton imaging.

G. PAWLIK and W.-D. HEISS ● Department of Neurology, Max Planck Institute of Neurological Research, Cologne University Clinics, D-5000 Cologne 41, Federal Republic of Germany.

Closely related to brain blood flow are the images obtained by single photon emission computed tomography (SPECT). However, its steady-state variants using [^{99m}Tc]HM-PAO or ^{123}I-labeled amphetamine yield qualitative results only, and quantitative dynamic SPECT of xenon-133 cannot provide reliable data on deep brain structures because of the low energy of the radionuclide.

Positron emission tomography (PET), by contrast, permits the quantitation and three-dimensional imaging of distinct physiological variables. This advanced nuclear medicine technology uniquely combines tracer kinetic principles with the specific advantages of coincidence counting and computed tomography, resulting in great methodological efficiency and flexibility. Depending on the physiochemical properties of the radiotracer, on the PET procedure, and on the applied biomathematical model, various major aspects of brain function can be investigated at satisfactory spatial and temporal resolution. Therefore, PET appears particularly well suited for studies of the physiological and anatomic basis of neuropsychology.

METHODS

Principles of PET

Certain biologically relevant elements, e.g., carbon, nitrogen, oxygen, and fluorine substituting for a hydrogen atom, have neutron-deficient radioisotopes of favorable lifetime that decay at a sufficiently high percentage by emission of positrons, i.e., of positively charged particles of the mass of an electron. Because of their short half-life of less than 110 min, most of these radionuclides must be produced close to the PET laboratory, using a low-energy particle accelerator with high beam current, e.g., a cyclotron. Commonly, isotope production is followed by a radiochemical multistep procedure for the synthesis of the final physiological tracer. The specific advantage of PET over all other nuclear medicine imaging techniques is founded essentially on the principle of coincidence detection. Following emission from the atomic nucleus, the positron takes a path marked by multiple collisions with ambient electrons. Approximately 1 to 3 mm from its origin, it has lost so much energy that it combines with an electron, resulting in the annihilation of the two oppositely charged particles by the emission at an angle of $180 \pm 0.5°$ of two 511-keV photons that are recorded as coincident events, using pairs of practically uncollimated detectors facing each other. The origin of the photons, therefore, can be localized directly to the straight line between these coincidence detectors. State-of-the-art PET scanners are equipped with thousands of detectors arranged in up to eight rings simultaneously scanning 15 slices of less than 1 cm thickness. Pseudocolor-coded tomographic images of the radioactivity distribution are then reconstructed by computer from the many projected coincidence counts, using CT-like algorithms and reliable scatter and attenuation corrections. Typical in-plane resolution (full width at half-maximum) is 5 mm.

Up to this point, there still is some similarity with conventional nuclear medicine imaging techniques: the radioactivity tomograms represent local tracer concentrations in units of the rate of nuclear decay that do not have much meaning in physiological terms. Only an appropriate biomathematical model describing the compartmental kinetics of the applied tracer makes it possible to transform these data into images of truly biological function. Several models require the collection of data by sequential PET scanning for dynamic analysis.

Brain Energy Metabolism

The biochemical energy storage capacity of the brain is extremely limited, and therefore, under physiological conditions, substrate supply according to demand is nearly instantaneous. Furthermore, cerebral energy delivery depends almost exclusively on aerobic glucose metabolism, with the synapse being the major location for consumption (Sokoloff, 1981).

Glucose Consumption

The cerebral metabolic rate for glucose (CMRglu) can be quantified with PET (Reivich, Kuhl, *et al.*, 1979; Phelps *et al.*, 1979) using 2-[^{18}F]fluoro-2-deoxyglucose (FDG) and a modification of the three-compartment model equation developed for autoradiography by Sokoloff *et al.* (1977). Like glucose, FDG is transported across the blood–brain barrier and further into brain cells, where it is phosphorylated by hexokinase activity. The FDG-6-phosphate, however, cannot be metabolized to the respective fructose-6-phosphate analogue, nor can it diffuse out of the cells in significant amounts. It rather must be dephosphorylated again by a phosphatase that is hidden inside the endoplasmic reticulum. This effect, however, is virtually nondetectable for approximately 45 min (Lucignani *et al.*, 1987), and the distribution of the radioactivity accumulated in the brain remains quite stable between 30 and 50 min after intravenous tracer injection, thus permitting multiple intercalated scans. Using (1) the local radioactivity concentration measured with PET during this steady-state period, (2) the concentration–time course of tracer in arterial plasma, (3) plasma glucose concentration, and (4) a lumped constant correcting for the differing behavior in brain of FDG and glucose, CMRglu can be computed pixel by pixel according to an optimized operational equation (Wienhard, Pawlik, Herholz, Wagner, & Heiss, 1985). The resulting pseudocolor-coded images then reflect all effects on cerebral glucose metabolism, weighted according to the individual, multiexponentially decreasing FDG plasma concentration–time curve. Because of its robustness with regard to procedure and model assumptions, the FDG method has been employed in most PET studies of interest for neuropsychology.

Oxygen Consumption

Various PET methods have been developed for determining the cerebral metabolic rate for oxygen (CMRO$_2$), using continuous (Frackowiak, Lenzi, Jones, & Heather, 1980) or single-breath inhalation (Mintun, Raichle, Martin, & Hersovitch, 1984) of air containing trace amounts of ^{15}O-labeled molecular oxygen. All require the concurrent estimation or paired measurement of CBF in order to convert the measured oxygen extraction fractions, corrected for local blood volume if necessary (Lammertsma *et al.*, 1983), into images of CMRO$_2$ as given by the product of arterial oxygen concentration, local oxygen extraction fraction, and local CBF. Because of the short half-life (123 s) of ^{15}O necessitating an on-site cyclotron, and for other methodological complexities, the use of CMRO$_2$ as a measure of brain function has been limited to very few neuropsychological PET studies.

Brain Blood Flow

Although in principle related primarily to brain vascular physiology rather than to neuronal function, CBF continues to be a target of major interest for neuropsychological research. This tradition comes from the close coupling of brain blood flow and energy metabolism in physiological conditions as well as from the procedural and analytical straightforwardness of CBF measurements. Almost all commonly applied methods for the quantitative imaging of CBF with PET are based on the principle of diffusible tracer exchange. Using ^{15}O-labeled water administered either directly by intravenous bolus injection (Herscovitch, Markham, & Raichle, 1983) or by the inhalation of ^{15}O-labeled carbon dioxide that is converted into water by carbonic anhydrase in the lungs (Frackowiak *et al.*, 1980), CBF can be estimated from steady-state distributions or from the radioactivity concentration–time curves in arterial plasma and brain. Alternatively, methyl fluoride, a freely diffusible, inert, gaseous tracer labeled with fluorine-18 (half-life, 109.7 min) or with carbon-11 (half-life, 20.4 min) can be used for dynamic PET studies (Koeppe *et al.*, 1985). Typical measuring times range between 40 s and 12 min, and because of the short biological half-life of the radiotracers, repeat studies can readily be performed.

PET Image Analysis

Detailed analysis of single PET images is performed best by visual interpretation by an experienced observer using an overlay of matching CT or MRI slices. For statistical assessment, however, comprehensive brain mapping into topographically meaningful regions of interest is well established. Of course, this approach entails some averaging error, but at the benefit of significantly improved anatomic comparability—especially when a similarity matching procedure is used (Pawlik, Herholz, Wienhard, Beil, & Heiss, 1986). Significance probability mapping can help explore single cases, whereas repeated-measures analysis of variance handles multiregional multiple-group designs most efficiently (Pawlik, 1988), treating hemispheres and regions as within factors and subject groups as a between factor.

RESULTS

Normal Functional States

So far, only a comparatively small number of typical neuropsychological functions have been investigated by PET in healthy human subjects. Many more remain to be studied. This relative lack of PET-based knowledge may be explained rather by the tremendous problems posed by the design of a valid, modality-specific psychophysical activation paradigm than by the often overstated high cost of this technology. Matters are complicated further by the fact that most PET methods, because of their temporal integrating characteristics, require a certain physiological state of interest to remain stable for at least several minutes—a condition barely met by repetitive stimulation as commonly applied, e.g., in electrophysiological studies.

Resting Wakefulness

Before tackling more complex activated states, a reproducible base-line condition had to be established that was applicable to normal volunteers and to critically ill patients alike. Therefore, Mazziotta, Phelps, Carson, and Kuhl (1982a) investigated the effect of progressive sensory deprivation on CMRglu in a series of 22 young, healthy right-handers. Global CMRglu showed a stepwise decrease, from the eyes-and-ears-open over the eyes-closed/ears-open to the eyes-open/ears-closed condition and further to the deprivation of both visual and auditory stimuli. This sequence of global effects was paralleled by progressively increasing relative hyperfrontality, i.e., frontal-to-occipital metabolic ratios. Only in the eyes-closed/ears-closed condition was functional asymmetry (left greater than right) observed. To date, partial sensory deprivation, most commonly with eyes closed and ears unoccluded, has been the most widely used standard condition for PET investigations. In a series of 44 normal resting subjects (16 women, 28 men; mean age 44.7 ± 19.60 years; eyes closed/ears open) studied in our laboratory, we found a whole-brain metabolic rate of 35.0 ± 3.98 μmole glucose/100 g tissue per min (mean $\pm SD$). There were not only significant differences among regions ($P < 0.0001$), with the highest values in basal ganglia, primary visual, cingulate, and frontal cortex, and the lowest in white matter, but also significant asymmetries ($P < 0.002$) with largely right-hemispheric predominance (Table 1). This mean functional right-greater-than-left asymmetry, although small in absolute terms and, therefore, detectable only in a homogeneous group of sufficient size (Pawlik, Heiss, Beil, Wienhard, et al., 1987), was largest in temporoparietal cortex. It may be explained either by the average subject's attentional tone and response to minimal auditory stimulation caused by uncharacteristic, monotonous background noise from technical equipment or by the difference in size of the cerebral hemispheres. Recent evidence suggests that global CMRglu may be inversely related to brain size, resulting in a fairly constant glucose consumption of the brain as a whole, independent of its weight (Hatazawa, Brooks, Di Chiro, & Bacharach, 1987).

TABLE 1. Regional Glucose Metabolic Rates and
Functional Asymmetries[a]

Region	Regional CMRglu (μmole/100 g per min)[b]	Asymmetry (%)[c]
Inferior frontal cortex	42.2 ± 5.79	L 0.24 ± 2.367
Midfrontal cortex	42.0 ± 5.91	R 0.20 ± 2.164
Superior frontal cortex	41.0 ± 5.96	R 0.23 ± 2.617
Broca's area	42.9 ± 5.87	R 1.58 ± 3.646
Inferior sensorimotor cortex	39.5 ± 4.84	R 1.83 ± 3.516
Temporal cortex	36.6 ± 4.37	R 2.39 ± 3.892
Wernicke's area	41.0 ± 4.76	R 3.03 ± 5.980
Superior somatosensory cortex	38.3 ± 4.63	R 1.26 ± 4.215
Supramarginal cortex	38.4 ± 4.95	R 3.92 ± 4.186
Occipitoparietal cortex	36.7 ± 4.53	R 1.24 ± 2.689
Primary visual cortex	42.5 ± 4.82	L 0.09 ± 5.044
Secondary visual cortex	36.5 ± 4.05	R 0.10 ± 2.963
Cingulate cortex	42.5 ± 6.81	L 0.67 ± 2.395
Hippocampal structures	31.6 ± 3.81	L 0.41 ± 3.551
Cerebral white matter	19.9 ± 2.35	R 0.94 ± 2.031
Caudate nucleus	43.3 ± 4.54	L 0.58 ± 5.076
Lenticular nucleus	44.2 ± 5.71	L 0.50 ± 6.006
Thalamus	40.3 ± 5.40	R 2.22 ± 5.790
Brainstem	29.5 ± 3.47	R 2.61 ± 4.262
Cerebellum	33.2 ± 3.95	R 0.16 ± 3.417

[a] Mean ± SD for healthy human subjects ($n = 44$) resting awake in relaxed supine position, eyes closed, ears open, at low ambient light and noise.
[b] Bilateral mean of homotopic regions.
[c] Right or left hemisphere predominance (R, L) of CMRglu, expressed as percentage difference between homotopic regions, normalized to the respective mean.

From the data reported by Frackowiak and co-workers (1980), who measured CBF and $CMRO_2$ with ^{15}O continuous-inhalation methods and PET in 14 resting, normal volunteers (11 men, 3 women) aged 26 to 74 (mean, 48 years), an average temporal gray matter blood flow of 62 ± 18.4 ml/100 g per min at an oxygen metabolic rate of 213 ± 61.3 μmole/100 g per min is obtained when the necessary corrections are applied; the corresponding values for cerebral white matter (centrum semiovale) are 20 ± 1.9 ml/100 g per min and 70 ± 9.0 μmole/100 g per min, respectively. In good agreement with the above observations on CMRglu, some functional right-greater-than-left asymmetry averaging 2.9% for CBF and $CMRO_2$ was found in various, largely cortical, cerebral regions of 27 healthy men (Lenzi et al., 1981). The apparent oxygen-to-glucose molar consumption ratios of less than 6, most conspicuous in white matter, probably must be ascribed to a combination of some Pasteur effect, alternative utilization pathways for glucose in the brain that do not provide substrates for oxidative metabolism, and technical discrepancies.

Age

Among the first problems addressed with PET was the effect of normal aging on physiological brain function. Of course, PET has not been around long enough to perform longitudinal studies, and therefore all studies published to date have been cross-sectional, thus leaving much room for controversy. It was only recently that CMRglu data reflecting brain functional development during infancy and childhood became available from a series of 29 children aged 5 days to 15 years, who were considered reasonably representative of normal (Chugani, Phelps, & Mazziotta, 1987). Compared to values found in young adults, glucose metabolism of various gray matter structures was low at birth (13 to 25 μmole/100 g per min), reached the level of adults (19 to 33 μmole/100 g per min) by 2 years, and continued to rise until the age of 3 to 4. Between 4 and 10 years, it was maintained at a high level (49 to 65 μmole/100 g per min), with the largest increases over adult values in cerebral cortex (185 to 224%), followed by cerebellar cortex (171%), lenticular nuclei (158%), thalamus (154%), caudate nuclei (152%), and brainstem (146%). At about 10 years, CMRglu began to decline. Kuhl, Metter, Riege, and Phelps (1982) found a rather uniform decrease by 26% of CMRglu in all investigated brain regions of 40 healthy, resting subjects between the ages of 18 and 78. In contrast, Duara et al. (1984) failed to observe statistically significant correlations with age of rCMRglu in 40 healthy men aged 21 to 83.

Chawluk and co-workers (1987) again demonstrated an age-related decline of frontal cortical CMRglu by 17% in 44 healthy subjects (age range, 18 to 83) studied in the eyes-open/ears-unoccluded condition. Furthermore, they noted some metabolic decrease in sensorimotor, superior temporal, and inferior parietal cortex, with little effect of minor accidental cardiovascular or systemic medical problems. Riege, Metter, Kuhl, and Phelps (1985), investigating resting CMRglu and multivariate memory test performance in relation to age (27 to 78 years) in 23 healthy, right-handed adults (9 men, 14 women), noted that particularly Broca's area showed a strong age-dependent metabolic decline corresponding with a progressive impairment of secondary memory for verbally processed material. In contrast, the cortical–subcortical interactions indicated by the close relationships among the relative CMRglu of parietal, temporal, posterior frontal, and thalamic regions, did not differ among age groups.

In a series of 42 normal subjects aged 15 to 85 and studied in the base-line resting state with eyes closed and ears open, we found a small (0.65 μmole/100 g per min per decade) but statistically significant ($P < 0.05$) age-dependent decrease in global CMRglu (Pawlik, Heiss, Beil, Wienhard, et al., 1987). However, as demonstrated in Figure 1, the various regions contributed differently to this overall effect: cingulate (-16.6%), frontal (-16.4%), parietal (-15.6%), insular (-14.6%), temporal (-12.1%), and sensorimotor cortex (-11.4%) as well as cerebral hemispheric white matter (-11.3%); virtually no change with age was observed, e.g., in cerebellum and primary visual cortex. These findings seem to be in line with the described loss with advancing age of the metabolic correlations between frontal and parietal areas (Horwitz, Duara, & Rapoport, 1986). Likewise, corroborating previous results obtained with ^{15}O steady-state methods (Frackowiak et al., 1980; Lenzi et al., 1981; Pantano et al., 1984), in a recent study of 34 healthy volunteers (16 women, 18 men; age range, 22 to 82 years), significant age-related declines of insular gray matter and frontal cortical CBF (-0.51% per year) were found (Leenders, Healy, et al., 1987). The $CMRO_2$ showed a reduction in those regions as well (-0.47% per year), but also in other cortical areas and in white matter.

Sleep

Hardly any normal functional state is as regularly associated with as dramatic changes of general behavior and shifting of attention as is sleep. Nevertheless, it was only recently that the conclusion of no effect of sleep on human cerebral hemodynamics and metabolism, derived from early Kety–Schmidt studies (Mangold et al., 1955), could be disproved with PET (Heiss, Pawlik, Herholz, Wagner, & Wienhard, 1985). As shown in Figure 2, during stage II–IV sleep, a significant ($P < 0.001$) global decrease of brain functional activity was observed, with the largest declines in orbitofrontal cortex and in thalamus. In dream sleep, by contrast, both a general metabolic increase and conspicuous regional activations of superior frontal, insular, inferior parietal, hippocampal, and visual association cortex were found.

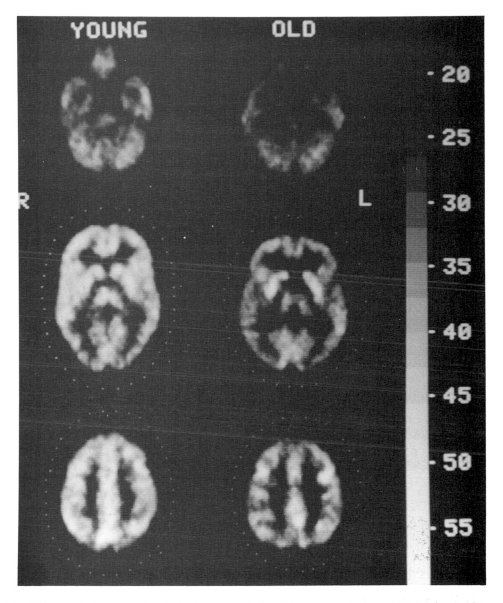

FIGURE 1. Typical corresponding brain glucose metabolic PET images of a 23-year-old (young) and a 67-year-old (old) healthy female volunteer, respectively. R (right) and L (left) indicate side of head. Reference scale is calibrated in micromoles glucose consumption per 100 g brain tissue per minute.

Attention and Cognition

Alertness, raised attention toward external stimuli, apprehensiveness, and anxiety are common behavioral responses to the experience of an unfamiliar situation. As illustrated in Figure 3, the degree of this emotional tension of individuals undergoing PET procedures quite obviously plays a major role for both global level and regional pattern of CMRglu. Buchsbaum, Holcomb, Johnson, King, and Kessler (1983), comparing CMRglu in 14 subjects who received mild to

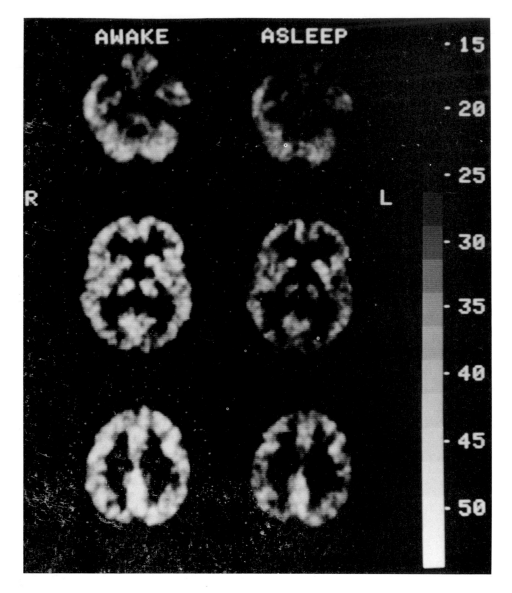

FIGURE 2. Matching CMRglu (μmol/100 g per min) PET images of a healthy 37-year-old man in standard resting condition (awake), eyes closed, ears unoccluded, and during non-rapid-eye-movement sleep, respectively.

painful electrical shocks to the skin of the right forearm to the values obtained in 7 normal volunteers in a resting state, reported not only the expected activation of the postcentral cortex contralateral to stimulation but also an increase in the anterior–posterior metabolic gradient. The latter observation may reflect the anxious expectation of pain. Accordingly, Reivich, Gur, and Alavi (1983), using Spielberger's state–trait anxiety inventory in 18 patients studied with PET, found that as anxiety increased, posterior fronto-orbital and midfrontal cortical activity increased to a point and then declined. Furthermore, in high anxiety, there was significantly ($P = 0.021$) greater right frontal involvement.

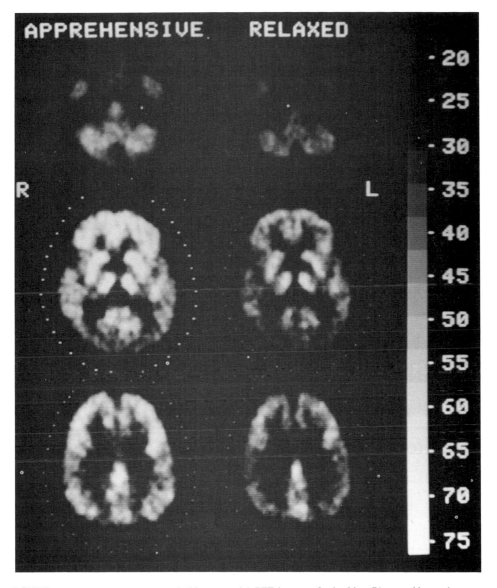

FIGURE 3. Matching CMRglu (μmol/100 g per min) PET images of a healthy, 71-year-old man in apprehensive versus relaxed resting state, eyes closed, ears unoccluded.

On factor analysis of various memory test scores and of the metabolic ratios (regional CMRglu divided by the average of all regions) in nine regions of interest per cerebral hemisphere of 23 healthy adults studied during rest with their eyes and ears unoccluded, close relationships between attentional processing and the relative functional activity of superior frontal, right parietal, and occipital cortex were demonstrated (Riege *et al.*, 1985). The cerebral glucose metabolic correlate of attention focused on a visual or auditory target stimulus was investigated in a series of 15 young, right-handed, male volunteers, whose PET findings were compared with the results obtained in 9 matched controls studied in an unstimulated base-line state, blindfolded and ears plugged (Reivich, Alavi, & Gur, 1984). Considering only cerebellum and superior and inferior parietal lobules, neither side nor modality of stimulation had significant effects on laterality of metabolism. However, there was a statistically significant ($P < 0.005$) right-greater-than-left functional asymmetry in the inferior parietal lobule of stimulated subjects as opposed to the controls. Besides, stimulated subjects manifested higher cerebellar metabolism.

These observations were taken as evidence suggesting right parietal predominance in directed attention. It is important to note, however, that neither stimulus (a specific utterance embedded in a verbal discourse presented in an unfamiliar language, the dimming of a light-emitting diode) carried any concrete, analyzable meaning, thus challenging intuition rather than rational thought. As described below, matters may be different in situations requiring a more analytical strategy.

In another PET investigation of CMRglu during concept formation and abstraction, two groups of 4 right-handed young men received, respectively, Benton's Judgment of Line-Orientation Test and verbal analogies from the Miller Analogies Test (Gur et al., 1983). Eight regions of interest in either cerebral hemisphere were analyzed. Both the verbal and the spatial task produced right hemispheric predominance ($P < 0.005$) that was significantly ($P = 0.002$) larger in the spatial-task group, especially in the frontal eye field and the inferior frontal cortex. The largest opposite lateralized effect was observed in superior temporal/inferior parietal cortex (Wernicke's area), showing left-greater-than-right functional asymmetry during verbal and inverse lateralization during spatial task performance. In addition to these regional side-to-side differences of Wernicke's area, consistent with expectation from clinical experience, the task-dependent activation of right frontal regions may reflect the degree of visuospatial attention and oculomotor readiness of the minor hemisphere. More detailed conclusions, however, cannot be drawn from this study because of the lack of an appropriate control state.

The effect on CMRglu of another frequently applied test of cognitive functions, memory, reasoning, and of the ability to adapt to changing conditions was investigated in 6 normal men (Metz, Yasillo, & Cooper, 1987). Each underwent PET scanning both during a simple match-to-sample control task and while performing the Wisconsin Card-Sorting Test. Forty-four brain regions were evaluated. Contrary to the expected primarily frontal cortical activation, a rather uniform, global metabolic increase was found, averaging during cognitive-task performance almost 30% above the control level. This result reemphasizes the fact that metabolic activation studies and pathological findings may not lead to the same conclusions regarding regional involvement in mental processes.

Tactile Processing

In a series of 23 healthy adults, a close relationship between the ability to recognize from memory tactual patterns and the relative CMRglu of caudate nuclei and right Wernicke's area was suggested by factor analysis (Riege et al., 1985). Roland, Eriksson, Widen, et al. (1987) studied the regional $CMRO_2$ changes associated with tactile learning and recognition in 12 subjects, each of whom was scanned at rest, during right-handed tactile examination of 10 geometrical nonsense objects, and while trying to recognize tactually those objects presented at random in a series of similar but unknown objects. The same distinct pattern of regional activations averaging 10–30% was found during the learning and recognition tasks. Aside from the expected stimulation of the left primary sensorimotor area (Reivich, Greenberg, et al., 1979), there were major metabolic increases in the middle portion of right prefrontal cortex and, bilaterally, in premotor and anterior parietal cortex, supplementary motor area, insula, cingulum, hippocampus, thalamus, striatum, and lateral cerebellum. Hence, evidence was shown for the participation in a seemingly simple task of numerous brain structures commonly implicated in connection with spatial attention, concept formation, voluntary movement and its fine tuning, tactile perception, sensorimotor coordination, secondary association, and memory storage and retrieval.

Visual Processing

Various PET studies of the metabolic response to stimulation of the visual system were reported (Greenberg *et al.*, 1981; Phelps *et al.*, 1981; Reivich, Greenberg, *et al.*, 1979; Reivich *et al.*, 1983). During full-field stimulation with white light, an 11% increase in CMRglu of the primary visual cortex was found bilaterally. The change rate was approximately 25% when the stimulus consisted of a slowly moving, high-contrast, black-and-white pattern of dots and small lines at various orientations, of abstract color images, or of a reversing checkerboard pattern. Significantly asymmetric activation was observed with neither monocular nor binocular full-field stimulation. Studies with partial-field stimulation (Schwartz, Christman, & Wolf, 1984) confirmed that macular stimulation results in selective activation of the occipital poles, whereas more peripheral stimulation activates more anterior striate cortical areas, and lateral half-field stimulation causes the expected functional asymmetry of primary visual cortex. Furthermore, by using the [^{15}O]water bolus method and patterned-flash stimulation, a frequency–response relationship was demonstrated for the primary visual cortex: CBF increased linearly between 0 and 7.8 Hz and then declined (Fox & Raichle, 1984). Compared with white light, increasing complexity of visual input was found to cause not only progressive activation of primary visual cortex but even more activation of associative visual cortex (Brodmann areas 18 and 19). Likewise, concurrent CBF increases in primary visual cortex and in lateral, extrastriate, occipital cortex were observed in 7 normal, right-handed subjects reading single nouns presented on a video monitor at a rate of one word per second (Fox, Petersen, Posner, & Raichle, 1987).

An inverse relationship between thalamic metabolic activity on one side and secondary memory for visual sequences, patterns, or ordered word lists on the other was suggested by factor analysis of multivariate memory test scores and regional CMRglu ratios determined with PET during rest in 23 normal subjects (Riege *et al.*, 1985). In a recent study of pure mental activity, CMRO$_2$ was measured with PET in 10 healthy male volunteers (9 right handers, 1 ambidextrous) aged 20 to 45 who were scanned with their eyes closed, both in a resting state and during visuo-spatial imagery without any immediate prior stimulation (Roland, Eriksson, Stone-Elander, & Widen, 1987). The task consisted of an imaginary walk starting at the subject's home and ending somewhere in the neighborhood after a few minutes of serial retrieval of stored images and continuous mental operation on those recalled images of familiar surroundings. Forty-six brain regions were evaluated in each hemisphere. Across subjects, some increase in CMRO$_2$ was found in almost any region during visual imagery, resulting in a significant ($P < 0.005$) increase in whole-brain oxygen consumption of approximately 15.8 ± 9.75%. However, regional activations were quite nonuniform, revealing a distinct response pattern. Bilateral activation was largest (22 to 30%) in posterior thalamus (left, 34.6%; right, 25.9%), superior and anterior intermediate prefrontal, anterior midfrontal, posterior superior and posterior intraparietal cortex, and in neostriatum. Major functional increases lateralized to the left were detected in posterior midfrontal (30.9%), posterior superior prefrontal (27.5%), and insular cortex (23.0%) as well as in the frontal eye field (18.0%). Predominantly right-sided activation occurred in middle midfrontal (29.1%), middle superior prefrontal (28.6%), and supramarginal cortex (23.2%), the head of the caudate nucleus (27.1%), and in the cerebellar hemisphere (22.5%). No metabolic change was observed in primary sensory or motor areas, especially not in striate cortex or in adjacent lateral and inferior cortical regions.

Despite some reservations concerning the PET procedure—the final oxygen metabolic images representing either the rest or the imagery condition resulted from three separate measurements (for oxygen extraction fraction, blood flow, blood volume) subdivided into six periods requiring identical physiological steady states—and the subjectivity of the paradigm (subject performance could not be monitored reliably), the large and consistent activations described are likely to reflect major functional relationships. As with other tasks requiring directed attention, the widespread midfrontal and superior prefrontal increases may indicate the participation of these areas in the intrinsic organization of brain work, while the cerebellar and anterior neostriatal activation may be considered secondary to the increased input from prefrontal cortex. Probably most task-specific were the functional activations of the posterior thalamic nuclei (visual image retrieval?), the superior posterior parietal areas (image reconstruction and holding?), right supramarginal (spatial attention and recognition?), and left insular (paralimbic connection?) cortex.

Auditory Processing

Several PET studies were aimed at elucidating the relationship between CMRglu and auditory stimulation. In general, irrespective of the stimulus type and side, the transverse superior temporal cortex was activated bilaterally, showing the well-known anatomic asymmetries for Heschl's gyri and the planum temporale (Mazziotta, Phelps, Carson, & Kuhl, 1982b). In addition, with monaural stimulation, there may be slight functional predominance of the contralateral primary auditory cortex (Greenberg et al., 1981; Kushner, Schwartz, et al., 1987) within which the anterior–posterior position of the area of maximal response may be different for different tone frequencies (Lauter, Formby, Fox, Herscovitch, & Raichle, 1983). Mazziotta and co-workers (1982b; Mazziotta, Phelps, & Carson, 1984) explored the effect of two nonverbal auditory stimulation paradigms in 4 and 8 normal subjects, respectively, studied both in a base-line resting state with their ears plugged and covered with soundproof headphones and in an activated state. When asked to identify differences in chord pairs (timbre test), the volunteers exhibited bilateral temporal–parietal increases in CMRglu as well as activation of the right superior posterior temporal cortex and diffuse right-greater-than-left metabolic asymmetry in frontal, posterior temporal, and temporal–occipital regions. In contrast, when stimuli consisted of pairs of tone sequences and subjects had to determine whether any of the tones in the second sequence differed in frequency from the tones in the first sequence (Seashore Tonal Memory Test), aside from activations of the right middle and posterior superior temporal cortices, metabolic responses were found to correlate with the individual analysis strategy: in subjects using highly structured visual imagery, activations and functional asymmetries occurred in the left posterior superior temporal cortex and in the head of the left caudate nucleus, whereas those subjects approaching the discrimination task more intuitively basically showed the same activation pattern as described above for stimulation with chord pairs. With neither paradigm did the side of (monaural) stimulation have any effect on the responses. Furthermore, even when measured during rest, CMRglu of both caudate nuclei and of the right Wernicke's area may correlate with test scores on nonverbal auditory (bird call series) recognition (Riege et al., 1985).

The PET findings pertaining to verbal auditory stimulation are somewhat contradictory. Using the paradigm of monaural presentation of a factual story, some authors (Reivich, Greenberg, et al., 1979) first noted 20–25% increases in CMRglu of the right temporal lobe regardless of the ear stimulated; they then reported a temporal cortical activation contralateral to the stimulated ear, with metabolic rates averaging 7% above the homotopic regional control level (Greenberg et al., 1981). Comparable results were obtained when the discourse was presented in an unfamiliar foreign language (Kushner et al., 1987b). Mazziotta and co-workers (1982b, 1984) again used the task paradigm of verbal auditory stimulation by a factual story in a base-line-controlled series of 4 right-handed volunteers. In addition to bilateral transverse and posterior temporal cortical activations, these authors consistently found significant metabolic increases and functional asymmetries in the left frontal, lateral occipital, and posterior superior temporal cortices as well as in both thalami (left greater than right). Listening to single nouns presented binaurally at a rate of one word per second produced bilateral CBF increases without lateralization in the primary auditory cortex and left angular gyrus activation in 7 right-handed normals (Fox et al., 1987). As with nonverbal auditory stimulation, these conflicting results may at least in part be explained by the subjects' different attitudes and verbal comprehension techniques. According to our experience with healthy volunteers binaurally receiving various verbal auditory tasks, highly motivated, alert, and bright individuals listening to complex, interesting news commonly exhibit hyperfrontality

and left-hemispheric functional cerebral predominance at a raised overall metabolic level; in con-
strast, poorly motivated, inattentive subjects bored by a lengthy tale primarily show asymmetric
(right-greater-than-left) activation of transverse and posterior superior temporal areas (Figure 4).

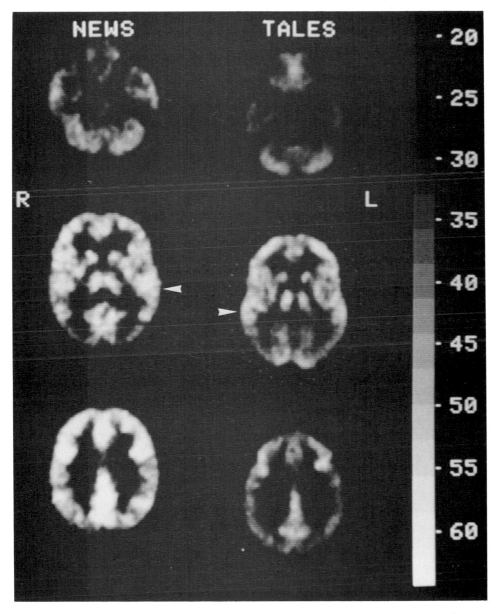

FIGURE 4. Typical corresponding CMRglu (μmol/100 g per min) PET images of normal right-handers during
verbal auditory stimulation, with one listening to interesting news and the other to boring tales, showing
differential activation and lateralization of Wernicke's area response (arrowheads).

Speech

Physiological correlates of spoken language were investigated with PET using elementary and global paradigms. In a CBF study of 7 healthy right-handers (Fox *et al.*, 1987) vocally repeating spoken cue words, symmetric activations were observed in primary auditory cortex and in the sensorimotor mouth area, while flow increases were lateralized to the left hemisphere in anterior superior premotor cortex, supplementary motor area, and angular gyrus, and to the right in the head of the caudate nucleus and cerebellum. In addition, left anterior inferior frontal lobe and anterior cingulate gyrus showed strong activation during semantic generation, i.e., when subjects said aloud a verb in response to a cue noun.

The CMRglu was measured with PET in 9 subjects performing the Word Fluency Test, i.e., they said as many words as possible beginning with a given letter, 1 min per letter, for 30 min (Duara, Chang, Barker, Yoshii, & Apicella, 1986). Interestingly, the correlations between word output per minute and regional metabolism were mainly negative, the strongest being in left inferior perirolandic ($r = -0.94$), right orbitofrontal cortex ($r = -0.90$), and left caudate nucleus ($r = -0.87$), suggesting an inverse relationship between the efficiency of word production and metabolic activity. The specific role of the striatum in overlearned, semiautomatic motor performance was demonstrated in a PET study of 11 healthy volunteers: while sequential finger movements produced significant activation only of contralateral sensorimotor cortex, additional bilateral striatal increases in CMRglu were observed during signature writing (Mazziotta, Phelps, & Wapenski, 1985). Similar metabolic behavior may be expected in "subcortical speech."

We used the FDG method and PET to investigate the regional CMRglu changes associated with spontaneous speech (Pawlik, Heiss, *et al.*, 1985; Pawlik, Heiss, Beil, Grünewald, *et al.*, 1987; Heiss *et al.*, 1987). Eight healthy male volunteers (age range, 24 to 32 years; 4 left-, 4 right-handers) were studied with their eyes closed in balanced random order both in a resting state and during spontaneous, meaningful speech on a social philosophical subject related to the individual's own biography. Speech caused a significant ($P < 0.005$), asymmetric ($P < 0.05$) increase in CMRglu, averaging 14% in the left and 12% in the right cerebral hemisphere and 19% in the left and 29% in the right cerebellar hemisphere, respectively. Furthermore, there was a distinct pattern of regional cerebral activations, with the largest changes in somatosensory cortex (32%), Wernicke's area (18%), prefrontal cortex (16%), thalamus (15%), sensorimotor mouth area (14%), and Broca's area (13%). Across subjects, these metabolic activations were quite similar for homotopic regions in either cerebral hemisphere. Only the prefrontal cortex exhibited increases that were consistently larger on the left.

As to the relationship between handedness and speech dominance, four response types could be distinguished by the degree of asymmetric functional activation: right-handed/left-dominant (Figure 5), right-handed/bilateral, left-handed/bilateral, and left-handed/right-dominant. However, right- and left-handers were distinguished most clearly by their respective metabolic asymmetry of Wernicke's area during speech, averaging 1.5% (L>R) in right-handers and 7.6% (R>L) in left-handers. These results demonstrate that the demanding task of continuously speaking about a rather abstract subject poses a major challenge to the brain as a whole and to some regions in particular. Although only a few of the observed regional activations, e.g., in Broca's and Wernicke's areas, may be considered language-specific and others may be related to self-awareness, memory retrieval, associative thinking, sensorimotor coordination, and vocalization, the largely bilateral representation of speech-related functions suggests a major supportive role for certain mirror regions in the minor hemisphere in the performance of tasks traditionally ascribed to the dominant hemisphere. Moreover, handedness may be more closely related to the understanding rather than to the production of language, and the consistently lateralized activation of left prefrontal cortex and right cerebellar hemisphere may reflect the participation of these structures in verbal memory processing, independent of handedness. Because of the widespread cerebral activation induced by spontaneous speech, this paradigm proved quite useful for the functional contrast enhancement between responsive and abnormal tissue (Heiss *et al.* 1987; Pawlik, Heiss,

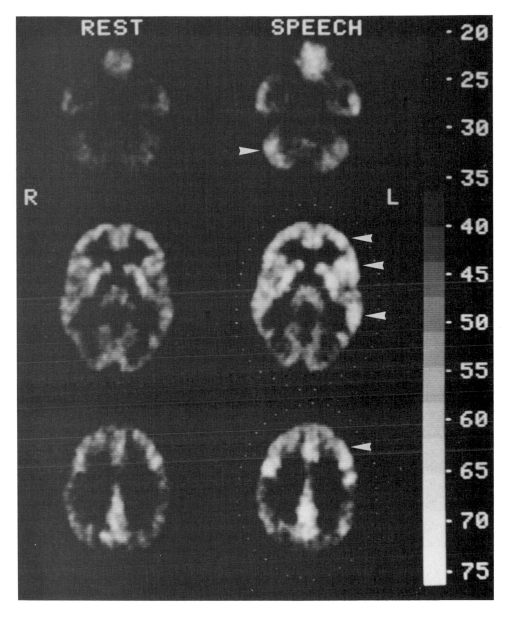

FIGURE 5. Characteristic CMRglu PET (μmol/100 g per min) images of a healthy 24-year-old strongly right-handed man during rest and spontaneous speech, respectively. Arrowheads point at regions of maximum lateralized activation.

Stefan, *et al.*, 1987). For example, in temporal lobe epilepsy, speech activation produces little effect in the interictally hypometabolic epileptogenic focus, concurrently increasing CMRglu of adjacent normal brain regions and thus improving the reference-to-focus metabolic ratio by more than 20% (Figure 6).

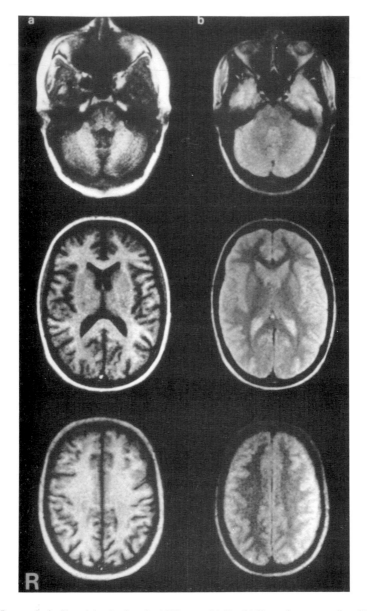

FIGURE 6. Progressively T$_2$-weighted spin-echo MRI scans (a) T$_1$, (b) T$_2$, and corresponding CMRglu (μmol/ 100 g per min) PET images of a typical right-handed patient with right temporal lobe epilepsy during rest (c) and spontaneous speech (d) activation, respectively. Arrows mark epileptogenic focus.

Memorizing

Learning, i.e., the formation of memory and experience, necessarily invokes some input-modality-specific regional activations. However, as described above in the section on tactile processing, the participation of limbic, paralimbic, and other deep gray matter structures in the

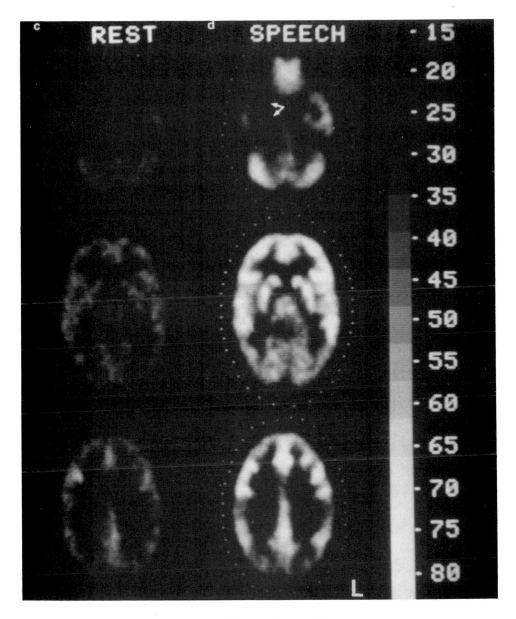

FIGURE 6. (*continued*)

memorizing process can also be visualized with PET. Accordingly, Mazziotta and co-workers (1982b) observed increased glucose consumption in the mesial temporal cortex (hippocampus, parahippocampal gyri) of subjects listening to a Sherlock Holmes story. The volunteers were instructed to remember specific phrases, and they were told that they would be paid in proportion to their performance on an examination following the PET session. The resulting activation, particularly of the amygdala, therefore, may reflect the strong affective coloring of their experience caused by a combination of suspense and greed. In contrast, Riege *et al.* (1985), using factor analysis of memory test scores and of regional CMRglu ratios determined with PET during rest in 23 normal adults, could demonstrate that the metabolic activity of Broca's area was closely related to the performance in tests that seemed to measure the elaborative encoding of concepts joined by meaning and images.

In a CMRglu study of 9 healthy volunteers memorizing historical dates and the respective events presented auditorily for a few seconds, a different pair of items every 30 s (Pawlik, Heiss, Beil, Wienhard, *et al.*, 1987), we found that the individual's metabolic response pattern was largely determined by his memorizing technique. Most subjects used maintenance rehearsal, i.e., silent speech repeating the items again and again, thus producing regional activations that were, except for the sensorimotor part, quite similar to the ones observed during spontaneous speech with vocalization, including left frontal and right cerebellar activation (Figure 7). Others, however, used creative global association, embedding the historical items in a short, vividly imagined scene appearing as a visual image. These subjects consistently showed the largest cortical acti-

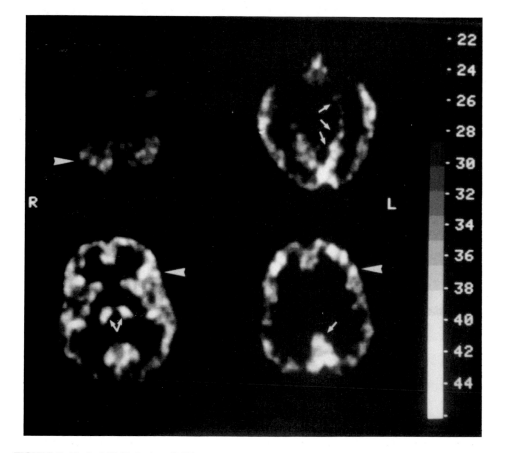

FIGURE 7. Typical CMRglu (μmol/100 g per min) PET images of a healthy right-handed volunteer during performance of a memorizing task, using maintenance rehearsal with silent speech. Arrows and arrowheads point to areas of maximum activation.

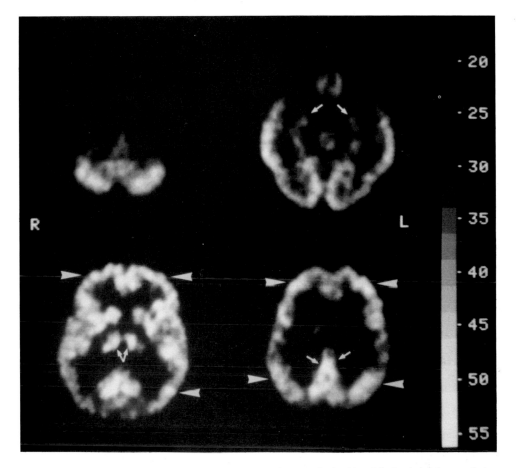

FIGURE 8. Typical CMRglu (μmol/100 g per min) PET images of a healthy right-handed volunteer during performance of a memorizing task using visual imagery and global emotional association. Arrows and arrowheads point at areas of maximum activation.

vations bilaterally in prefrontal, intraparietal, and posterior superior parietal regions (Figure 8). In addition, all subjects exhibited some thalamic, neostriatal, insular, cingulate, and hippocampal activation.

Other task paradigms calling primarily for short-term memory are conceivable, e.g., the continuous mental calculation of primes in ascending order, requiring repeated operations according to a limited set of rules on some abstract numbers that must be kept in mind for a short period of time. Figure 9 illustrates the brain work associated with this specific task: somewhat increased CMRglu is found bilaterally in posterior superior temporal and inferior parietal cortex; major activations lateralized to the left occur in prefrontal, posterior and middle inferior frontal, insular, hippocampal, and posterior superior parietal cortex as well as in the frontal eye field, neostriatum, and dorsomedial thalamus; only the cerebellum exhibits some right-hemispheric predominance. This regional activation pattern bears resemblance with both silent-speech memorizing and visual imagery, but the functional left-greater-than-right asymmetry obviously is much stronger in neostriatum and insular and hippocampal cortex, while intraparietal areas show lesser involvement,

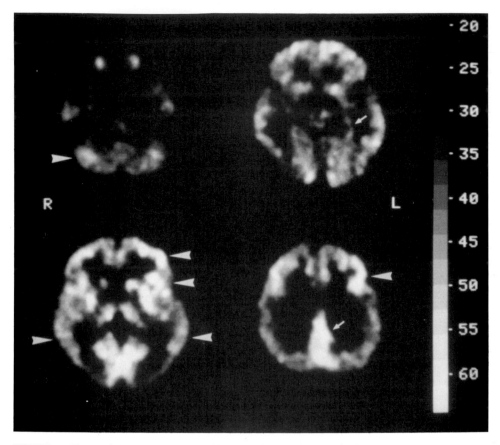

FIGURE 9. Characteristic CMRglu (μmol/100 g per min) PET images of a healthy right-handed volunteer performing mental arithmetic. Arrows and arrowheads point at areas of maximum activation.

and the moderate bilateral temporoparietal activations may be related to the task-specific computing operations.

Clinical Syndromes

It should be clear from the observations of normal functional states described in the previous sections that neurobehavioral activations practically never elicit unilobular cerebral responses. Therefore, it is of little surprise that distinct neuropsychological signs or symptoms rarely occur in isolation. Common focal brain lesions, e.g., infarcts and tumors, in general impair more than just one functional network because each brain region may serve as a module participating in various cognitive or behavioral tasks, and lesions do not respect functional boundaries. Thus, in the following, syndromes are grouped pragmatically according to the most prominent clinical abnormality, and no claim is made as to the physiological significance of this structuring. Furthermore, it should be noted that, to date, reports on PET findings in distinct neuropsychological disorders other than the aphasias and dementias have been anecdotal at best, although PET holds great promise for their elucidation as well.

Focal brain lesions commonly give rise to metabolic disturbances extending far beyond the site of morphological damage. Of course, this secondary deactivation (diaschisis, deafferentation) caused by the disruption of established connections between normally interacting neuronal modules cannot be detected by CT or MRI. It may underlie not only the nonspecific organic brain syndrome frequently accompanying acute brain injury but also the various neuropsychological deficits that remain unexplained on anatomic grounds alone, and it can be quantitatively visualized only with PET. Using multivariate analysis of regional CMRglu in a series of 62 stroke patients, we could demonstrate that focal ischemic lesions cause global and region-specific remote effects in varying proportions, depending on the residual functional activity (or size) of the infarcted area and on its location (Pawlik, Herholz, *et al.*, 1985). Perhaps it is the great variability of effective lesion parameters on the one side and the stochastic nature of neuropsychological network operations on the other that make the deduction of general neuropsychological principles from the results of correlative clinical deficit studies so exceedingly difficult.

Aphasias and Related Disorders

Infarcts in the supply territory of the middle cerebral artery (MCA) in the dominant hemisphere are the most common cause of aphasia. Such ischemic lesions can be detected by metabolic PET imaging at an earlier stage and with greater sensitivity than by CT or MRI of protons (Pawlik, Heiss, Wienhard, *et al.*, 1987). Furthermore, the severity of aphasia and the spatial extent of the metabolic lesion appear closely correlated (Kushner, Reivich, *et al.*, 1987), and therefore the aphasia resulting from typical MCA infarction (Figure 10) often is global, permitting no detailed localization of language functions.

Even with smaller lesions, localization may be quite ambiguous when PET studies are performed in the resting condition only. Using factor analysis of regional metabolic asymmetries during rest, and of subtest scores from the Boston Diagnostic Aphasia Examination and the Porch Index of Communicative Ability, Metter *et al.* (1984) found close relationships between speaking and Broca's area and left caudate and frontal activity, as well as between auditory comprehension, naming, oral reading, repetition, and automatic speech and Wernicke's area, left posterior middle-inferior temporal, and parietal CMRglu. When the metabolic patterns in distinct aphasia types as determined by the Western Aphasia Battery were compared, similar functional asymmetry was described for temporal and parietal regions of patients with Wernicke's, conduction, and Broca's aphasia; the three syndromes differed with respect to asymmetric left frontal deactivation, which was marked in Broca's, moderate in Wernicke's, and essentially absent in conduction aphasia (Metter *et al.*, 1986). However, the cerebral localization, particularly of Broca's and of conduction aphasia, may be rather elusive, as illustrated in Figure 11 showing no major abnormality of left posterior inferior frontal and caudate metabolism in a right-handed patient with moderate Broca's aphasia and distinct hypometabolism in left parietal and anterior temporal cortex.

The focal dysfunction underlying mild conduction aphasia may be barely detectable on resting PET images (Kempler *et al.*, 1987) but obvious during speech activation (Heiss *et al.*, 1987; Pawlik, Heiss, Beil, Grünewald, *et al.*, 1987). Mostly, there is significant hypometabolism of left superior temporal, insular, parietal opercular, and supramarginal cortex concurrent with mild deactivation of Broca's area and ipsilateral caudate nucleus (Figure 12). In a case of pure word mutism (aphemia), significant metabolic depression was observed subcortically in the space lateral to the left caudothalamic line (Marie's quadrilateral space), with transient and less severe dysfunction of the surrounding frontal cortex (Kushner *et al.*, 1982).

Among the more common types of aphasia, Wernicke's aphasia seems to have the most consistent metabolic pattern characterized by the functional loss of posterior middle and superior temporal, parietal opercular, and inferior and intraparietal cortex in the dominant hemisphere, generally in conjunction with marked secondary deactivation of ipsilateral middle and posterior inferior frontal regions and striatum (Figure 13), although the correlation with morphological abnormalities may be poor (Metter *et al.*, 1984).

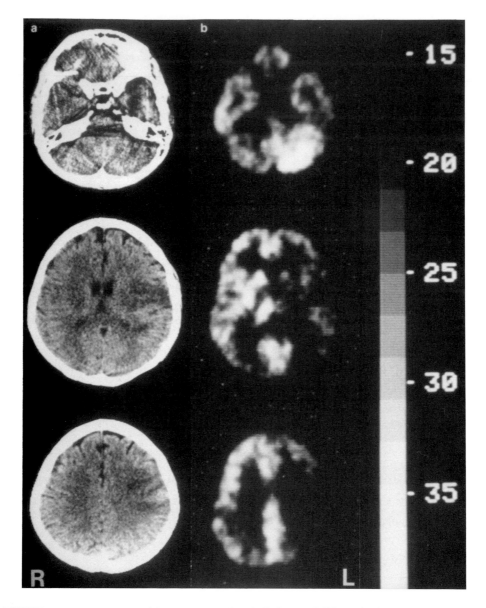

FIGURE 10. Characteristic CT (a) and corresponding CMRglu (μmol/100 g per min) PET (b) images of a right-handed, globally aphasic patient 12 days after left MCA infarction.

In a patient presenting with mild dementia and mixed transcortical aphasia consequent to prolonged hypoglycemia, whose only remaining language competence was repetition, CT showed moderate, diffuse brain atrophy but provided no clue as to the pathoanatomic basis of this rare clinical syndrome (Figure 14a). The metabolic images, however, revealed moderately decreased global CMRglu as well as distinct focal dysfunction in the supply territory of the left anterior cerebral artery and its MCA border zone, in left inferior temporal cortex, and in most of the left parietal lobe including the posterior watershed area. In addition, there was marked deactivation

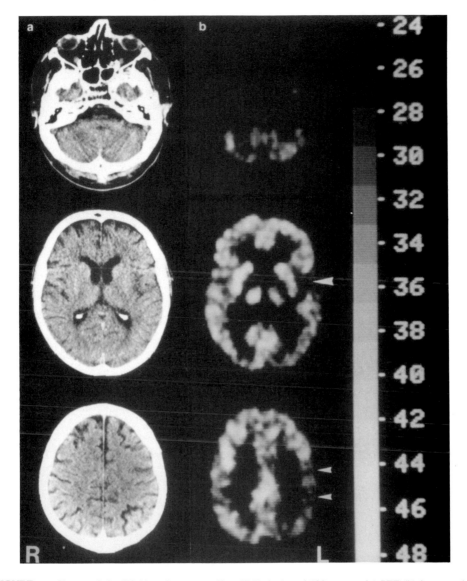

FIGURE 11. Characteristic CT (a) and corresponding CMRglu (μmol/100 g per min) PET (b) images of a right-handed stroke patient with moderate Broca's aphasia, 9 days after minor left MCA infarction. Arrowheads point at most severely affected regions.

of ipsilateral anterior insular and frontal opercular cortex, thalamus, hippocampus, and contralateral cerebellum, while the left frontal eye field and Wernicke's and Broca's areas exhibited relatively increased activity (Fig. 14b), suggesting an isolation of the sensorimotor speech areas depending entirely on intracortical conduction. We observed a similar metabolic lesion pattern, albeit with complete functional loss of Wernicke's area and with sparing of the anterior cerebral artery territory and the anterior watershed area, in the resting PET study of a right-handed stroke victim presenting with a Wernicke-like aphasia and dramatically preserved repetition characterizing the syndrome as transcortical sensory aphasia. Only from the speech activation study did it

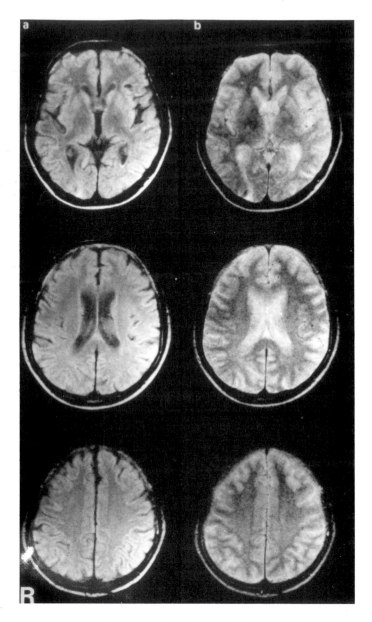

FIGURE 12. Characteristic, progressively T_2-weighted spin echo MRI scans (a) T_1, (b) T_2, and corresponding CMRglu (μmol/ 100 g per min) PET images during rest (c) and spontaneous speech (d), respectively, of a right-handed stroke patient with mild conduction aphasia 2 weeks after minor left MCA infarction. Arrowheads point at dysfunctional area.

become clear that this patient primarily used his right-hemispheric homologues of Wernicke's and Broca's areas.

Recently, interest has focused on the role of subcortical structures in aphasia. In a PET study of 3 right-handed, mildly aphasic patients who had sustained vascular lesions of the left thalamus, putamen, or caudate head and internal capsule 6 weeks to 6 months previously, the only consistent

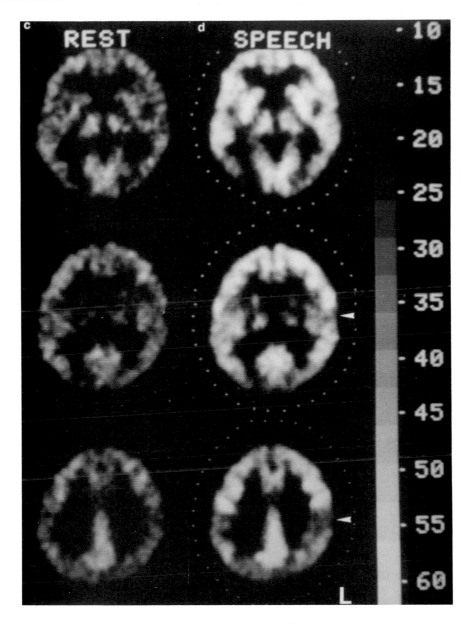

FIGURE 12. (*continued*)

metabolic finding was a depression of CMRglu in the left caudate and thalamus, while cortical deactivation was mild and varied among subjects (Metter *et al.*, 1983). Their clinical syndrome was characterized by word-finding, animal-naming, body-part-naming, and writing difficulties, particular problems with auditory comprehension of complex material, but relatively preserved repetition. Although the authors ascribed the aphasia to cortical dysfunction, they explained the observed verbal memory deficits by thalamic hypometabolism as we reported on a series of 8

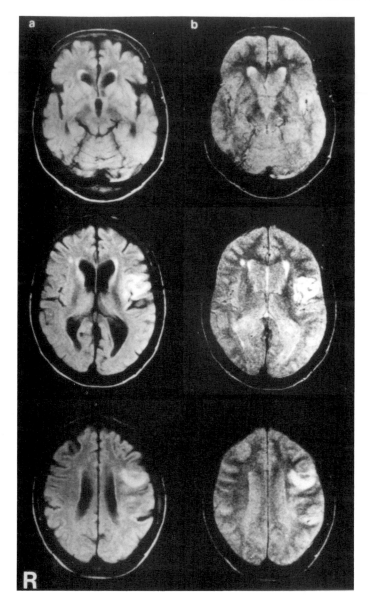

FIGURE 13. Characteristic, progressively T_2-weighted spin echo MRI scans (a) T_1, (b) T_2, and corresponding CMRglu (μmol/100 g per min) PET images (c) of a right-handed stroke patient with moderate Wernicke's aphasia 9 days after left MCA infarction. Arrowheads point at area of most severe hypometabolism.

patients with small ischemic lesions of the left anterior, medial, and dorsal thalamic nuclei (Pawlik, Beil, *et al.*, 1985a). Differences in the aphasia test profiles between the subcortical aphasics and a comparable group of age-matched patients with cortical aphasia (2 Wernicke, 1 Broca, 1 mixed) corresponded with lower left-to-right metabolic ratios in parietal, Wernicke's, posterior middle-inferior temporal, and caudate areas of the latter group.

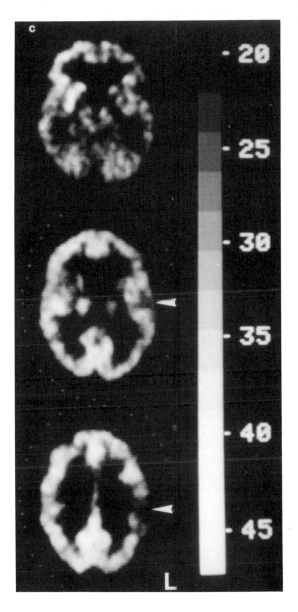

FIGURE 13. (*continued*)

As symptoms of subcortical aphasia often are transient despite the persistence of morphological damage, and the blood flow–metabolism relationships in the various involved structures are
unclear, we performed paired CBF and CMRglu PET studies in a 50-year-old right-hander 1 week
and again 1 year after left capsulostriatal infarction (Figure 15). The patient had acutely developed
very mild right brachial and facial paresis concurrent with mutism. A week later, her neuropsychological syndrome consisted of a nonfluent aphasia with hypophonic, paraphasic verbal output
and relatively preserved auditory comprehension, naming, and repetition. At this time, global

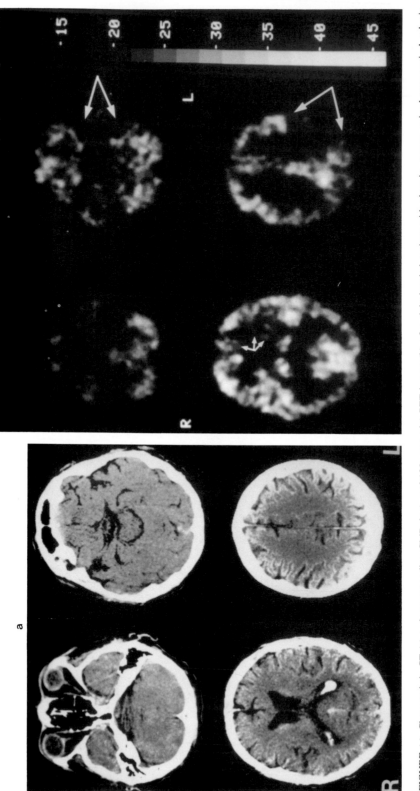

FIGURE 14. Characteristic CT (a) and corresponding CMRglu (μmol/100g per min) PET images (b) of a right-handed patient with chronic, severe, mixed transcortical aphasia and mild dementia caused by hypoglycemic–hypoxic–ischemic brain damage. Arrows mark areas of most severe metabolic dysfunction.

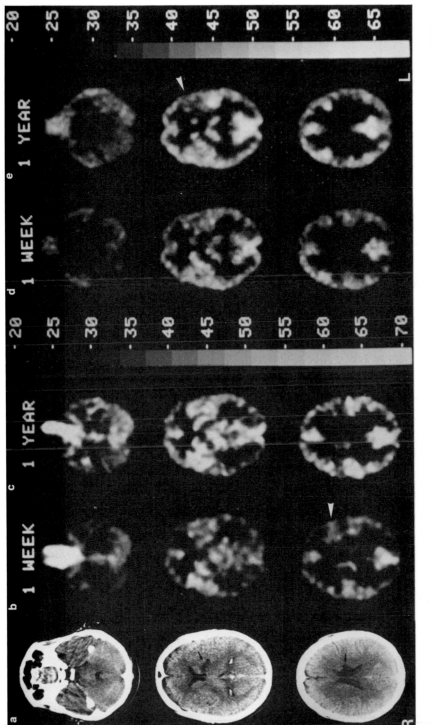

FIGURE 15. Characteristic CT (a) (early and late scans showed lesion virtually unchanged) and corresponding early and late CBF (ml/100 g per min) (b, c, respectively,) and CMRglu (μmol/100 g per min) (d, e, respectively,) PET images of a right-handed stroke patient with subcortical aphasia consequent to left capsulostriatal infarction (black arrows). White arrowhead in CBF image points at an area of early luxury perfusion, and that in the CMRglu image at a region of persistent deactivation.

CBF was more severely decreased than global CMRglu, with maximum ischemia in left frontal cortex. In contrast, both in the infarcted area showing almost zero metabolic activity, and in nearby directly deactivated regions (rest of left striatum, left thalamus, Broca's area), CBF was reduced to a lesser degree than was CMRglu, and at the upper outer border of the infarct there was obvious luxury perfusion. After 1 year the patient had made an almost full recovery except for some word-finding difficulties and a slight decrease in the fluency of spontaneous speech as compared with the premorbid state. Global CBF and CMRglu had returned to normal levels, however, despite normal regional CBF, distinct metabolic dysfunction was observed in the morphologically intact Broca's area, left frontal operculum, anterior insular and temporal cortex, putamen, and anterior thalamus. Other regions (left frontal and parietal cortex, right cerebellum) showed minor functional decreases that were proportional for CBF and CMRglu. This case clearly demonstrates that lack of activating neuronal input rather than persistent ischemia is the cause of symptoms in subcortical aphasia. It further suggests that this dysfunctional state can be overcome, albeit imperfectly, by spontaneous functional readjustment within the cortical speech network.

The compensatory neuronal mechanisms can come into effect even repeatedly, as we observed in a 44-year-old right-handed patient with cerebral microangiopathy and her fifth serious ischemic episode in 4 years (Figure 16). Small, circular ischemic lesions, old and recent, were scattered along the deep border zone between the left anterior cerebral artery (ACA) and MCA territories, involving capsulostriatum, ventrolateral thalamus, and frontal and parietal white matter, but nowhere extending into cortex. As during previous attacks, she presented with near-mutism, syntactic alexia, acalculia, and difficulties with auditory comprehension as well as with the recognition of body parts; in addition, there was severe right hemiparesis, hemisensory loss, apathy, and psychomotor retardation. Again, as after the previous insults, her neuropsychological symptoms receded largely within a few days, and, as in the previous case, particularly in left frontal and temporal cortex, CBF exceeded the metabolic demand in functionally deactivated regions.

Neuropsychological recovery may not always be complete and rapid, yet it is possible even with large cortical defects. Figure 17 shows the cerebral metabolic changes with time observed in a right-handed patient with acute infarction of the left posterior MCA territory. For weeks after the ictus, her clinical syndrome was rather stable, consisting of severe Wernicke's aphasia, attention deficit and psychomotor retardation, ideomotor apraxia (including the inability to perform, on written or gestural command, a motor activity that was easily carried out spontaneously), acalculia, right–left disorientation (tested nonverbally), and agraphia without alexia. Early PET images revealed severe global hypometabolism, functional loss of almost the entire left parietal and superior temporal lobes, as well as marked regional deactivation of left cuneus and precuneus, frontal eye field, thalamus, and right cerebellar hemisphere, and bilaterally of basal ganglia and midfrontal and prefrontal cortex. Gradually over years, the patient regained full neuropsychological competence except for minor word-finding difficulties and a continued prolongation of reaction time. Her late PET images showed major metabolic abnormalities only in the area of morphological damage, in immediately adjacent structures (superior temporal, insular, occipital cortex; putamen, anterior thalamus), and in contralateral cerebellum, suggesting excellent functional compensation.

Of particular interest with regard to the cerebral localization of written language processing may be the case of a young right-handed diabetic who had suffered left posterior cerebral artery infarction during diabetic coma. Her clinical picture was dominated by right homonymous hemianopia, posterior alexia, and copying agraphia; i.e., she was unable to read despite normal vision in the left hemifield, spontaneous writing and writing on dictation were unimpaired, but she could not copy even a single letter. Furthermore, there was mild transcortical sensory aphasia and an isolated memorizing disturbance for semantic context. In addition to the ischemic lesion seen on CT and MRI, PET scanning revealed marked focal hypometabolism of the splenium of the corpus callosum, of left hippocampus, fusiform and parahippocampal gyri, amygdaloid nucleus, poste-

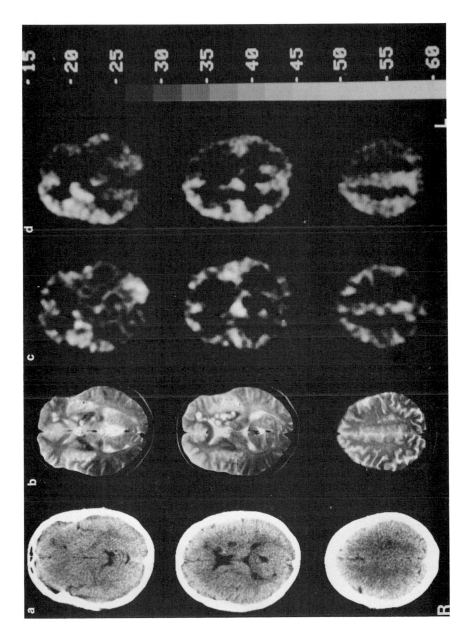

FIGURE 16. Characteristic CT (a), T_2-weighted MRI (b), CBF (ml/100 g per min) (c), and CMRglu (μmol/100 g per min) (d) PET images of a right-handed patient with subcortical aphasia consequent to recurrent ischemic attacks in deep left arterial border zone.

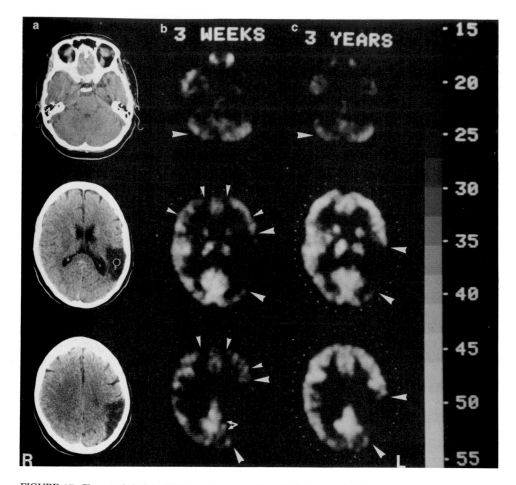

FIGURE 17. Characteristic late CT (a) and corresponding CMRglu (μmol/100 g per min) PET (b) images of a right-handed stroke patient with severe dominant temporoparietal syndrome at postacute (b) and at chronic stage (c) of left posterior MCA infarction. Arrows and arrowheads mark areas of maximum hypometabolic dysfunction.

rior, ventrolateral, medial, and dorsal nuclei of the left thalamus, as well as minor deactivation of most of the left cerebral cortex (Figure 18). Obviously, the patient's alexia and input-modality-specific agraphia must be attributed to disruption of the transfer of visual information from the intact right hemisphere to the left parietal cortex, and her partial amnestic syndrome may be explained by concurrent dominant thalamic, limbic, and paralimbic dysfunction (see below).

Discorders of Complex Visual Processing

Cortical processing of sensory information involves essentially three more or less distinct steps: primary perception, recognition (i.e., neuronal image synthesis and comparison with similar modality-specific images from memory), and heteromodal association (i.e., interaction with information from other input channels and formation of a response macro). In the human brain, the visual association areas are particularly well developed, and therefore damage to these structures produces conspicuous clinical deficits of great localizing value.

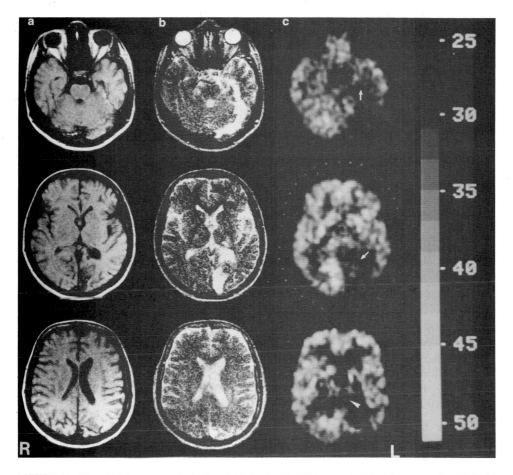

FIGURE 18. Characteristic, progressively T_2-weighted spin-echo MRI scans (a, b) and corresponding CMRglu (μmol/100 g per min) PET (c) images of a right-handed patient at chronic stage after left PCA infarction that caused posterior alexia, copying agraphia, mild transcortical sensory aphasia, and semantic memory disorder. Arrows and arrowheads point at lesions presumably corresponding to prominent clinical signs.

One of the most remarkable properties of the primate visual system is its large capacity for simultaneous perception, recognition, and spatial analysis of highly complex input, resulting in a gaze shift to bring a target into optimal central vision and guiding motor output. These functions are typically impaired in Balint's syndrome, which we observed in a patient who had survived a severe chest trauma necessitating long-term artificial respiration complicated by prolonged hypoxic episodes. At the chronic stage, his clinical syndrome consisted of visual disorientation (simultagnosia), optic ataxia, ocular apraxia, and achromatopsia. His visual fields were intact, and relative visual acuity was 0.8. Morphological imaging failed to detect any abnormality. However, PET demonstrated severe bilateral occipital hypometabolism extending from the lingual gyrus, occipital pole, and adjacent lateral cortex to the mesial occipitoparietal junction and reaching farther into Brodmann's area 7 but sparing most of the calcarine cortex (Figure 19). Functional activity also was somewhat decreased bilaterally in the anterior temporal lobe, dorsolateral putamen, and insular and sensorimotor cortex, while the frontal eye fields exhibited increased activity.

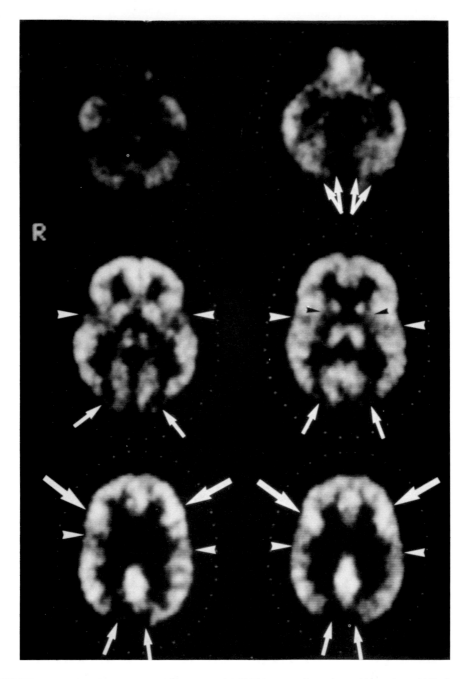

FIGURE 19. Various CMRglu (μmol/100 g per min) PET images of a patient with persistent Balint's syndrome and achromatopsia years after posttraumatic hypoxia. Small arrows point at areas of maximum hypometabolic dysfunction, arrowheads mark regions showing deactivation of lesser degree, and large arrows point at areas of increased activity.

Quite similar with regard to disease mechanism and lesion pattern was the case of a patient with hypoxic brain damage whom we studied years after cardiopulmonary resuscitation because of postsurgical hemorrhagic shock. She had a central scotoma measuring 5° and homonymous visual field defects comprising both lower quadrants. Behaviorally, she presented with visual

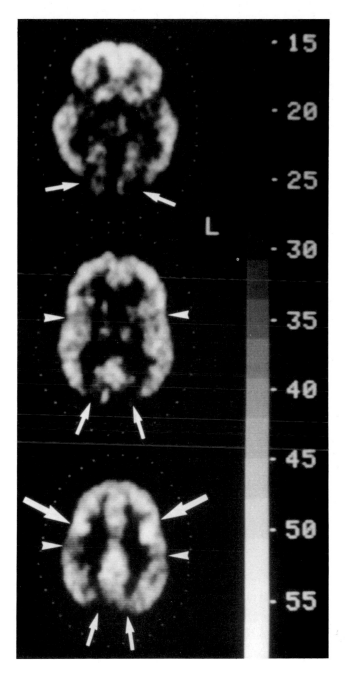

FIGURE 19. (*continued*)

object agnosia, astereopsis, and achromatopsia. There was no gross structural damage, but PET revealed a symmetrical circular area of severe cortical hypometabolism centered around the occipital poles, comprising the upper lip of the calcarine cortex and Brodmann's area 18 (Figure 20). Hypometabolism was mild in perisylvian and sensorimotor cortex, but posterior cingulate gyrus,

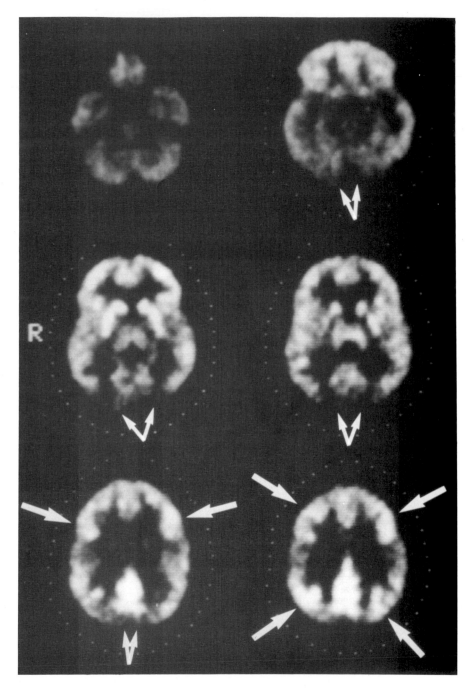

FIGURE 20. Various CMRglu (μmol/100 g per min) PET images of a patient with visual object agnosia, astereopsis, and achromatopsia years after hypoxic brain injury. Small arrows mark areas of most severe hypometabolism, and large arrows point at regions of increased activity.

anterior cuneus, precuncus, intraparietal cortex, and the fontal eye field appeared to be activated bilaterally. Compared with the previous case, calcarine, polar, and infrapolar occipital cortex were more severely affected, but the lesion did not extend into the occipitoparietal junction area, which may account for the clinical differences.

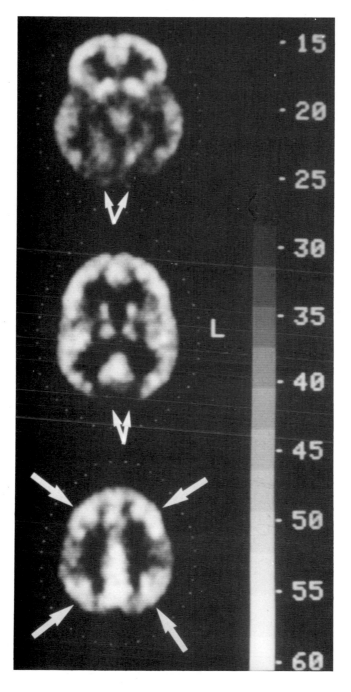

FIGURE 20. (*continued*)

The most common cause of focal brain injury involving the visual system is posterior cerebral artery (PCA) infarction, typically producing homonymous hemianopias or quadrantanopias
that may be transient when the primary lesion is subcortical (Bosley *et al.*, 1987). Not so seldom

though, infarcts are bilateral, and it is particularly in these cases that various neuropsychological deficits prevail. Figure 21 shows the CT and CMRglu images of a patient who, 5 weeks before, had suffered asymmetric (right-greater-than-left) bilateral PCA infarctions involving the hippo-campal formation, fusiform and lingual gyrus, optic radiation, visual cortex, and thalamus. The initial cortical blindness accompanied by Anton's syndrome soon receded, with vision returning homonymously in the lower right quadrant and, for the left eye, also in the lower left quadrant.

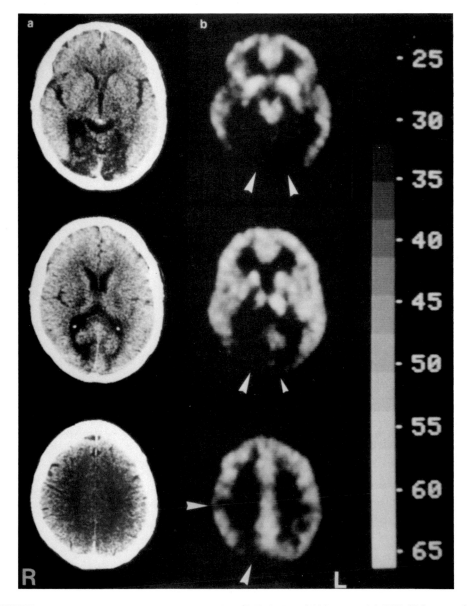

FIGURE 21. Characteristic CT (a) and corresponding CMRglu (μmol/100 g per min) PET (b) images of a stroke patient with prosopagnosia, static visual object agnosia, achromatopsia, visual hemineglect, and memory disorder weeks after bilateral PCA infarction. Arrowheads point at most severely deactivated regions.

Concurrently, prosopagnosia, static visual object agnosia, visual neglect to the left, achromatopsia, and serious memorizing difficulties become apparent. In addition to the morphological lesions documented by CT, PET studies demonstrated severely decreased functional metabolic activity bilaterally in the entire infracalcarine occipital lobe including the temporoccipital junction area and hippocampal structures (greater dysfunction on the right side)), in right thalamus, right posterior upper calcarine cortex, right upper half of Brodmann's area 18, and right supramarginal and intraparietal cortex, while CMRglu in the frontal eye fields was relatively increased. Comparison with the previous cases suggests that achromatopsia may be related to infracalcarine damage, whereas visual agnosia may require larger lingual gyrus lesions. Furthermore, the frontal eye fields may attempt to compensate for the dysfunction of visual association areas.

Another patient with prosopagnosia, moderate global amnesia with prominent loss of factual knowledge, semantic and word-selection anomia mostly for animate objects, mild Wernicke's aphasia, and ideational apraxia was studied 18 months after repeated cardiopulmonary resuscitation because of cardiac arrest. His CT (Figure 22a) and MRI (Figure 22b) showed the morphological sequelae of cerebral hypoxia/ischemia: moderate external and severe internal atrophy that was most marked in the temporal and lower occipital lobes as well as small demyelination spots in parietal white matter. The PET, however, revealed not only a moderate decrease in the overall metabolic level commensurate with the atrophy but also (Figure 22c) severe bilateral hypometabolism in Brodmann's area 37, fusiform, parahippocampal, and hippocampal gyri, Wernicke's area and adjacent insular cortex, left (and right) anterior and dorsolateral thalamus, and (predominantly posterior superior) parietal cortex (more severe dysfunction on the right side).

The clinical similarities in the latter two cases comprise prosopagnosia and certain amnestic symptoms, and their bilateral metabolic lesions overlap in the basal occipitotemporal junction area, inferior limbic and paralimbic structures, and in right parietal cortex. As outlined in the following section, serious memorizing deficits most frequently are related to hippocampal and thalamic dysfunction, thus leaving only the observed bilateral occipitotemporal and right parietal hypometabolism to account for the patients' inability to recognize and associate a specific meaning with familiar faces.

Amnesias

Memory disorders frequently are part of more generalized neuropsychological syndromes of various etiologies, especially dementive processes, and pure amnestic syndromes are comparatively rare. Although clinical evidence suggests that there are multiple dissociable memory systems, they may be particularly difficult to localize in mixed syndromes because of possible overlap of functional modules (see sections on other clinical syndromes). The pure amnesias, by contrast, are characterized by fairly constant patterns of symptoms, suggesting distinct sites of pathology that are selectively vulnerable to specific types of injury, but only a small number of patients with isolated memory defects have been studies with PET.

For example, Volpe, Herscovitch, Raichle, Hirst, and Gazzaniga (1983) investigated CBF and $CMRO_2$ in a single patient with transient global amnesia both in the amnestic state and following resolution of symptoms. The CT was normal, but PET demonstrated a global reduction of blood flow and, even more so, of oxygen consumption during amnesia; in both mesial temporal lobes as compared with the rest of the cerebrum, the oxygen extraction fraction was further decreased by some 19%. One day after the patient's recovery, the global depression of CBF and $CMRO_2$ as well as the focal mesial temporal depression of oxygen extraction had resolved.

In contrast, persistent bilateral oxygen hypometabolism of mesial temporal structures, with regional oxygen extraction fractions decreased by 19 to 32%, was found in chronic amnesias caused by global hypoxic–ischemic brain injury and characterized by intact short-term memory, severely impaired free recall, and less depressed recognition of visual and verbal material (Volpe & Hirst, 1983; Volpe, Herscovitch, & Raichle, 1984).

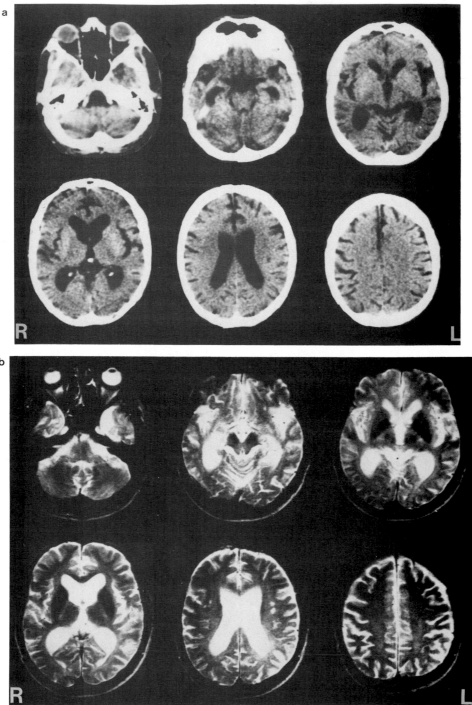

FIGURE 22. Characteristic CT (a), T_2-weighted MRI scans (b), and CMRglu (μmol/100 g per min) PET images (c) of a right-handed patient with chronic prosopagnosia, global amnesia, anomia, Wernicke's aphasia, and ideational apraxia consequent to hypoxic–ischemic brain damage. Arrows and arrowheads point at areas of most severe hypometabolic dysfunction.

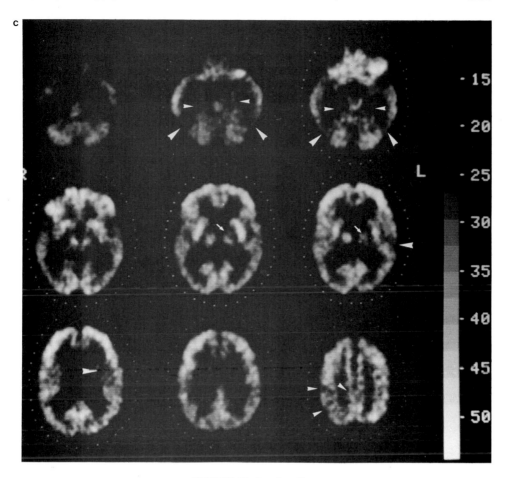

FIGURE 22. (*continued*)

We studied CMRglu in several patients with transient global amnesia, mostly during the early recovery phase, and usually found global metabolic decreases of varying degree, occasionally with marked bilateral functional depression of the hippocampal formation. A 60-year-old patient underwent PET scanning toward the end of his amnestic episode, approximately 20 hr after onset of total anterograde and lacunar retrograde amnesia. Typically, CT was unremarkable, and the overall brain glucose consumption rate was moderately decreased. This functional deactivation was more marked in the posterior circulation territory than in the carotid supply area, with the largest depression bilaterally in polar and mesial temporal gray matter and, to a lesser degree, in intraparietal cortex (Figure 23). Concurrently, there was a significant relative increase in left thalamic activity, perhaps reflecting some functional compensation.

These observations emphasize the critical role of mesial temporal lobe structures in the formation of lasting memory traces and in the retrieval of stored information. Bilateral lesions seem to be required for the manifestation of a severe amnestic syndrome. Although the CA1 field of the hippocampus may be most susceptible to hypoxic ischemia (Zola-Morgan, Squire, & Amarel, 1986), damage to this area may cause functional deactivation of closely related limbic and para-limbic structures as well, because in none of the described patients was hypometabolism limited

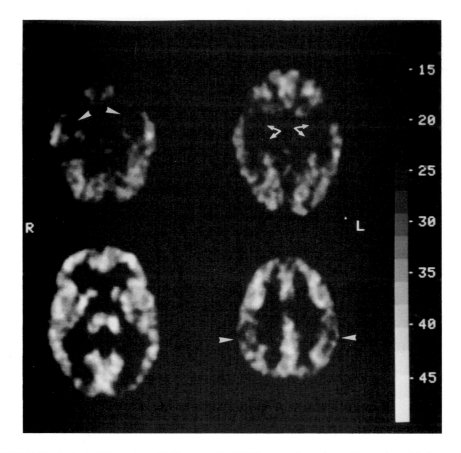

FIGURE 23. Typical CMRglu (μmol/100 g per min) PET images of a patient with transient global amnesia. Arrows and arrowheads mark deactivated regions.

to the hippocampus proper. Other well-documented cases with histologically verified bilateral chronic ischemic lesions exceeding the CA1 sector, e.g., also showing neuronal loss and gliosis in the subiculum and amygdala, presented clinically with dementia rather than with pure amnesia (Volpe & Petito, 1985).

Korsakoff's syndrome is another example of a relatively pure, at least partially reversible memory disorder. Its clinical characteristics generally include retrograde amnesia according to Ribot's law, a deficit in acquiring new information, and frequently confabulation. Thiamine deficiency as a result of alcoholism is the most common etiology, but there are other possible causes as well, e.g., thalamic hemorrhage rupturing into the third ventricle, temporarily producing mass effect on, and depression of CBF and $CMRO_2$ of, both thalami (Babikian et al., 1985). In a series of patients with alcoholic Korsakoff's syndrome and confabulation, we found no gross morphological abnormality other than mild brain atrophy, but PET, as in the study of Kessler et al. (1985), consistently demonstrated moderate decreases of whole-brain glucose consumption and severe, more or less symmetrical, bilateral focal hypometabolism of the hippocampal formation, cingulate cortex, hypothalamus, and ventrolateral, anterior, and dorsal thalamic nuclei (Figure 24).

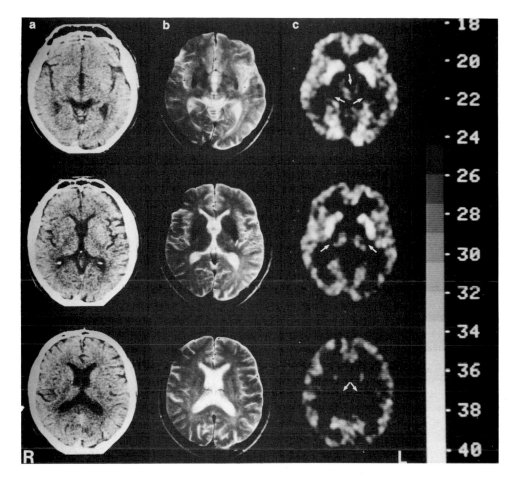

FIGURE 24. Characteristic CT (a), T_2-weighted MRI scans (b), and corresponding CMRglu (μmol/100 g per min) PET (c) images of a confabulating patient with alcoholic Korsakoff's syndrome. Arrows point at most dysfunctional regions.

Even small, lacunar infarcts in the territory of the polar thalamic artery in the dominant hemisphere, involving the mamillothalamic tract, produce complex neuropsychological syndromes that may be dominated by amnesia (Pawlik, Beil, *et al.*, 1985). Figure 25 illustrates the case of a 61-year-old right-hander presenting with severe anterograde and moderate lacunar retrograde amnesia, mild dementia with loss of judgment and intitiative, attention and verbal memory deficits, and moderate Wernicke's aphasia. Correlative clinical testing and neuroimaging studies were performed 6 weeks after acute onset of symptoms. Both CT and MRI showed mild bilateral temporal atrophy, slight left ventricular enlargement, and two lacunar infarcts in the left anterior and ventrolateral thalamic nuclei. In addition, small white matter lesions (ventricular capping and lining) and a spotlike lesion with increased signal intensity in the right posterior putamen were detected by MRI. Global CBF was moderately, and CMRglu mildly decreased. Regional dysfunction was proportionate for blood flow and glucose utilization in the left anterior thalamus, left frontal cortex, and right cerebellum. A depression of CBF in excess of CMRglu was found in right occipital cortex, while perfusion was relatively increased in Wernicke's area. However, the majority of deactivated regions, e.g., the rest of the left thalamus, right posterior putamen and insula, both inferior temporal lobes including the hippocampal formation, and left insular, pari-

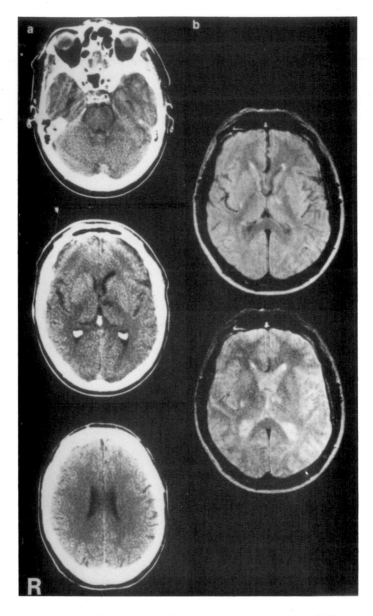

FIGURE 25. Characteristic CT (a), corresponding CBF (ml/100 g per min) (c) and CMRglu (μmol/100 g per min) (d) PET images, and progressively T_2-weighted spin-echo MRI scans (b) of center slice of a right-handed stroke patient with thalamic amnesia, Wernicke's aphasia, and mild dementia 6 weeks after left anterior thalamic infarction. Arrowheads mark most severely deactivated areas.

etal, and middle and superior temporal cortex, exhibited predominantly metabolic depression. This case clearly demonstrates the pitfalls of functional localization by morphological methods. It further supports the view that metabolic rather than blood flow imaging provides the closest correlations with the various neurobehavioral deficits: only the CMRglu images showed the full extent of left thalamic and bilateral mesial temporal involvement that may explain the amnestic

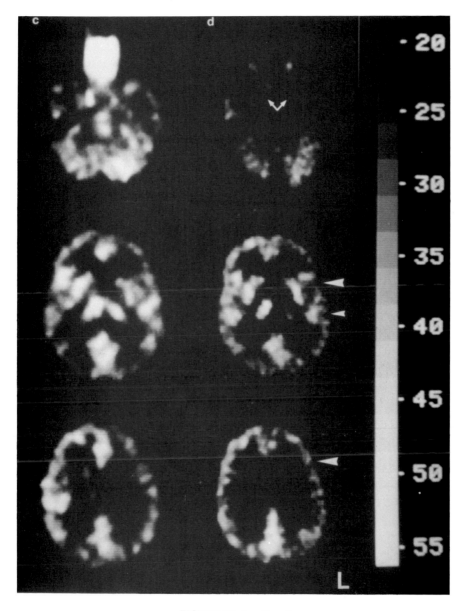

FIGURE 25. (*continued*)

syndrome, and the patient's cognitive impairment may be related to left frontal, temporal, and parietal cortical hypometabolism secondary to the lack of activating input from thalamus.

Dementias

In contrast with the amnestic syndromes, dementias are characterized by a more pervasive deterioration in intellectual function and personality in excess of the symptoms of normal aging, including benign senile forgetfulness and the progressive substitution of crystallized for fluid intelligence. Clinically, learning and memory are impaired in proportion to deficits of attention, orientation, insight, and judgment. Apraxias and agnosias as well as disorders of visuospatial abilities and of language-related functions may be found in addition. The dementias are a very heterogeneous group of diseases, and the diagnostic yield of modern neuroimaging techniques is as varied as are the etiologies. In general, however, PET is superior to morphological imaging because of its greater sensitivity and specificity with regard to the detection of abnormal patterns or absolute changes of the overall level of brain function (DeLeon, George, et al., 1983; McGeer, Kamo, Harrop, Li, et al., 1986). In the chronic progressive course of degenerative processes, PET abnormalities may even precede the manifestation or deterioration of clinical signs and symptoms. Despite some partial volume error of regional data because of brain atrophy, and although PET scanning during rest cannot distinguish decreased neuronal function from loss of synapses and neurons, PET often can provide essential information that is specific enough to permit the differential diagnosis among the various dementia types under consideration.

Alzheimer's disease and senile dementia of the Alzheimer type (DAT) not only account for the majority of dementias, they also are among the diseases most extensively studied with PET. An inverse relationship between the severity of the neuropsychological deficit and global measures of cerebral function, e.g., CBF and $CMRO_2$ (Frackowiak et al., 1981), CMRglu (Alavi, Fazekas, Chawluk, & Zimmerman, 1980; Benson, Kuhl, Phelps, Cummings, & Tsai, 1981; De Leon, Ferris, et al., 1983; Foster et al., 1983; Friedland et al., 1983; Kuhl et al., 1983; Chase, Fedio, et al., 1984; Duara, Grady, et al., 1986), and protein synthesis as determined by the cerebral uptake of L-[^{11}C]methionine (Bustany et al., 1983), was consistently reported. Another common finding was the pattern of bilateral regional dysfunction, with the most severe coupled decrease in metabolism and blood flow observed in posterior parietotemporal (up to 50% reduction as compared to age-matched normals) and lateral frontal (up to 40% reduction) association cortices, relatively sparing cerebellum and brainstem, subcortical gray matter structures, and primary visual and sensorimotor cortex (Benson et al., 1981; Friedland et al., 1983; Kuhl et al., 1983; Duara, Grady et al., 1986; McGeer, Kamo, Harrop, Li, et al., 1986; Jamieson et al., 1987). In autopsy-proven AD, the degree of regional CMRglu deficit was shown to parallel the extent of gliosis (McGeer, Kamo, Harrop, McGeer, et al., 1986).

As illustrated in Figure 26, the depression of regional energy metabolism generally is most profound in the temporoparietal junction area, affecting the frontal lobe primarily in more advanced stages of the disease (Frackowiak et al., 1981; Friedland et al., 1983; Duara, Grady, et al., 1986). Some authors also reported a significant negative correlation between left frontal CMRglu and age at onset of symptoms (Friedland et al., 1987). However, in early DAT presenting only with memory impairment or mild dementia and showing little or no brain atrophy, the slight functional abnormalities may readily escape detection by PET (Jamieson et al., 1987) unless discriminating regional metabolic ratios are used, e.g., temporoparietal/frontal cortex (Friedland et al., 1983), parietal cortex/caudate–thalamus (Kuhl et al., 1983), parietal/cerebellar cortex (Kuhl et al., 1985), parietal/sensorimotor and temporal/occipital cortex, or left-right asymmetries (Haxby et al., 1986).

Recent longitudinal PET studies of patients with possible DAT, or with probable DAT and mild dementia, with observation periods ranging between 6 and 40 months, consistently demonstrated that the distinct abnormalities of the above metabolic ratios remained essentially unchanged despite the clear-cut deterioration of dementia, which occurred in the majority of patients (Haxby, Grady, Koss, Friedland, & Rapoport, 1987; Kuhl et al., 1987); significant progression was noted only for the metabolic asymmetries of prefrontal, parietal, and lateral temporal association cortices (Haxby et al., 1987). In contrast, when the dementia was moderate initially and then became severe, the impairment of temporoparietal versus frontal cortex and the asymmetries of frontal and occipital cortex were larger on follow-up (Haxby et al., 1987; Jagust et al., 1987). These pattern changes may reflect a progressive functional disintegration within association areas of the temporal, parietal, and frontal lobes and between homologous association regions in the right and left hemispheres, as demonstrated by Horwitz, Grady, Schlageter, Duara, and Rapoport (1987), who used computer-intensive methods to identify reliable partial intercorrelations among the regional CMRglu values obtained in 21 mildly to moderately demented DAT patients and in 21 age-matched controls.

Several studies focused on the relationship between neuropsychological test performance and regional cerebral metabolism during rest. In general, verbal competency correlated with the metabolic activity in left parasylvian frontal, temporal, and parietal cortex, whereas visuoconstructive performance mainly localized to the right posterior parietal lobe (Foster et al., 1983; Chase, Fedio, et al., 1984). Likewise, apraxia scores on tests of imitation related most closely with CMRglu of right parietal cortex, and scores on tests to command correlated best with cortical metabolism in the left inferior cerebral hemisphere, especially in frontal regions (Foster, Chase, Patronas, Gillespie, & Fedio, 1986). Furthermore, significant positive correlations were obtained for the Token test and left temporoparietal cortex CMRglu and for the Drawing test and the glucose consumption of right temporal cortex; left temporal and occipital cortex-to-basal ganglia ratios correlated with the Naming subtest of the Boston Diagnostic Aphasia Examination, whereas the Map test related to the respective indices for the right hemisphere (Friedland et al., 1987). Although both mildly and moderately demented DAT patients exhibited significant metabolic asymmetry in associative neocortical regions, correspondence with appropriate discrepancies between language and visuoconstructive test scores was found only in patients with moderate dementia but not in mild cases, suggesting that metabolic dysfunction may precede neocortically mediated cognitive deficits in the course of DAT (Haxby et al., 1986).

Loewenstein et al. (1987), investigating hemispheric CMRglu normalized to occipital lobe metabolism in 18 patients with memory disorder or with mild dementia who also were given a battery of neuropsychological tests selected to distinguish between left and right hemisphere impairment, failed to detect statistically significant correlations between neuropsychological and metabolic measures related to the right cerebral hemisphere. However, they obtained significant positive correlations with left-hemisphere metabolic indices not only for putative left-hemisphere tasks but also for Visual Reproduction, the Picture Arrangement and Block Design subtests of the Wechsler Adult Intelligence Scale Revised, and Design Fluency. These findings were interpreted as suggesting that substantial left hemisphere impairment may be required for the clinical expression of right hemisphere dysfunction in mild dementia.

Age-dependent changes in cognition and behavior, as well as neuropathological abnormalities, in Down's syndrome largely correspond with manifestations found in DAT. Therefore, Schapiro et al. (1987) compared language, visuospatial ability, attention, and visual recognition memory test performance with CMRglu in 14 young (19 to 33 years old) and 6 old (47 to 63 years of age) subjects with Down's syndrome. Both the test scores and the metabolic activity in all investigated cerebral regions were significantly decreased in the old group (including 2 dements), with the most prominent dysfunction in right parietal and left temporal association areas as opposed to sensorimotor and occipital regions. Thus, the remarkable similarity between advanced Down's syndrome and DAT was essentially confirmed by PET.

The effect of simple neuropsychological task paradigms on regional CMRglu in early stages of dementia was investigated in a few activation PET studies. Miller et al. (1987) performed paired measurements in 2 mildly and 5 moderately impaired DAT patients and in 7 age-matched controls (all right-handed) during rest and during recognition from memory of written words (grocery items) presented one every 5 s on a computer screen. As expected, the controls showed good task performance, whereas the patients responded to most trials but performed at a chance level. For the five regions analyzed in either cerebral hemisphere, group differences in absolute metabolic rates were similar in the base-line and activated conditions, essentially replicating previously described reductions in DAT (temporal and parietal, -27%, frontal, -22%). However, patients and normals were distinguished even more clearly by their response patterns of metabolic lateral asymmetry in the temporal region (small portion of superior temporal gyrus, continuing posteriorly to include part of the inferior temporal gyrus, all at basal ganglia level): the controls

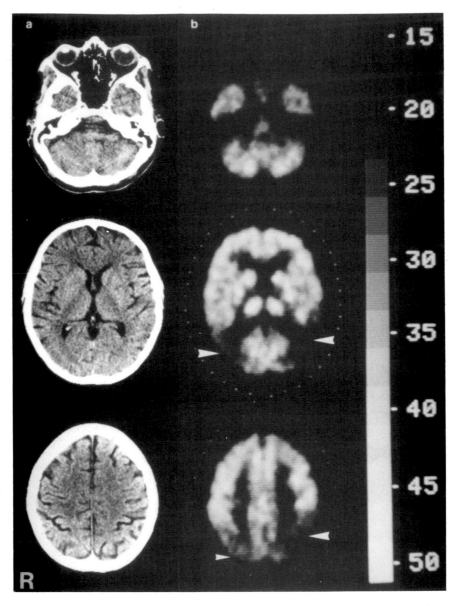

FIGURE 26. Characteristic CT (a) and corresponding CMRglu (μmol/100 g per min) PET images of a patient with Alzheimer's disease at an early stage (b); severe DAT (c) for comparison. Arrowheads point at most hypometabolic areas.

exhibited right predominance during both conditions (R 2.5 $\pm$ 5.8% and R 1.7 $\pm$ 6.8%, respectively), whereas in the DAT group left predominance was found during rest and right-greater-than-left CMRglu during task performance (L 5.6 $\pm$ 15.6% and R 4.7 $\pm$ 17.0%, respectively). Furthermore, the change in temporal percentage asymmetry from the base-line condition to the activated state was significantly correlated ($r = 0.66$, $P < 0.02$) with the score on the 7-point Global Deterioration Scale.

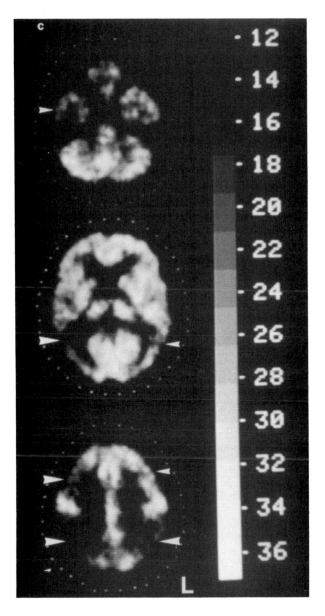

FIGURE 26. (*continued*)

In another paired rest/activation study of 6 patients with mild DAT and 3 normal controls, also using a verbal memory task, global CMRglu increased by approximately 11% in normals, with maximum activation in orbitofrontal and inferior temporal regions, whereas the patients activated only left inferior frontal cortex and cerebellum (Duara, Barker, Chang, Yoshii, & Apicella, 1986). Concurrently, significant increases in metabolic asymmetry during activation relative to rest were observed only in the dements, e.g., in medial temporal regions, where the percentage asymmetry was R 4.6 ± 4.0% at base line and R 8.6 ± 0.6% during task performance. When

13 aged normal subjects and 14 memory-impaired subjects with or without mild dementia read passages similar to those in the Wechsler Memory Scale and immediately thereafter recalled what was read, repeating this procedure approximately 15 times during the 30-min FDG uptake period, global CMRglu increased compared to the individual resting values by 15.6% in the controls and by 21.5% in the patients (Duara et al., 1987). Both groups showed the greatest activation in occipital regions (26.5% and 31.0%, respectively), and percentage increases were equivalent in those brain regions where the patients had significant hypometabolism during rest (left premotor, left superior and inferior parietal, left superior temporal). But the bilaterally least responsive areas differed: superior parietal (approximately 11% activation) in the controls, orbitofrontal (19.5%) in the patients.

Yoshii et al. (1987) analyzed the ratios of regional to whole-brain CMRglu determined with PET either during rest or while subjects viewed color slides of complex scenes, faces, animals, or flowers, projected during FDG uptake for 4 s each. In primary and associative visual cortex of young and old stimulated normals alike, relative metabolic activity was significantly greater than in age-matched, resting controls. However, no significant regional difference was found between two heterogeneous groups of mildly demented patients of comparable age (72.2 ± 5.9 years), with one studied at rest (5 DAT, 11 multiinfarct, and 4 mixed dementias) and the other (5 DAT, 4 multiinfarct, and 1 mixed) during behaviorally effective visual activation.

Although the activation studies described were largely exploratory, and the ideal stimulation paradigm may not have been found yet, reported results suggest that physiological abnormalities in early stages of dementia can be enhanced by various stress tests. In this regard, pattern shifts of functional asymmetries rather than of absolute or normalized metabolic rates may provide a sensitive measure of impaired associative regional interaction, unless too nonspecific a task is chosen. So far, the only consistent finding has been an abnormal shift in functional lateralization toward the right temporal lobe in patients with mild to moderate DAT performing a continuous verbal recognition task. This result may reflect the attempt of cooperative DAT patients to substitute inadequate right hemisphere function, e.g., increased visuospatial memory, attention, and global intuition, for the lacking analytic left hemisphere capabilities required to solve the task properly. Overall, because of the more global nature of the disease process and the great variability of maximum regional involvement, it may be concluded that PET studies of energy metabolism in DAT are not very likely to advance current knowledge of the cerebral localization of specific neuropsychological functions substantially, no matter how much they actually contribute to clinical differential diagnosis and to the monitoring of treatment effects (Heiss et al., 1988).

Pick's disease, a rarer form of degenerative dementia, is not always easy to distinguish from DAT on clinical grounds alone or by morphological neuroimaging because signs and symptoms often are quite similar. However, PET can demonstrate distinct differences in the pattern of regional CMRglu abnormalities as illustrated in Figure 27. Typically, whole-brain glucose consumption is moderately decreased, and, in contrast with other dementias, there is profound asymmetric depression of resting CMRglu in large, confined areas of atrophic frontal and inferior temporal cortex. Less severe asymmetric dysfunction may also occur in the parietal lobe, basal ganglia, thalamus, and cerebellum, with the latter structure largely reflecting a mirror image of frontal functional asymmetry. Except for the cerebellum showing no histomorphological abnormality, good agreement was found between the severity of regional hypometabolism and the degree of gliosis and neuronal loss (Kamo et al., 1987).

Progressive supranuclear palsy (Steele–Richardson–Olszewski syndrome, SRO) is characterized clinically by vertical gaze palsy, dysarthria, nuchal and limb rigidity, bradykinesia, and toppling gait, generally accompanied by mild to moderate dementia. Neuropathological changes are of the degenerative type (neurofibrillary tangles, neuronal loss, and gliosis). They commonly affect several subcortical gray matter structures (e.g., reticular formation, substantia nigra, raphe system, nucleus basalis of Meynert, locus coeruleus, pallidum, striatum, and the mesocortico-limbic system), all of which project either directly or indirectly via thalamic nuclei onto frontal cortex, but they usually spare cerebral neocortex. Accordingly, CT often is unremarkable, show-

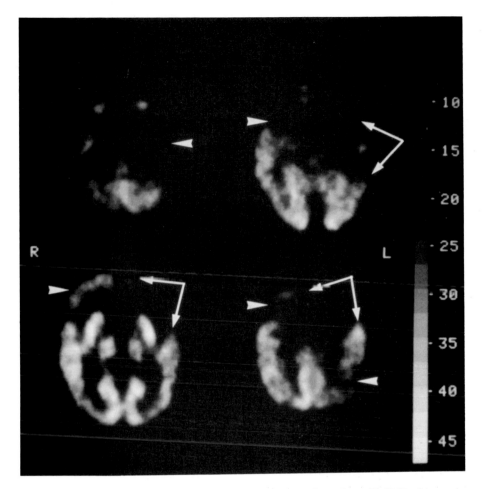

FIGURE 27. Typical CMRglu (μmol/100 g per min) PET images of a patient with Pick's disease. Arrows
and arrowheads mark regions showing most severe hypometabolism.

ing at the most some diffuse brain atrophy. In contrast, PET studies of CMRglu in 6 patients
(D'Antona *et al.*, 1985) and of CBF and $CMRO_2$ in 4 patients with probable SRO (Leenders,
Frackowiak, *et al.*, 1987) concurrently demonstrated a reduction of global energy metabolism by
approximately 20% and of CBF by approximately 40%, with the largest metabolic decreases
(-21 to -30%) in almost symmetric regions of frontal and temporal cortex. Furthermore, we
observed distinct bilateral CMRglu deficits in thalamus, brainstem, and cerebellum. The degree
of frontal metabolic depression was shown to correlate with the duration of symptoms but not
with their severity; i.e., pathophysiological changes may precede clinical manifestations.

Despite obvious similarity as to the areas of maximum abnormality, some PET criteria clearly
distinguish SRO from Pick's disease: (1) there is no major functional asymmetry, (2) frontal and
temporal deactivation is less severe, and (3) the dysfunctional area is not as well demarcated.
Leenders, Frackowiak, *et al.* (1987a) also measured with PET the L-[^{18}F] fluoro-DOPA uptake in
5 SRO patients and found striatal dopamine metabolism to be significantly reduced, thus supporting the concept of degeneration of the nigrostriatal dopaminergic system. However, using the

high-affinity D_2-receptor ligand 3-N-[^{18}F]fluoroethylspiperone (Wienhard et al., 1988), we failed to observe any decrease in the striatum-to-cerebellum ratio (>8) of tracer concentrations in SRO 3 hr after injection that might have indicated reduced striatal dopamine receptor density, as reported by Mazière, Baron, Loc'h, Cambon, and Agid (1987), who applied the comparatively low-affinity ligand [^{76}Br]bromospiperone (striatum-to-cerebellum ratio <2 after 4.5 hr) in 7 SRO patients. According to the PET findings described, the symptoms of dementia characteristic of SRO (apathy and slow cerebration, loss of abstracting ability and judgment, perseveration, and personality changes) probably must be attributed to frontal cortical deactivation resulting from the lack of largely thalamus-mediated excitatory input, which in turn may require functional integrity of the basal dopaminergic system.

Huntington's disease (HD), another degenerative disorder affecting the striatal dopaminergic system, usually manifests with chorea, personality change, and depression, but at least in more advanced cases dementia is quite common. Pathoanatomically, there is neuronal loss primarily in caudate nucleus and putamen, eventually leading to bulk loss of striatal tissue and central atrophy. However, long before any morphological changes can be detected by CT or MRI, PET measures of caudate (and usually also of putamen) CMRglu are significantly decreased (Kuhl, Metter, Riege, & Markham, 1984), even in clinically normal subjects at high genetic risk who are going to develop symptoms of HD within a couple of years (Hayden et al., 1987). Although the severity of dementia was found closely associated with the degree of central atrophy, extrastriatal CMRglu remains generally unchanged (Kuhl et al., 1984). In patients with advanced HD and prominent dementia, though, in addition to severe bilateral striatal hypometabolism, we repeatedly observed (Figure 28) moderate to severe decreases in whole-brain glucose consumption as well as asymmetric regional dysfunction of thalamus and parietal, temporal, and frontal cortex, bearing some resemblance to the metabolic changes in DAT. The relative observation frequence of such cases makes comorbidity with HD and DAT an unlikely explanation.

Similar results were obtained in 14 patients with Parkinson's disease, the most common disorder of the nigrostriatal dopaminergic system, presenting with dementia in addition to the typical motor symptoms at a higher rate than expected by chance alone (Kuhl et al., 1985): global CMRglu was significantly below normal when the dementia was moderate to severe; in contrast with DAT, bilateral cortical hypometabolism also involved the occipital calcarine cortex, but otherwise the metabolic patterns were similar again, showing maximum dysfunction in posterior parietal cortex and minimum impairment of cerebellum, caudate, and thalamus; the parietal-cortex/cerebellar cortex metabolic ratio correlated negatively with both the severity and the duration of dementia. Therefore, the findings described in huntingtonian and parkinsonian dementia suggest that prolonged corticostriatal functional imbalance may cause permanent abnormalities in various regions of cerebral cortex.

Frackowiak, Herold, Petty, and Morgan-Hughes (1987) investigated $CMRO_2$ and CMRglu in 4 demented and in 4 nondemented patients with mitochondrial cytopathies. Although these adult-onset syndromes of respiratory chain dysfuntion are characterized primarily by distinct enzyme abnormalities in skeletal muscle mitochondria, neural tissue may be affected secondarily. The patients without dementia had normal glucose and mildly increased oxygen consumption, suggestive of some ketone use. In the demented patients, by contrast, a 50% decrease in $CMRO_2$ and a 26% fall in CMRglu were found, with a molar oxygen-to glucose utilization ratio of 3.8 $\pm$ 0.8, reflecting profound cerebral degeneration.

Among the infectious etiologies of dementia, herpes simplex encephalitis plays a special role because a frequent site of the focal disease process is the inferior temporal lobe, often showing bilateral involvement. When the inflammation is limited to mesial temporal structures, a pure amnestic syndrome may result, whereas patients with more extensive bilateral lesions may develop additional cognitive and behavioral changes reminiscent of the Klüver–Bucy syndrome observed in monkeys after experimental bilateral temporal lobotomy with ablation or isolation of the amygdala.

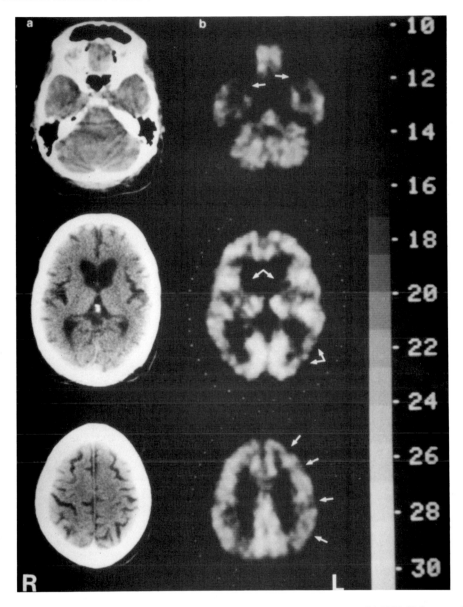

FIGURE 28. Characteristic CT (a) and corresponding CMRglu (μmol/100 g per min) PET (b) images of a right-handed patient with advanced Huntington's disease and prominent dementia. Arrows mark most hypometabolic regions.

The case illustrated in Figure 29 was characterized clinically by a combination of amnesia, prosopagnosia, loss of orientation, judgment, insight, and affective adequacy, disinhibition of responses to external stimuli, and lack of restraint of sexual and mouthing impulses. Both CT and MRI showed moderate brain atrophy, accentuated in temporal and inferior frontal regions, as well as large, nearly symmetrical areas of morphological damage comprising the inferior temporal lobe, insular cortex, and hippocampal gyrus. Severe depression of global CMRglu sparing only

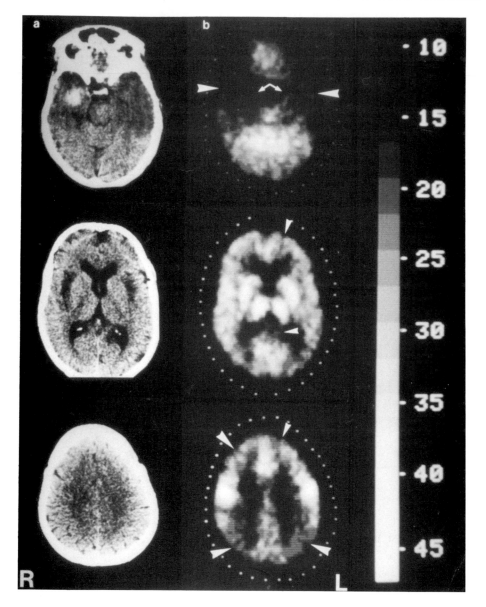

FIGURE 29. Characteristic CT (a) and corresponding CMRglu (μmol/100 g per min) PET (b) images of a patient with a Klüver–Bucy-like syndrome 1 month after bitemporal herpes simplex encephalitis. Arrows and arrowheads mark areas of structural damage and regions showing severe functional deactivation.

infratentorial structures, thalamus, basal ganglia, and superior sensorimotor cortex was demonstrated by PET. Deactivation of morphologically intact cerebral cortex was largest bilaterally in the basal occipitotemporal junction, posterior parietal, and pre- and midfrontal regions. Therefore, functional rather than morphological neuroimaging again revealed the disorder of limbic, paralimbic, and associative cortex interactions presumably underlying this peculiar type of dementia.

The pattern of cerebral glucose metabolic dysfunction may be more varied in Creutzfeldt–Jakob disease, a subacute spongiform encephalitis caused by a slow infectious agent, leading to progressive global dementia, pyramidal and extrapyramidal findings, myoclonus, and early death. Using the FDG method and PET, Horowitz, Benson, Kuhl, and Cummings (1982) observed patchy metabolic hypofunction with loss of definition of subcortical structures that was similar in the 2 patients studied. As in DAT, Friedland, Prusiner, Jagust, Budinger, and Davis (1984) found relative FDG uptake to be asymmetrically decreased in both posterior temporal cortices of a severely demented patient, whose CT was normal at that time, 7 months prior to his death.

A correlative study of CMRglu and neuropsychological test results in 12 patients with the AIDS dementia complex occurring in more than two-thirds of all AIDS cases, failed to demonstrate significant relationships between regional metabolic rates on the one side and the poor performance on tests of verbal fluency and problem solving on the other (Navia et al., 1987). However, significant decreases as compared to 18 controls were noted in the relative metabolic activity of a small number of brain regions (e.g., basal ganglia, thalamus, and frontal cortex) also showing significant correlations with motor test scores.

Nowadays a rarity dementia paralytica, a late form of cerebral syphilis, used to be a major differential diagnosis of progressive mental disability. Typical CSF abnormalities including positive serological tests establish the diagnosis. Morphological imaging may show focal changes commensurate with Lissauer's cerebral sclerosis, but usually there is no direct correspondence with dementia. We studied a patient with only discrete focal neurological signs and prominent dementia dominated by impaired memory, lack of orientation, insight, and judgment, and inappropriate jocosity. The MRI demonstrated moderate bilateral temporal atrophy and signal hyperintensity as well as a small left prefrontal area of increased signal intensity (Figure 30). Overall CMRglu was within the lower range of normal, but there was a distinct pattern of decreased regional cortical glucose consumption, showing maximum bilateral involvement of inferior anterior and mesial temporal lobe structures as well as of insular cortex. Furthermore, slightly asymmetric dysfunction was observed in anterior intraparietal and superior pre- and midfrontal regions. This particular topographical distribution of metabolic abnormalities distinguishes dementia paralytica from the other forms of dementia described above, and it may explain the characteristic clinical deficit. There also is little similarity to the dementia caused by cerebral vasculitis with prominent arteriolar involvement, where we found moderate to severe symmetrical metabolic depression bilaterally in the vascular watershed areas, which appeared normal on CT or MRI.

Multiinfarct dementias (MIDs), together with mixed forms of vascular and degenerative origin accounting for some 30%, constitute the second largest group of all dementias. Although the clinical appearance of MID and DAT often is quite similar on cross-sectional examination, distinction between the two generally can be achieved when the patient's history and focal neurological and neuroimaging findings are put into proper perspective (Hachinski et al., 1975; Rosen, Terry, Fuld, Katzman, & Peck, 1980). In this regard, PET is superior to CT (Kuhl et al., 1983; Heiss et al., 1986), and MRI can demonstrate white matter lesions with maximum sensitivity (Alavi et al., 1987). In a series of 9 patients with vascular dementia (4 moderate, 5 severe), 5 of whom had typical MID and 4 met clinical and CT criteria of subcortical arteriosclerotic encephalopathy (Binswanger's disease), a proportional decrease by approximately 23%, as compared to 14 controls, was found for whole-brain CBF and $CMRO_2$ (Frackowiak et al., 1981). As in DAT, this functional depression paralleled the severity of dementia, and the principal effect occurred in parietal regions. Frontal regions, however, showed consistently less involvement than in DAT. Kuhl et al. (1983), investigating CMRglu in 6 MID patients, noted a 17% reduction of global cerebral metabolism as well as scattered metabolic defects in cortex, caudate, thalamus, white matter, and cerebellum. Compared with the DAT pattern, we found greater functional asymmetry in MID, the defects were more sharply demarcated from surrounding tissue, and their primary location was in the border zones of the MCA territory or in the terminal PCA supply area, i.e., in typical sites of infarction.

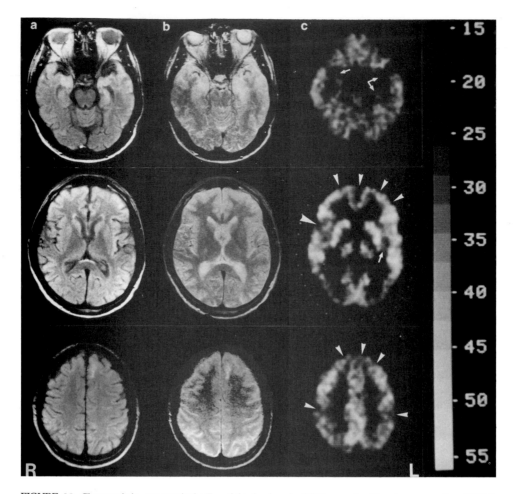

FIGURE 30. Characteristic, progressively T_2-weighted spin-echo MRI scans (a, b) and corresponding CMRglu (μmol/100 g per min) PET images (c) of a patient with dementia paralytica. Arrows and arrowheads mark most dysfunctional areas.

Occasionally, even two small "strategic" infarcts can produce a severe dementia, as we observed in a 38-year-old woman with bilateral, nearly symmetrical, paramedian thalamic infarction involving dorsomedial, centromedian, anterior, and lateral ventral nuclei and the mamillo-thalamic tract (Figure 31). Following acute onset, the clinical picture was dominated for months by extreme hypersomnia, with an attention span of a few seconds only, thus precluding any formal testing. There also was persistent, complete, bilateral ophthalmoplegia and severe thalamic dysarthria. At this stage, PET demonstrated a 36% global decrease in CMRglu as compared to 12 age-matched controls, with the most severe deficits, aside from the lesions, in noninfarcted thalamic nuclei (-44%), in frontal (-42%), cingulate (-39%), parietal (-38%), and insular (-38%) cortices, and in the caudate nuclei (-38%). Eleven months after the ictus, the patient was still hypersomnic, but she was no longer bedridden, and she would stay awake while receiving personal attention, Orientation to time was completely lost, and there was a severe memory disorder with prominent anterograde amnesia, global cognitive impairment, marked psychomotor retardation, lack of insight and judgment. Moreover, compared with the premorbid state, a pro-

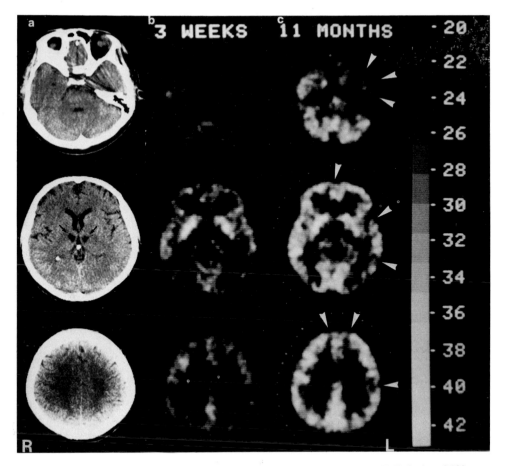

FIGURE 31. Characteristic late CT (a) and corresponding early (b) and late (c) CMRglu (μmol/100 g per min) PET images of a right-handed patient with severe hypersomnia and thalamic dementia consequent to bilateral paramedian thalamic infarction (black arrows). Arrowheads point at cortical regions showing severe, persistent functional deactivation.

found personality change had occurred, manifesting itself with naivete, infantile curiosity, and emotional instability with occasional aggressive outbursts.

Overall, the patient's personality and mental capacity were comparable to a developmental age of 3. Corresponding with the clinical course, the repeat PET study showed markedly improved metabolic activity, although whole-brain CMRglu still was 15% below normal. The degree of functional improvement, however, differed considerably among regions, with the largest relative increases in visual cortex, white matter, and cerebellum. Moderate to severe deactivation persisted bilaterally in noninfarcted thalamic nuclei (-34%), frontal cortex (-27%), brainstem, (-22%), and striatum (-22%). Prominent asymmetric dysfunction was found in the left inferior temporal lobe and intraparietal cortex. These PET findings suggest that a complex disturbance of the interactions among distinct limbic, paralimbic, thalamic, and heteromodal associative cortical areas, primarily in the dominant hemisphere, constitutes the pathophysiological basis of thalamic dementia.

Normal-pressure hydrocephalus (NPH) is generally accepted as a not so rare cause of treatable dementia. It may be suspected in patients presenting with a generalized decline in cognitive function associated with gait difficulty and urinary (or double) incontinence, showing dilated ventricles with comparatively little cortical atrophy as well as typical findings in radionuclide cisternography and/or CSF pressure monitoring. However, the diagnosis is certain only when the patient's clinical condition improves following CSF shunting, thus emphasizing the need for discriminating preoperative selection criteria. Preliminary PET results obtained in 3 patients, whose diagnosis was established by other technical findings and clinical course, suggest that cerebral glucose metabolism may be typically altered in NPH (Jagust, Friedland, & Budinger, 1985). Compared to 7 healthy elderly controls, cortical CMRglu at the midventricular level was diffusely decreased by approximately 32%. Major focal changes or functional asymmetries caused by NPH were not observed, but there was a trend toward hypofrontality relative to temporoparietal cortex activity, thus contrasting, e.g., with the metabolic pattern characteristic of DAT.

Traumatic dementia may be observed in the wake of prolonged coma and stupor consequent to severe craniocerebral injury. It commonly is a transient state of cognitive and affective deficiency, eventually leading to more or less complete recovery of mental functions. As in the case illustrated in Figure 32, morphological neuroimaging may show some diffuse brain atrophy of minor degree, aside from possible contusional defects. However, PET typically reveals marked global cerebral hypometabolism and a chracteristic pattern of largely symmetrical regional dysfunction, with the most severe CMRglu deficits in orbito- and prefrontal, parietal, and inferior temporal cortices. Other regions, by contrast, may exhibit only minor metabolic depression or none: primary sensory and motor areas, basal ganglia, thalamus, and infratentorial structures. Overall, hypometabolic dysfunction of heteromodal association areas seems to be the common denominator of the various described forms of dementia, whereas the characteristic symptomatology of a specific type of dementia appears related to additional differential involvement of certain basal and thalamic nuclei, of limbic, paralimbic, and occasionally also of unimodal association areas.

Disorders of Attention, Judgment, Affect, Psychomotor Activity, and Social Behavior

Perhaps the most important prerequisite for the manifestation of all higher mental functions, attention, with its level and channel components, regulates both the overall cerebral information-processing capacity and its instantaneous focus. Likewise, the experience and expression of affect fundamentally influence all cognitive processes and human behavior. In conjunction with memory, these qualities constitute the functional basis of proper judgment and appropriate social interaction. They can be seriously impaired in a great variety of neurological and psychiatric syndromes, but only disorders likely to be manifestations of primarily abnormal brain physiology are considered here, as far as they have been studied with PET. Therefore, the vast and rather contradictory literature on PET abnormalities associated with typical psychiatric illness, e.g., CMRglu (Farkas *et al.*, 1984; DeLisi *et al.*, 1985; Kling, Metter, Riege, & Kuhl, 1986; Volkow *et al.*, 1986; Buchsbaum *et al.*, 1987; Gur *et al.*, 1987; Wiesel *et al.*, 1987), CBF (Early, Reiman, Raichle, & Spitznagel, 1987), cerebral amino acid pools (Kishimoto *et al.*, 1987a), cerebral methionine metabolism (Bustany *et al.*, 1983), and striatal D_2 dopamine receptor density (Wong *et al.*, 1986; Farde *et al.*, 1987) in schizophrenia, are not discussed. Likewise, for initial advances with PET in depression research, the interested reader is referred to a review article (Pawlik, Beil, *et al.*, 1986); the latest developments in major affective disorders or in obsessive–compulsive disorder were reported by Kling *et al.* (1986), Post *et al.* (1987), Kishimoto, Takazu, *et al.* (1987), and Baxter *et al.* (1987); regional CMRglu in anorexia nervosa was investigated by Herholz *et al.* (1987).

The special role of the frontal lobes and the limbic system in the maintenance of attention, psychomotor activity, social judgment, and in the control of drive and affect is widely acknowledged. It may be illustrated by the case of a 30-year-old right-handed male tax consultant, who

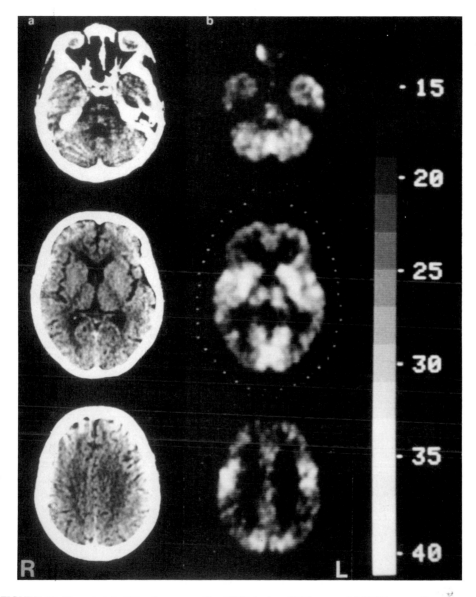

FIGURE 32. Characteristic CT and corresponding CMRglu (μmol/100 g per min) PET images of an 18-year-old patient with traumatic dementia 6 weeks after severe closed craniocerebral injury.

came to medical attention because of acute, irresistable sleepiness, apathy, mild memory deficits, and slurred speech. In contrast with his previously flawless legal record and personal and professional achievements, during the previous couple of months he had been arrested repeatedly for drunk driving, speeding, and driving without a license, and he had lost several customers because of unprofessional conduct and mismanagement of client's affairs. Despite the persistence of mental slowing and deficient attention, his apathy soon gave way to grandiose ideas about his professional future that were readily advanced; his affect was labile but largely dull, and he showed little concern for his family, his fellow patients, or his impending court trials. At this time, CT revealed a large low-density area in the right orbitofrontal to midfrontal ACA territory (Figure 33) that was diagnosed by angiography as an infarct resulting from moya-moya disease. The PET demonstrated severe regional ischemia not only in the infarcted area but also in right superior

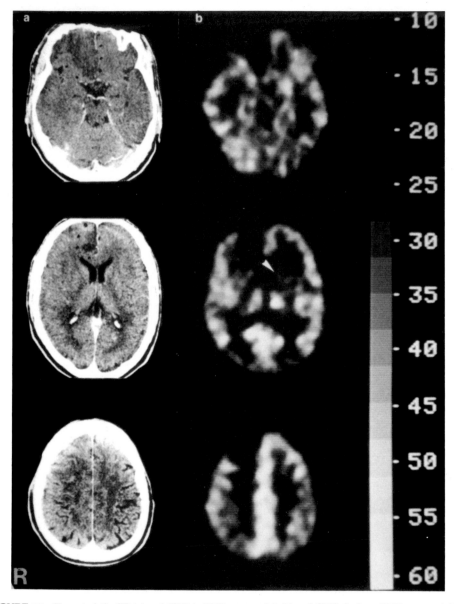

FIGURE 33. Characteristic CT (a) and CBF (ml/100 g per min) (b) and CMRglu (μmol/100 g per min) (c) PET images of a right-handed patient with right frontal ACA infarction resulting from moya-moya disease, causing disorders of attention, judgment, and affect. Arrowhead points at left striatal region, where flow deficit exceeds metabolic decrease. Arrows mark regions showing more severe metabolic than hemodynamic impairment.

prefrontal cortex and bilaterally in anterior insula and striatum, showing the larger CBF deficit in the left hemisphere. The CMRglu was significantly reduced in the same structures, but the lateral asymmetry of striatum and insular cortex was reversed, and additional hypometabolic dysfunction was found in both amygdala regions, in the right thalamus, and in right posterior superior prefrontal cortex. Because moya-moya disease is a more chronic, progressing–remitting type of cerebrovascular disease, it may be assumed that both the past behavioral changes and the present

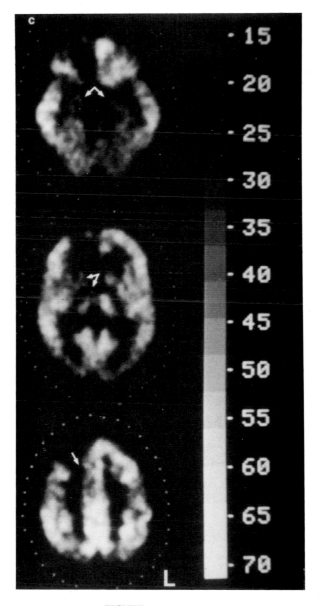

FIGURE 33. (*continued*)

complex neuropsychological deficits were produced by the loss of regional brain functions seen on the metabolic PET images, thus characterizing the syndrome as one of severe frontothalamo-limbic disintegration.

    The strong, globally modulating effect of even unilateral thalamic activity on brain metabolism and fundamental aspects of behavior was quite obvious in a study of patients with typical thalamic pain syndrome consequent to lacunar infarction in the posterolateral supply territory of the thalamogeniculate artery who had had a stimulation electrode implanted in the posterolateral ventral thalamus for pain control (Pawlik, Beil, *et al.*, 1985). Patients were remarkably similar with respect to infarct size and location, behavior, and response to electrical stimulation. A few hours after the stimulator was switched off, i.e., when the painful deep sensations in the ataxic, hypalgesic, and mildly paretic limbs reappeared, they became depressed and irritated, affect was labile, there was marked psychomotor retardation and perseveration and loss of attention, judgment, and initiative; these behavioral abnormalities largely receded within minutes when stimulation was resumed. In the unstimulated state, whole-brain CMRglu was decreased by 7% as compared to 10 age-matched controls. More conspicuous, though, was the asymmetric pattern of regional deactivations and of abnormal increases in relative functional activity (Figure 34). With-

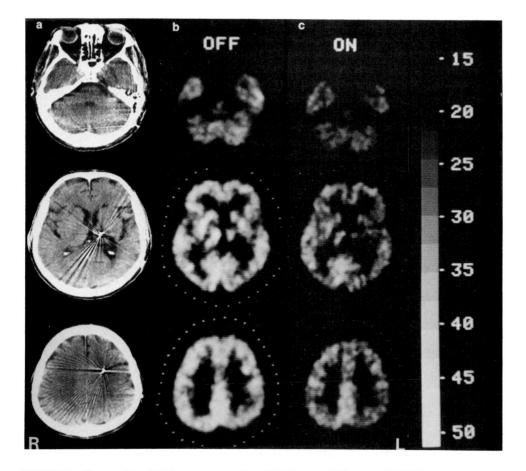

FIGURE 34. Characteristic CT (a) and corresponding CMRglu (μmol/100 g per min) PET images of a right-handed stroke patient having a stimulation electrode implanted for control of thalamic pain syndrome consequent to left posterolateral thalamic infarction. (b) Metabolic images were obtained after stimulator had been switched off for 12 hr; (c) images show CMRglu shortly after stimulator was turned on again.

out stimulation, metabolic depression was largest in noninfarcted nuclei of the affected thalamus (-26%), in ipsilateral caudate (-16%), in temporal limbic structures (-15%), as well as in contralateral thalamus (-13%) and ipsilateral (-13%) and contralateral cerebellum (-15%), while major relative activation was apparent primarily in contralateral frontal, parietal, and occipital cortex ($+5$ to 13%). During stimulation, global CMRglu decreased by another 18%, with the previously more active regions showing the larger and the directly deactivated areas the lesser decline, thus reducing functional asymmetry. Therefore, the described behavioral changes in the electrode-off condition may represent release phenomena related to gradually developing, lateralized, regional cortical overstimulation caused by the lack of thalamic filtering and integrating activity that can be partially replaced by appropriate electrical stimulation.

Gilles de la Tourette syndrome is a rare, likely neurological disorder of unknown etiology, characterized by multiple motor and vocal tics that are commonly associated with hyperactivity, troublesome sexual and aggressive impulses, echopraxia, palilalia, involuntary cursing, coprolalia, and deficits of attention and perception. In a PET study of 5 such patients aged 30 ± 8.9, all of whom were drug-free right-handers, Chase, Foster, et al. (1984) found no significant abnormality of overall CMRglu or of side-to-side differences. However, compared with 7 older (59 ± 5.3 years) normal subjects, glucose utilization in the patients' basal ganglia averaged 16% above control level, and relative hypermetabolism was noted in certain portions of the frontal and temporal lobes bilaterally. The severity of vocal tics appeared to be inversely related to the metabolic activity of middle and inferior frontal regions extending from the frontal pole to the postcentral gyrus, whereas coprolalia correlated with left parasylvian CMRglu. These preliminary findings suggest that some dyssynergy of subcortical nuclei and frontal and temporal cortex may underlie this complex behavioral and movement disorder.

Mountz and co-workers (1986) performed repeated PET studies of CBF in 5 right-handed women with simple phobia (of small animals) both during rest and while confronted with the stimulus that produced intense anxiety. In the latter condition, there was a bilateral, symmetrical flow increase in a region near the anterior inferomedial aspect of the temporal lobe. This observation probably does not provide any clue to the pathogenesis of phobia, but it suggests significant involvement of mesial temporal structures in producing irrational anxiety responses.

At least in some patients with panic disorder, there may be a strong organic factor predisposing to anxiety attacks, as indicated by the results of a comprehensive PET study of CBF, cerebral blood volume, and $CMRO_2$ (Reiman et al., 1986). Sixteen patients with panic disorder (7 without, 9 with limited phobic avoidance; 5 men, 11 women; 12 right-, 4 left-handed; age range, 22–52 years), divided into two equal groups according to their separately determined response (subjective report of typical panic attack, group I) or nonresponse (group II) to lactate infusion, and 25 healthy volunteers (16 men, 9 women; 23 right-, 2 left-handed; age range, 19–74 years) were scanned in a nonpanic, resting state with their eyes closed and ears unoccluded. Both patient groups had a significantly higher number of anxiety symptoms than the controls, but only those patients who were vulnerable to lactate-induced panic had a mild uncompensated respiratory alkalosis, apparently in response to the experimental situation. As to the PET measurements, only the data for whole brain and for the left and right parahippocampal regions were analyzed.

Both the global CBF corrected for differences in Pco_2 and whole brain $CMRO_2$ were significantly higher ($+22.3\%$ both) in patients vulnerable to lactate-induced panic than in nonvulnerable patients and in controls (54.8 ± 10.06 ml/100 g per min and 120 ± 17.6 μmole/100 g per min, respectively), but differences in the Pco_2-adjusted blood volume (4.8 ± 0.92 ml/100 g) and in oxygen extraction fraction ($37 \pm 9.6\%$) were not statistically significant. Likewise, adjusted CBF (left, 66.2 ± 20.10; right, 72.4 ± 18.70 ml/100 per min) and $CMRO_2$ (left, 135 ± 21.9; right, 146 ± 25.0 μmole/100 g per min) in both parahippocampal regions and adjusted right parahippocampal blood volume (5.7 ± 1.22 ml/100 g) of group-I patients were asymmetrically elevated above the values measured in the other subjects (adjusted CBF: left, 54.7 ± 10.04; right, 56.4 ± 10.82 ml/100 g per min; $CMRO_2$: left, 125 ± 23.6; right, 126 ± 24.2 μmole/100 g per min; adjusted right blood volume: 5.3 ± 1.02 ml/100 g). The right parahippocampal predominance associated with lactate-vulnerable panic disorder was most obvious in the significant decreases in the left-to-right ratios of parahippocampal CBF (-10%), $CMRO_2$ (-7.5%), and blood volume (-15.7%).

According to the authors' interpretation, these findings suggest that there are at least two discrete populations of individuals with panic disorder that can best be distinguished by either response to lactate or functional parahippocampal asymmetry. Considering the known anatomic and chemoarchitectural relationships between the parahippocampal gyrus on the one side and hippocampus, subiculum, entorhinal cortex, heteromodal association areas, amygdala, raphe nuclei, and locus coeruleus on the other, the described parahippocampal abnormalities may well be explained as unbalanced increases in the regional vascular surface permeability product, caused by some hyperactivity of the central noradrenergic system arising in the locus coeruleus. In the same vein, the respiratory alkalosis observed in the patients vulnerable to lactate-induced panic may reflect their abnormal susceptibility to sympathetic arousal.

The glucose metabolic correlates of mental retardation were determined with PET in 4 children (age range, 2–6 years) with severe birth asphyxia resulting in profound psychomotor retardation, microcephaly, spastic quadriparesis, and seizures and in 8 children with normal birth who developed seizures and mild to moderate mental retardation between 2 months and 6 years of age (Chugani, Phelps, Light, & Mazziotta 1987). Global CMRglu in all 12 retarded children was well below normal values for age, showing an apparent correlation with the degree of mental impairment. Furthermore, glucose utilization rates were generally lower in frontal, parietal, and temporal cortices compared to basal ganglia, thalamic, and primary sensory areas. Two of the severely retarded children also displayed bizarre metabolic patterns, with patches of active tissue surrounded by extremely hypometabolic areas. These findings probably are quite representative of the majority of cases of mental retardation.

However, there remains a comparatively small group that has attracted particular interest: patients with obvious mental defects but with an unremarkable family history, uneventful pregnancy and perinatal period, and without signs of any other developmental abnormality or neurological disorder. It comprises various forms of pervasive developmental disorders of childhood in which a neurological basis often can only be suspected. The core symptom of impaired relatedness cuts across widely varying levels of neuropsychological functioning and commonly persists into adulthood.

Rumsey et al. (1985) investigated CMRglu in 10 men aged 18 to 36, with well-documented histories of infantile autism. Eight had a residual state without substantial language deficit or deviance, and 2 presented the full syndrome with language disturbances. All were drug-free and had normal CT scans. In the resting state, with eyes and ears occluded, whole-brain CMRglu was raised significantly above the control values determined in 15 healthy male volunteers between the ages of 20 and 37 years. This hypermetabolism involved most of the 59 regions analyzed. Relative metabolic decreases were found only in superior frontal, postcentral, and superior parietal cortex as well as in the paracentral lobule. Although the autistic group as a whole appeared to display fairly normal functional asymmetry, significant lateralization to the left was detected in superior frontal cortex, where the controls showed some right predominance. These findings are at variance with the results obtained by Herold, Frackowiak, Rutter, and Howlin (1985), who measured CBF, $CMRO_2$, cerebral blood volume, and CMRglu in a single plane across basal ganglia and temporal gray matter of 6 unmedicated men (age range, 21–25 years) with an unequivocal diagnosis of infantile autism. Three had normal CT scans, and 3 showed definite signs of atrophy. Compared with 6 older healthy men and 2 women, or with 6 age-matched normal volunteers, no significant difference was found for any of the physiological variables or any of the regions evaluated, and functional asymmetries were similar in all groups. Only when the autistic subjects were compared to the young controls did a trend toward lower $CMRO_2$ emerge that reached significance in the left anterior temporal region and in the right basal ganglia. Because of these conflicting results, further PET studies must be awaited before valid conclusions can be drawn on the pattern of cerebral functional abnormalities in autism.

Childhood-onset pervasive developmental disorder is an even rarer syndrome than infantile autism, characterized by a later onset ($>2\frac{1}{2}$ years of age) of severe impairment in social relationships and of a great variety of affective, language, motor, and other cognitive and behavioral abnormalities. We studied an 18-year-old male with an unremarkable family history (parents and 2 elder sisters in good health) and normal pregnancy and birth who, at the age of 5, after years of timely, normal development (except for some lack of initiative and insensitivity to pain), without any obvious external cause, became rapidly withdrawn, ceased to play with his friends at preschool, and would follow only his own impulses. His affective behavior was marked by silliness and aggression alternating with apathy, his attention span was dramatically reduced, and there were recurrent episodes of psychomotor agitation with vocal tics and inappropriate laughter or shrill crying. Despite intensive educational and rehabilitative efforts, his motor coordination deteriorated, and verbal output soon was reduced to gibberish and yes/no answers; he learned to read fluently but did not understand the meaning of what he read, and he required constant charge in his basic activities of daily living. Comprehensive blood chemistry and urinalyses were negative for any metabolic disorder worth considering, and repeated electroencephalographic studies were unremarkable. Both CT scans and MRI (Figure 35a) consistently showed some enlargement of the great cistern and of the lateral ventricles (right $>$ left), but radionuclide cisternography demonstrated normal CSF circulation. The PET revealed a mild decrease in overall CMRglu and a largely symmetrical pattern of regional functional abnormalities (Figure 35b). Hypometabolic dysfunction was accentuated in prefrontal, anterior cingulate, anterior temporal, and intraparietal cortex, and relatively increased CMRglu was found in striatum, posterior inferior frontal, midfrontal, and premotor cortex. Although this singular observation does not warrant any generalization, the metabolic changes described are suggestive of a profound imbalance between frontostriatal activating systems and paralimbic and heteromodal association areas that may explain the patient's communicative difficulties with his environment.

CONCLUSIONS

In his famous introductory textbook on experimental biomedicine, the great physiologist Claude Bernard (1865) noted that whenever a novel and reliable research tool becomes available, scientific progress is invariably made in those fields to which this technique can be applied. His insightful statement fittingly defines the relationship between behavioral neurology or neuropsychology and PET. However, like any technology, PET has its idiosyncrasies that one must appreciate in order to avoid spurious interpretations of image data. First, the radioisotopic tracer has a long way to go from the accelerator to a test subject's brain, and along this line it can be mishandled chemically and otherwise. Second, high procedural standards must be maintained at all times to assure the interpretability of results. Third, only very few tracer kinetic models have been sufficiently validated to permit quantitation in a wide range of physiological and abnormal conditions. Fourth, because of the limited spatial resolution of all PET scanners, quantitative results can be relative at best, with each anatomic structure having its own isotope- and machine-dependent recovery coefficient. Fifth, anatomic localization on functional images is anything but standardized.

Further complexities are added by the elusive nature of the subject of behavioral neurology. Although tremendous knowledge has been gained during the past three decades, current concepts of how the brain works still are crude oversimplifications, and much creative imagination will be needed to generate the proper physiological hypotheses to be tested. Because of its rapid tracer detection characteristics in three dimensions, PET certainly is better suited for the elucidation of gross structure–function relationships than any other presently available technology, but it is nei-

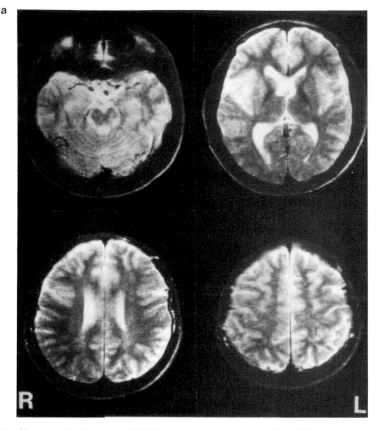

FIGURE 35. Characteristic T_2-weighted MRI scans (a) and corresponding CMRglu (μmol/100 g per min) PET images (b) of an 18-year-old patient with childhood-onset pervasive developmental disorder. Arrows mark areas of hypometabolic dysfunction; arrowheads point at activated cortical regions.

ther a cheap, single-button routine nor an answering machine responding explicitly to topistic questions. As illustrated in the previous sections, each PET study rather produces evidence for the concerted action of the brain as a whole and of varying regional sets of closely connected modules temporarily joining in the performance of a specific operation, with the among-subject differences in strategy (i.e., modular set configuration) being most obvious in nonprimary areas. Thus, PET not only provides solutions to well-known problems, it also raises many questions that previously could not be asked in the rigid framework of established localization schemes and that have no satisfactory correspondence in common neuropsychological classifications, as demonstrated most clearly by the results of the various cited activation PET studies in health and disease. Considering the intrinsically multivariate nature of PET data and the comparatively small series investigated so far, the vague and perhaps somewhat speculative interpretations offered in this chapter may be excused. They were attempted nevertheless in order to provide a reference set of hypothetical views that are widely held among PET workers.

Moreover, it must be clearly understood that "brain function" is quite an ambiguous term, having different meanings to different people. In the present context, it was largely represented by measures of energy metabolism and blood flow because these physiological variables have been the target of most PET investigations that are likely to have a similar impact on neuropsy-

b

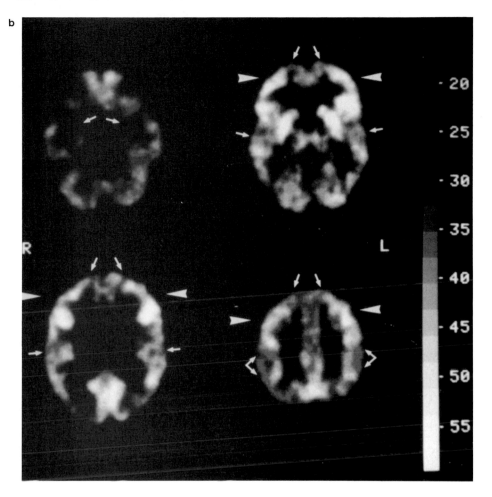

FIGURE 35. (*continued*)

chology as two-dimensional CBF studies once had. However, although they will remain the PET workhorses for years to come, other PET methods are dawning already. In view of the current pace at which tracer techniques for the *in vivo* investigation of various transmitter systems are being developed, it seems fair to predict the beginning of a transsynaptic era of behavioral neurology in the near future.

REFERENCES

Alavi, A., Ferris, S., Wolf, A., Reivich, M., Farkas, T., Dann, R., Christman, D., Mac-Gregor, R. R., & Fowler, J. (1980). Determination of cerebral metabolism in senile dementia using F-18-deoxyglucose and positron emission tomography. *The Journal of Nuclear Medicine, 21,* P21.
Alavi, A., Fazekas, F., Chawluk, J., & Zimmerman, R. (1987). Magnetic resonance imaging of the brain in normal aging and dementia. In J. S. Meyer, H. Lechner, M. Reivich, & E. O. Ott (Eds.), *Cerebral*

vascular disease 6 (pp. 191–195). Amsterdam, New York, Oxford: Excerpta Medica.

Babikian, V. L., Ackerman, R. H., Kelley, R. E., Correia, J. A., Alpert, N. M., Feinberg, W. M., & Taveras, J. M. (1985). CT and PET findings after onset and resolution of Korsakoff's syndrome due to thalamic hemorrhage. *Annals of Neurology, 18,* 128.

Baxter, L. R., Phelps, M. E., Mazziotta, J. C., Guze, B. H., Schwartz, J. M., & Selin, C. E. (1987). Local cerebral glucose metabolic rates in obsessive–compulsive disorder—a comparison with rates in unipolar depression and in normal controls. *Archives of General Psychiatry, 44,* 211–218.

Benson, D. F., Kuhl, D. E., Phelps, M. E., Cummings, J. L., & Tsai, S. Y. (1981). Positron emission computed tomography in the diagnosis of dementia. *Transactions of the American Neurological Association, 106,* 68–71.

Bernard, C. (1865). *Introduction à l'étude de la médicine expérimentale.* Paris: Baillière.

Bosley, T. M., Dann, R., Silver, F. L., Alavi, A., Kushner, M., Chawluk, J. B., Savino, P. J., Sergott, R. C., Schatz, N. J., & Reivich, M. (1987). Recovery of vision after ischemic lesions: Positron emission tomography. *Annals of Neurology, 21,* 444–450.

Buchsbaum, M. S., Holcomb, H. H., Johnson, J., King, A. C., & Kessler, R. (1983). Cerebral metabolic consequences of electrical cutaneous stimulation in normal individuals. *Human Neurobiology, 2,* 35–38.

Buchsbaum, M. S., Wu, J. C., DeLisi, L. E., Holcomb, H. H., Hazlett, E., Cooper-Langston, K., & Kessler, R. (1987). Positron emission tomography studies of basal ganglia and somatosensory cortex neuroleptic drug effects: Differences between normal controls and schizophrenic patients. *Biological Psychiatry, 22,* 479–494.

Bustany, P., Henry, J. F., Sargent, T., Zarifian, E., Cabanis, E., Collard, P., & Comar, D. (1983). Local brain protein metabolism in dementia and schizophrenia: *In vivo* studies with ^{11}C-L-methionine and positron emission tomography. In W.-D. Heiss & M. E. Phelps (Eds.), *Positron emission tomography of the brain* (pp. 208–211). Berlin, Heidelberg, New York: Springer.

Chase, T. N., Fedio, P., Foster, N. L., Brooks, R., Di Chiro, G., & Mansi, L. (1984). Wechsler Adult Intelligence Scale performance—cortical localization by fluorodeoxyglucose F^{18}-positron emission tomography. *Archives of Neurology, 41,* 1244–1247.

Chase, T. N., Foster, N. L., Fedio, P., Brooks, R., Mansi, L., Kessler, R., & Di Chiro, G. (1984). Gilles de la Tourette syndrome: Studies with the fluorine-18-labeled fluorodeoxyglucose positron emission tomographic method. *Annals of Neurology, 15 (Suppl.),* S175.

Chawluk, J. B., Alavi, A., Jamieson, D. G., Hurtig, H. I., Gur, R. E., Resnick, S. M., Rosen, M., & Reivich, M. (1987). Changes in local cerebral glucose metabolism with normal aging: The effect of cardiovascular and systemic health factors. *Journal of Cerebral Blood Flow and Metabolism, 7 (Suppl. 1),* S411.

Chugani, H. T., Phelps, M. E., Light, R. K., & Mazziotta, J. C. (1987). Metabolic correlates of mental retardation in childhood determined with FDG positron emission tomography (PET). *Journal of Cerebral Blood Flow and Metabolism, 7 (Suppl. 1),* S533.

Chugani, H. T., Phelps, M. E., & Mazziotta, J. C. (1987). Positron emission tomography study of human brain functional development. *Annals of Neurology, 22,* 487–497.

D'Antona, R., Baron, J. C., Samson, Y., Serdaru, M., Viader, F., Agid, Y., & Cambier, J. (1985). Subcortical dementia—frontal cortex hypometabolism detected by positron tomography in patients with progressive supranuclear palsy. *Brain, 108,* 785–799.

De Leon, M. J., Ferris, S. H., George, A. E., Christman, D. R., Fowler, J. S., Gentes, C., Reisberg, B., Gee, B., Emmerich, M., Yonekura, Y., Brodie, J., Kricheff, I. I., & Wolf, A. P. (1983). Positron emission tomographic studies of aging and Alzheimer disease. *American Journal of Neuroradiology, 4,* 568–571.

De Leon, M. J., George, A. E., Ferris, S. H., Rosenbloom, S., Christman, D. R., Gentes, C. I., Reisberg, B., Kricheff, I. I., & Wolf, A. P. (1983). Regional correlation of PET and CT in senile dementia of the Alzheimer type. *American Journal of Neuroradiology, 4,* 553–556.

DeLisi, L. E., Buchsbaum, M. S., Holcomb, H. H., Dowling-Zimmerman, S., Pickar, D., Boronow, J., Morihisa, J. M., van Kammen, D. P., Carpenter, W., Kessler, R., & Cohen, R. M. (1985). Clinical correlates of decreased anteroposterior metabolic gradients in positron emission tomography (PET) of schizophrenic patients. *American Journal of Psychiatry, 142,* 78–81.

Duara, R., Grady, C., Haxby, J., Ingvar, D., Sokoloff, L., Margolin, R. A., Manning, R. G., Cutler, N. R., & Rapoport, S. I. (1984). Human brain glucose utilization and cognitive function in relation to age. *Annals of Neurology, 16,* 702–713.

Duara, R., Barker, W., Chang, J. Y., Yoshii, F., & Apicella, A. (1986). The use of a behavioral "stress" test to enhance metabolic abnormalities in Alzheimer's disease (AD). *The Journal of Nuclear Medicine, 27*, 1025.

Duara, R., Chang, J., Barker, W., Yoshii, F., & Apicella, A. (1986). Correlation of regional cerebral metabolic activation to performance in activating tasks. *Neurology, 36 (Suppl. 1)*, 349.

Duara, R., Grady, C., Haxby, J., Sundaram, M., Cutler, N. R., Heston, L., Moore, A., Schlageter, N., Larson, S., & Rapoport, S. I. (1986c). Positron emission tomography in Alzheimer's disease. *Neurology, 36*, 879–887.

Duara, R., Yoshii, F., Chang, J., Barker, W., Apicella, A., Parks, R., & Emran, A. (1987). Complex reading memory task during PET in normal and memory-impaired subjects. *Journal of Cerebral Blood Flow and Metabolism, 7 (Suppl. 1)*, S312.

Early, T. S., Reiman, E. M., Raichle, M. E., & Spitznagel, E. L. (1987). Left globus pallidus abnormality in never-medicated patients with schizophrenia (cerebral blood flow/positron emission tomography). *Proceedings of the National Academy of Sciences of the United States of America, 84*, 561–563.

Farde, L., Wiesel, F. A., Hall, H., Halldin, C., Stone-Elander, S., & Sedvall, G. (1987). No D$_2$ receptor increase in PET study of schizophrenia. *Archives of General Psychiatry, 44*, 671–672.

Farkas, T., Wolf, A. P., Jaeger, J., Brodie, J. D. Christman, D. R., & Fowler, J. S. (1984). Regional brain glucose metabolism in chronic schizophrenia—a positron emission transaxial tomographic study. *Archives of General Psychiatry, 41*, 293–300.

Foster, N. L., Chase, T. N., Fedio, P., Patronas, N. J., Brooks, R. A., & Di Chiro, G. (1983). Alzheimer's disease: Focal cortical changes shown by positron emission tomography. *Neurology, 33*, 961–965.

Foster, N. L., Chase, T. N., Patronas, N. J., Gillespie, M. M., & Fedio, P. (1986). Cerebral mapping of apraxia in Alzheimer's disease by positron emission tomography. *Annals of Neurology, 19*, 139–143.

Fox, P. T., & Raichle, M. E. (1984). Stimulus rate dependence of regional cerebral blood flow in human striate cortex demonstrated by positron emission tomography. *Journal of Neurophysiology, 51*, 1109–1121.

Fox, P. T., Petersen, S. E., Posner, M. I., & Raichle, M. E. (1987). Language-related brain activation measured with PET: Comparison of auditory and visual word presentations. *Journal of Cerebral Blood Flow and Metabolism, 7 (Suppl. 1)*, S294.

Frackowiak, R. S. J., Lenzi, G.-L., Jones, T., & Heather, J. D. (1980). Quantitative measurement of regional cerebral blood flow and oxygen metabolism in man using ^{15}O and positron emission tomography: Theory, procedure and normal values. *Journal of Computer Assisted Tomography, 4*, 727–736.

Frackowiak, R. S. J., Pozzilli, C., Legg, N. J., Du Boulay, G. H., Marshall, J., Lenzi, G. L., & Jones, T. (1981). Regional cerebral oxygen supply and utilization in dementia—a clinical and physiological study with oxygen-15 and positron tomography. *Brain, 104*, 753–778.

Frackowiak, R. S. J., Herold, S., Petty, R. K. H., & Morgan-Hughes, J. A. (1987). Dementia in mitochondrial cytopathy: Clinical heterogeneity explained by altered cerebral oxygen : glucose stoichiometry. *Journal of Cerebral Blood Flow and Metabolism, 7 (Suppl. 1)*, S384.

Friedland, R. P., Budinger, T. F., Ganz, E., Yano, Y., Mathis, C. A., Koss, B., Ober, B. A., Huesman, R. H., & Derenzo, S. E. (1983). Regional cerebral metabolic alterations in dementia of the Alzheimer type: Positron emission tomography with [^{18}F]fluorodeoxyglucose. *Journal of Computer Assisted Tomography, 7*, 590–598.

Friedland, R. P., Prusiner, S. B., Jagust, W. J., Budinger, T. F., & Davis, R. L. (1984). Bitemporal hypometabolism in Creutzfeldt–Jakob disease measured by positron emission tomography with [^{18}F]-2-fluorodeoxyglucose. *Journal of Computer Assisted Tomography, 8*, 978–981.

Friedland, R. P., Jagust, W. J., Budinger, T. F., Koss, E., Ober, B. A., Dronkers, N. F., & Swain, B. E. (1987). Consistency of temporal–parietal cortex hypometabolism in probable Alzheimer's disease (AD): Relationships to cognitive decline. *Journal of Cerebral Blood Flow and Metabolism, 7 (Suppl. 1)*, S403.

Gooddy, W., & McKissock, W. (1951). The theory of cerebral localisation. *The Lancet, 260*, 481–483.

Greenberg, J. H., Reivich, M., Alavi, A., Hand, P., Rosenquist, A., Rintelmann, W., Stein, A., Tusa, R., Dann, R., Christman, D., Fowler, J., MacGregor, B., & Wolf, A. (1981). Metabolic mapping of functional activity in human subjects with the [^{18}F]fluorodeoxyglucose technique. *Science, 212*, 678–680.

Gur, R. C., Gur, R. E., Rosen, A. D., Warach, S., Alavi, A., Greenberg, J., & Reivich, M. (1983). A cognitive–motor network demonstrated by positron emission tomography. *Neuropsychologia, 21*, 601–606.

Gur, R. E., Resnick, S. M., Alavi, A., Gur, R. C., Caroff, S., Dann, R., Silver, F. C., Saykin, A. J.,

Chawluk, J. B., & Kushner, M. (1987). Regional brain function in schizophrenia—1. A positron emission tomography study. *Archives of General Psychiatry, 44,* 119–125.

Hachinski, V. C., Iliff, L. D., Zilhka, E., Du Boulay, G. H., McAllister, V. L., Marshall, J., Ross Russell, R. W., & Symon, L. (1975). Cerebral blood flow in dementia. *Archives of Neurology, 32,* 632–637.

Hatazawa, J., Brooks, R. A., Di Chiro, G., & Bacharach, S. L. (1987). Glucose metabolic rate versus brain size in humans. *Journal of Cerebral Blood Flow and Metabolism, 7 (Suppl. 1),* S301.

Haxby, J. V., Grady, C. L., Duara, R., Schlageter, N., Berg, G., & Rapoport, S. I. (1986). Neocortical metabolic abnormalities precede nonmemory cognitive defects in early Alzheimer's-type dementia. *Archives of Neurology, 43,* 882–885.

Haxby, J. V., Grady, C. L., Koss, E., Friedland, R. P., & Rapoport, S. I. (1987). Longitudinal study of brain metabolic and neuropsychological heterogeneity in dementia of the Alzheimer type: Evidence for subtypes. *Journal of Cerebral Blood Flow and Metabolism, 7 (Suppl. 1),* S377.

Hayden, M. R., Hewitt, J., Stoessl, A. J., Clark, C., Ammann, W., & Martin, W. R. W. (1987). The combined use of positron emission tomography and DNA polymorphisms for preclinical detection of Huntington's disease. *Neurology, 37,* 1441–1447.

Heiss, W.-D., Pawlik, G., Herholz, K., Wagner, R., & Wienhard, K. (1985). Regional cerebral glucose metabolism in man during wakefulness, sleep, and dreaming. *Brain Research, 327,* 362–366.

Heiss, W.-D., Herholz, K., Böcher-Schwarz, H. G., Pawlik, G., Wienhard, K., Steinbrich, W., & Friedman, G. (1986). PET, CT, and MR imaging in cerebrovascular disease. *Journal of Computer Assisted Tomography, 10,* 903–911.

Heiss, W.-D., Pawlik, G., Hebold, I., Herholz, K., Wagner, R., & Wienhard, K. (1987). Metabolic pattern of speech activation in healthy volunteers, aphasics, and focal epileptics. *Journal of Cerebral Blood Flow and Metabolism, 7 (Suppl. 1),* S299.

Heiss, W.-D., Hebold, I., Klinkhammer, P., Ziffling, P., Szelies, B., Pawlik, G., & Herholz, K. (1988). Effect of piracetam on cerebral glucose metabolism in Alzheimer's disease as measured by positron emission tomography. *Journal of Cerebral Blood Flow and Metabolism, 8,* 613–617.

Herholz, K., Krieg, J. C., Emrich, H. M., Pawlik, G., Beil, C., Pirke, K. M., Pahl, J. J., Wagner, R., Wienhard, K. Ploog, D., & Heiss, W.-D. (1987). Regional cerebral glucose metabolism in anorexia nervosa measured by positron emission tomography. *Biological Psychiatry, 22,* 43–51.

Herold, S., Frackowiak, R. S. J., Rutter, M., & Howlin, P. (1985). Regional cerebral blood flow, oxygen metabolism, and glucose metabolism in young autistic adults. *Journal of Cerebral Blood Flow and Metabolism, 5 (Suppl. 1),* S189–S190.

Herscovitch, P., Markham, J., & Raichle, M. E. (1983). Brain blood flow measured with intravenous $H_2^{15}O$. I. Theory and error analysis. *The Journal of Nuclear Medicine, 24,* 782–789.

Horowitz, S., Benson, D. F., Kuhl, D. E., & Cummings, J. L. (1982). FDG scan to confirm Creutzfeldt–Jakob diagnosis. *Neurology, 32,* A167.

Horwitz, B., Duara, R., & Rapoport, S. I. (1986). Age differences in intercorrelations between regional cerebral metabolic rates for glucose. *Annals of Neurology, 19,* 60–67.

Horwitz, B., Grady, C. L., Schlageter, N. L., Duara, R., & Rapoport S. I. (1987). Intercorrelations of regional cerebral glucose metabolic rates in Alzheimer's disease. *Brain Research, 407,* 294–306.

Ingvar, D. H., & Lassen, N. A. (Eds.). (1975). *Brain work: The coupling of function, metabolism and blood flow in the brain.* Copenhagen: Munksgaard.

Jagust, W. J., Friedland, R. P., & Budinger, T. F. (1985). Positron emission tomography with [^{18}F]fluorodeoxyglucose differentiates normal pressure hydrocephalus from Alzheimer-type dementia. *Journal of Neurology, Neurosurgery, and Psychiatry, 48,* 1091–1096.

Jagust, W. J., Friedland, R. P., Koss, E., Ober, B. A., Mathis, C. A., Huesman, R. H., & Budinger, T. F. (1987). Sequential studies of regional cerebral glucose metabolism in Alzheimer's disease. *Journal of Cerebral Blood Flow and Metabolism, 7 (Suppl. 1),* S414.

Jamieson, D. G., Chawluk, J. B., Alavi, A., Hurtig, H. I., Rosen, M., Bais, S., Dann, R., Kushner, M., & Reivich, M. (1987). The effect of disease severity on local cerebral glucose metabolism in Alzheimer's disease. *Journal of Cerebral Blood Flow and Metabolism, 7 (Suppl 1),* S410.

Kamo, H., McGeer, P. L., Harrop, R., McGeer, E. G., Calne, D. B., Martin, W. R. W., & Pate, B. D. (1987). Positron emission tomography and histopathology in Pick's disease. *Neurology, 37,* 439–445.

Kempler, D., Metter, E. J., Jackson, C. A., Hanson, W. R., Mazziotta, J. C., & Phelps, M. E. (1987). A metabolic investigation of a disconnection syndrome: Conduction aphasia. *Annals of Neurology, 22,* 134–135.

Kessler, R. M., Parker, E. S., Clark, C. M., Martin, P. R., George, D. T., Weingartner, H., Sokoloff, L., Ebert, M. H., & Mishkin, M. (1985). Regional cerebral glucose metabolism in patients with alcoholic Korsakoff's syndrome. *The Journal of Nuclear Medicine, 26,* P46.

Kishimoto, H., Kuwahara, H., Ohno, S., Takazu, O., Hama, Y., Sato, C., Ishii, T., Nomura, Y., Fujita, H., Miyauchi, T., Matsushita, M., Yokoi, S., & Iio, M. (1987). Three subtypes of chronic schizophrenia identified using ^{11}C-glucose positron emission tomography. *Psychiatry Research, 21,* 285–292.

Kishimoto, H., Takazu, O., Ohno, S., Yamaguchi, T., Fujita, H., Kuwahara, H., Takayoshi, I., Matsushita, M., Yokoi, S., & Iio, M. (1987). ^{11}C-Glucose metabolism in manic and depressed patients. *Psychiatry Research, 22,* 81–88.

Kleist, K. (1937). Bericht über die Gehirnpathologie in ihrer Bedeutung für Neurologie und Psychiatrie. *Zeitschrift für die gesamte Neurologie und Psychiatrie, 158,* 159–193.

Kling, A. S., Metter, E. J., Riege, W. H., & Kuhl, D. E. (1986). Comparison of PET measurement of local brain glucose metabolism and CAT measurement of brain atrophy in chronic schizophrenia and depression. *American Journal of Psychiatry, 143,* 175–180.

Koeppe, R. A., Holden, J. E., Polcyn, R. E., Nickles, R. J., Hutchins, G. D., & Weese, J. L. (1985). Quantitation of local cerebral blood flow and partition coefficient without arterial sampling: Theory and validation. *Journal of Cerebral Blood Flow and Metabolism, 5,* 214–223.

Kuhl, D. E., Metter, E. J., Riege, W. H., & Phelps, M. E. (1982). Effects of human aging on patterns of local cerebral glucose utilization determined by the (^{18}F)fluorodeoxyglucose method. *Journal of Cerebral Blood Flow and Metabolism, 2,* 163–171.

Kuhl, D. F., Metter, E. J., Riege, W. H., Hawkins, R. A., Mazziotta, J. C., Phelps, M. E., & Kling, A. S. (1983). Local cerebral glucose utilization in elderly patients with depression, multiple infarct dementia, and Alzheimer's disease. *Journal of Cerebral Blood Flow and Metabolism, 3 (Suppl. 1),* S494–S495.

Kuhl, D. E., Metter, E. J., Riege, W. H., & Markham, C. H. (1984). Patterns of cerebral glucose utilization in Parkinson's disease and Huntington's disease. *Annals of Neurology, 15 (Suppl.),* S119–S125.

Kuhl, D. E., Metter, E. J., Benson, D. F., Ashford, J. W., Riege, W. H., Fujikawa, D. G., Markham, C. H., Mazziotta, J. C., Maltese, A., & Dorsey, D. A. (1985). Similarities of cerebral glucose metabolism in Alzheimer's and parkinsonian dementia. *Journal of Cerebral Blood Flow and Metabolism, 5 (Suppl. 1),* S169–S170.

Kuhl, D. E., Small, G. W., Riege, W. H., Fujikawa, D. G., Metter, E. J., Benson, D. F., Ashford, J. W., Mazziotta, J. C., Maltese, A., & Dorsey, D. A. (1987). Cerebral metabolic patterns before the diagnosis of probable Alzheimer's disease. *Journal of Cerebral Blood Flow and Metabolism, 7 (Suppl. 1),* S406.

Kushner, M. J., Reivich, M., Alavi, A., Greenberg, J., Stern, M., & Dann, R. (1982). A PET study of the physiopathology of pure word mutism. *Neurology, 32,* A125.

Kushner, M., Reivich, M., Fazekas, F., McElhany, K., Rosen, M., Chawluk, J., & Alavi, A. (1987). Studies of cerebral metabolism and language dysfunction after acute ischemic aphasia. *Annals of Neurology, 22,* 135.

Kushner, M. J., Schwartz, R., Alavi, A., Dann, R., Rosen, M., Silver, F., & Reivich, M. (1987). Cerebral glucose consumption following verbal auditory stimulation. *Brain Research, 409,* 79–87.

Lammertsma, A. A., Wise, R. J. S., Heather, J. D., Gibbs, J. M., Leenders, K. L., Frackowiak, R. S. J., Rhodes, C. G., & Jones, T. (1983). Correction for the presence of intravascular oxygen-15 in the steady-state technique for measuring regional oxygen extraction ratio in the brain: 2. Results in normal subjects and brain tumour and stroke patients. *Journal of Cerebral Blood Flow and Metabolism, 3, 425*–431.

Lauter, J. L., Formby, C., Fox, P., Herscovitch, P., & Raichle, M. E. (1983). Tonotopic organization in human auditory cortex as revealed by regional changes in cerebral blood flow. *Journal of Cerebral Blood Flow and Metabolism, 3 (Suppl. 1),* S248–S249.

Leenders, K. L., Frackowiak, R. S. J., & Lees, A. J. (1987). Steele–Richardson–Olszewski (SRO) syndrome (supranuclear palsy) studied with positron emission tomography (PET). *Journal of Cerebral Blood Flow and Metabolism, 7 (Suppl. 1),* S397.

Leenders, K. L., Healy, M., Frackowiak, R., Buchingham, P., Lammertsma, A., & Jones, T. (1987). Effect of age on oxygen metabolism and CBF in healthy volunteers studied with positron emission tomography (PET). *Journal of Cerebral Blood Flow and Metabolism, 7 (Suppl. 1),* S398.

Lenzi, G. L., Frackowiak, R. S. J., Jones, T., Heather, J. D., Lammertsma, A. A., Rhodes, C. G., & Pozzilli, C. (1981). $CMRO_2$ and CBF by the oxygen-15 inhalation technique: Results in normal volunteers and cerebrovascular patients. *European Neurology, 20,* 285–290.

Loewenstein, D., Yoshii, F., Barker, W. W., Apicella, A., Emran, A., Chang, J. Y., & Duara, R. (1987). Predominant left hemisphere metabolic deficit predicts early manifestation of dementia. *Journal of Cerebral Blood Flow and Metabolism, 7 (Suppl. 1)*, S416.

Lucignani, G., Mori, K., Jay, T., Palombo, E., Nelson, T., Schmidt, K., & Sokoloff, L. (1987). Refinement of the kinetic model of the 2-[14C]deoxyglucose method to conform with known cell biology of G6Pase activity in brain. *Journal of Cerebral Blood Flow and Metabolism, 7 (Suppl. 1)*, S478.

Mangold, R., Sokoloff, L., Conner, E., Kleinerman, J., Therman, P.-O. G., & Kety, S. S. (1955). The effects of sleep and lack of sleep on the cerebral circulation and metabolism of normal young men. *Journal of Clinical Investigation, 34*, 1092–1100.

Mazière, B., Baron, J. C., Loc'h, C., Cambon, H., & Agid, Y. (1987). Progressive supranuclear palsy studied by PET. In W.-D. Heiss, G. Pawlik, K., Herholz, & K. Wienhard (Eds.), *Clinical efficacy of positron emission tomography* (pp. 101–109). Dordrecht, Boston, Lancaster: Martinus Nijhoff.

Mazziotta, J. C., Phelps, M. E., Carson, R. E., & Kuhl, D. E. (1982a). Tomographic mapping of human cerebral metabolism: Sensory deprivation. *Annals of Neurology, 12*, 435–444.

Mazziotta, J. C., Phelps, M. E., Carson, R. E., & Kuhl, D. E. (1982b). Tomographic mapping of human cerebral metabolism: Auditory stimulation. *Neurology, 32*, 921–937.

Mazziotta, J. C., Phelps, M. E., & Carson, R. E. (1984). Tomographic mapping of human cerebral metabolism: Subcortical responses to auditory and visual stimulation. *Neurology, 34*, 825–828.

Mazziotta, J. C., Phelps, M. E., & Wapenski, J. A. (1985). Human cerebral motor system metabolic responses in health and disease. *Journal of Cerebral Blood Flow and Metabolism, 5 (Suppl. 1)*, S213–S214.

McGeer, P. L., Kamo, H., Harrop, R., Li, D. K. B., Tuokko, H., McGeer, E. G., Adam, M. J., Ammann, W., Beattie, B. L., Calne, D. B., Martin, W. R. W., Pate, B. D., Rogers, J. G., Ruth, T. J., Sayre, C. I., & Stoessl, A. J. (1986). Positron emission tomography in patients with clinically diagnosed Alzheimer's disease. *Canadian Medical Association Journal, 134*, 597–607.

McGeer, P. L., Kamo, H., Harrop, R., McGeer, E. G., Martin, W. R. W., Pate, B. D., & Li, D. K. B. (1986). Comparison of PET, MRI, and CT with pathology in a proven case of Alzheimer's disease. *Neurology, 36*, 1569–1574.

Metter, J. E., Riege, W. H., Hanson, W. R., Kuhl, D. E., Phelps, M. E., Squire, L. R., Wasterlain, C. G., & Benson, D. F. (1983). Comparison of metabolic rates, language, and memory in subcortical aphasias. *Brain and Language, 19*, 33–47.

Metter, J. E., Riege, W. H., Hanson, W. R., Camras, L. R., Phelps, M. E., & Kuhl, D. E. (1984). Correlations of glucose metabolism and structural damage to language function in aphasia. *Brain and Language, 21*, 187–207.

Metter, J. E., Jackson, C. A., Kempler, D., Camras, L., Hanson, W. R., Mazziotta, J. C., & Phelps, M. E. (1986). Glucose metabolic asymmetries in chronic Wernicke's, Broca's and conduction aphasias. *Neurology, 36 (Suppl. 1)*, 317.

Metz, J. T., Yasillo, N. J., & Cooper, M. (1987). Relationship between cognitive functioning and cerebral metabolism. *Journal of Cerebral Blood Flow and Metabolism, 7 (Suppl. 1)*, S305.

Meyer, J. S., Haymann, L. A., Yamamoto, M., Sakai, F., & Nakajima, S. (1980). Local cerebral blood flow measured by CT after stable xenon inhalation. *American Journal of Neuroradiology, 1*, 213–225.

Miller, J. D., De Leon, M. J., Ferris, S. H., Kluger, A., George, A. E., Reisberg, B., Sachs, H. J., & Wolf, A. P. (1987). Abnormal temporal lobe response in Alzheimer's disease during cognitive processing as measured by 11C-2-deoxy-D-glucose and PET. *Journal of Cerebral Blood Flow and Metabolism, 7*, 248–251.

Mintun, M. A., Raichle, M. E., Martin, W. R. W., & Herscovitch, P. (1984). Brain oxygen utilization measured with O-15 radiotracers and positron emission tomography. *The Journal of Nuclear Medicine, 25*, 177–187.

Mountz, J., Curtis, G., Santa, C., Gebarski, S., Berent, S., Ehrenkaufer, R., & Koeppe, R. (1986). Alteration in regional cerebral blood flow in simple phobic anxiety demonstrated by O-15-H2O positron emission tomography (PET). *Journal of Nuclear Medicine, 27*, 901–902.

Navia, B., Sidtis, J. J., Moeller, J. R., Strother, S. C., Dhawan, V., Ginos, J. Z., Price, R. W., & Rottenberg, D. A. (1987). Metabolic anatomy of the AIDS dementia complex. *Journal of Cerebral Blood Flow and Metabolism, 7 (Suppl. 1)*, S388.

Pantano, P., Baron, J.-C., Lebrun-Grandie, P., Duquesnoy, N., Bousser, M.-G., & Comar, D. (1984). Regional cerebral blood flow and oxygen consumption in human aging. *Stroke, 15*, 635–641.

Pawlik, G. (1988). Positron emission tomography and multiregional statistical analysis of brain function: From exploratory methods for single cases to inferential tests for multiple group designs. In J. L. Willems, J. H. van Bemmel, & J. Michel (Eds.), *Progress in computer-assisted function analysis* (pp. 401–408). Amsterdam, New York, Oxford: Elsevier/North-Holland.

Pawlik, G., Beil, C., Herholz, K., Szelies, B., Wienhard, K., & Heiss, W.-D. (1985). Comparative dynamic FDG–PET study of functional deactivation in thalamic versus extrathalamic focal ischemic brain lesions. *Journal of Cerebral Blood Flow and Metabolism, 5 (Suppl. 1),* S9–S10.

Pawlik, G., Heiss, W.-D., Herholz, K., Beil, C., Szelies, B., Wienhard, K., & Grünewald, G. (1985). Positron emission tomographic study of variations in brain glucose metabolism related to spontaneous speech. *Journal of Neurology, 232 (Suppl.),* 227.

Pawlik, G., Herholz, K., Beil, C., Wagner, R., Wienhard, K., & Heiss, W.-D. (1985). Remote effects of focal lesions on cerebral flow and metabolism. In W.-D. Heiss (Ed.), *Functional mapping of the brain in vascular disorders* (pp. 59–83). Berlin, Heidelberg, New York, Tokyo: Springer.

Pawlik, G., Beil, C., Hebold, I., Herholz, K., Wienhard, K., & Heiss, W.-D. (1986). Positron emission tomography in depression research: Principles—results—perspectives. *Psychopathology, 19 (Suppl. 2),* 85–93.

Pawlik, G., Herholz, K., Wienhard, K., Beil, C., & Heiss, W.-D. (1986). Some maximum likelihood methods useful for the regional analysis of dynamic PET data on brain glucose metabolism. In S. L. Bacharach (Ed.), *Information processing in medical imaging* (pp. 298–309). Dordrecht, Boston, Lancaster: Martinus Nijhoff.

Pawlik, G., Heiss, W.-D., Beil, C., Grünewald, G., Herholz, K., Wienhard, K., & Wagner, R. (1987). Three-dimensional patterns of speech-induced cerebral and cerebellar activation in healthy volunteers and in aphasic stroke patients studied by positron emission tomography of $2(^{18}F)$-fluorodeoxyglucose. In J. S. Meyer, H. Lechner, M. Reivich, & E. O. Ott (Eds.), *Cerebral vascular disease 6* (pp. 207–210). Amsterdam, New York, Oxford: Excerpta Medica.

Pawlik, G., Heiss, W.-D., Beil, C., Wienhard, K., Herholz, K., & Wagner, R. (1987). PET demonstrates differential age dependence, asymmetry, and response to various stimuli of regional brain glucose metabolism in healthy volunteers. *Journal of Cerebral Blood Flow and Metabolism, 7 (Suppl. 1),* S376.

Pawlik, G., Heiss, W.-D., Stefan, H., Hebold, I. R., Herholz, K., & Wienhard, K. (1987). Refractory temporal lobe epilepsy: Diagnostic yield of interictal positron emission tomography improved by functional activation. In E. Bental & J. Manelis (Eds.), *17th epilepsy international congress* (p. 91). Jerusalem: The Israel League against Epilepsy.

Pawlik, G., Heiss, W.-D., Wienhard, K., Hebold, I. R., Ziffling, P., Staffen, W., Herholz, K., & Wagner, R. (1987). Brain glucose metabolism and blood flow in ischemic stroke. In W.-D. Heiss, G. Pawlik, K. Herholz, & K. Wienhard (Eds.), *Clinical efficacy of positron emission tomography* (pp. 37–53). Dordrecht, Boston, Lancaster: Martinus Nijhoff.

Penfield, W., & Rasmussen, T. (1950). *The cerebral cortex of man.* New York: Macmillan.

Phelps, M. E., Huang, S. C., Hoffman, E. J., Selin, C., Sokoloff, L., & Kuhl, D. E. (1979). Tomographic measurement of local cerebral glucose metabolic rate in humans with (F-18)2-fluoro-2-deoxyglucose: Validation of method. *Annals of Neurology, 6,* 371–388.

Phelps, M. E., Mazziotta, J. C., Kuhl, D. E., Nuwer, M., Packwood, J., Metter, J., & Engel, J. (1981). Tomographic mapping of human cerebral metabolism: Visual stimulation and deprivation. *Neurology, 31,* 517–529.

Post, R. M., DeLisi, L. E., Holcomb, H. H., Uhde, T. W., Cohen, R., & Buchsbaum, M. S. (1987). Glucose utilization in the temporal cortex of affectively ill patients: Positron emission tomography. *Biological Psychiatry, 22,* 545–553.

Reiman, E. M., Raichle, M. E., Robins, E., Butler, F. K., Herscovitch, P., Fox, P., & Perlmutter, J. (1986). The application of positron emission tomography to the study of panic disorder. *American Journal of Psychiatry, 143,* 469–477.

Reivich, M., Greenberg, J., Alavi, A., Christman, D., Fowler, J., Hand, P., Rosenquist, A., Rintelmann, W., & Wolf, A. (1979). The use of the ^{18}F-fluorodeoxyglucose technique for mapping of functional neural pathways in man. *Acta Neurologica Scandinavica, 60 (Suppl. 72),* 198–199.

Reivich, M., Kuhl, D., Wolf, A., Greenberg, J., Phelps, M. E., Ido, T., Casella, V., Fowler, J., Hoffman, E., Alavi, A., Som, P., & Sokoloff, L. (1979). The (^{18}F)fluorodeoxyglucose method for the measurement of local cerebral glucose utilization in man. *Circulation Research, 44,* 127–137.

Reivich, M., Gur, R. C., & Alavi, A. (1983). Positron emission tomographic studies of sensory stimuli, cognitive processes and anxiety. *Human Neurobiology, 2,* 25–33.

Reivich, M., Alavi, A., & Gur, R. C. (1984). Positron emission tomographic studies of perceptual tasks.

Annals of Neurology, 15 (Suppl.), S61–S65.

Riege, W. H., Metter, E. J., Kuhl, D. E., & Phelps, M. E. (1985). Brain glucose metabolism and memory functions: Age decrease in factor scores. *Journal of Gerontology, 40*, 459–467.

Roland, P. E., Eriksson, L., Stone-Elander, S., & Widen, L. (1987). Does mental activity change the oxidative metabolism of the brain? *The Journal of Neuroscience, 7*, 2373–2389.

Roland, P. E., Eriksson, L., Widen, L., & Stone-Elander, S. (1987). Increases of the regional cerebral oxidative metabolism during tactile learning and recognition. *Journal of Cerebral Blood Flow and Metabolism, 7 (Suppl. 1)*, S315.

Rosen, W. G., Terry, R. D., Fuld, P. A., Katzman, R., & Peck, A. (1980). Pathological verification of ischemic score in differentiation of dementias. *Annals of Neurology, 7*, 486–488.

Rumsey, J. M., Duara, R., Grady, C., Rapoport, J. L., Margolin, R. A., & Rapoport, S. I. (1985). Brain metabolism in autism. *Archives of General Psychiatry, 42*, 448–455.

Schapiro, M. B., Haxby, J. V., Grady, C. L., Duara, R., Schlageter, N. L., White B., Moore, A., Sundaram, M., Larson, S. M., & Rapoport, S. I. (1987). Decline in cerebral glucose utilisation and cognitive function with aging in Down's syndrome. *Journal of Neurology, Neurosurgery, and Psychiatry, 50*, 766–774.

Schwartz, E. L., Christman, D. R., & Wolf, A. P. (1984). Human primary visual cortex topography imaged via positron tomography. *Brain Research, 294*, 225–230.

Sokoloff, L. (1981). The relationship between function and energy metabolism: Its use in the localization of functional activity in the nervous system. *Neurosciences Research Program Bulletin, 19*, 159–210.

Sokoloff, L., Reivich, M., Kennedy, C., Des Rosiers, M. H., Patlak, C. S., Pettigrew, K. D., Sakurada, O., & Shinohara, M. (1977). The [^{14}C]deoxyglucose method for the measurement of local cerebral glucose utilization: Theory, procedure, and normal values in the conscious and anesthetized albino rat. *Journal of Neurochemistry, 28*, 897–916.

Volkow, N. D., Brodie, J. D., Wolf, A. P., Gomez-Mont, F., Cancro, R., Van Gelder, P., Russell, J. A. G., & Overall, J. (1986). Brain organization in schizophrenia. *Journal of Cerebral Blood Flow and Metabolism, 6*, 441–446.

Volpe, B. T., & Hirst, W. (1983). The characterization of an amnesic syndrome following hypoxic ischemic injury. *Archives of Neurology, 40*, 436–440.

Volpe, B. T., & Petito, C. K. (1985). Dementia with bilateral medial temporal lobe ischemia. *Neurology, 35*, 1793–1797.

Volpe, B. T., Herscovitch, P., Raichle, M. E., Hirst, W., & Gazzaniga, M. S. (1983). Cerebral blood flow and metabolism in human amnesia. *Journal of Cerebral Blood Flow and Metabolism, 3 (Suppl. 1)*, S5–S6.

Volpe, B. T., Herscovitch, P., & Raichle, M. E. (1984). PET evaluation of patients with amnesia after cardiac arrest. *Stroke, 15*, 196.

Wienhard, K., Pawlik, G., Herholz, K., Wagner, R., & Heiss, W.-D. (1985). Estimation of local cerebral glucose utilization by positron emission tomography of [^{18}F]2-fluoro-2-deoxy-D-glucose: A critical appraisal of optimization procedures. *Journal of Cerebral Blood Flow and Metabolism, 5*, 115–125.

Wienhard, K., Coenen, H. H., Pawlik, G., Hebold, I., Jovkar, S., Stöcklin, G., & Heiss, W.-D. (1988). PET investigation of the dopaminergic system in neurological patients, using 3-N-(F-18)fluoroethylspiperone. *The Journal of Nuclear Medicine, 29(1)*, 821.

Wiesel, F.-A., Wik, G., Sjögren, I., Blomqvist, G., Greitz, T., & Stone-Elander, S. (1987). Regional brain metabolism in drugfree schizophrenic patients as measured by positron emission tomography. In W.-D. Heiss, G. Pawlik, K. Herholz, & K. Wienhard (Eds.), *Clinical efficacy of positron emission tomography* (pp. 203–211). Dordrecht, Boston, Lancaster: Martinus Nijhoff.

Wong, D. F., Wagner, H. N., Jr., Tune, L. E., Dannals, R. F., Pearlson, G. D., Links, J. M., Tamminga, C. A., Broussolle, E. P., Ravert, H. T., Wilson, A. A., Toung, J. K. T., Malat, J., Williams, J. A., O'Tuama, L. A., Snyder, S. H., Kuhar, M. J., & Gjedde, A. (1986). Positron emission tomography reveals elevated D$_2$ dopamine receptors in drug-naive schizophrenics. *Science, 234*, 1558–1563.

Yoshii, F., Baker, W. W., Chang, J. Y., Kothari, P., Apicella, A., & Duara, R. (1987). Cerebral metabolism during visual activation: Effects of aging and dementia. *Journal of Cerebral Blood Flow and Metabolism, 7 (Suppl. 1)*, S313.

Zola-Morgan, S., Squire, L. R., & Amaral, D. G. (1986). Human amnesia and the medial temporal region: Enduring memory impairment following a bilateral lesion limited to field CA1 of the hippocampus. *The Journal of Neuroscience, 6*, 2950–2967.

5

Computed Tomographic Scanning and New Perspectives in Aphasia

DAVID S. KNOPMAN, OLA A. SELNES, and ALAN B. RUBENS

INTRODUCTION

Classical theories of aphasia are based on detailed studies of very small numbers of patients whose brains came to postmortem examination. Following the introduction of computed tomography (CT), a wealth of clinical-radiographic observations became available and prompted a major reexamination of how language is represented in the brain. Although early CT studies attempted primarily to confirm classical theory (Mazzocchi & Vignolo, 1979; Kertesz, Harlock, & Coates, 1979; Naeser & Hayward, 1978; Damasio & Damasio, 1980), more recent studies have begun to revise concepts of language representation in the brain (Selnes, Knopman, Niccum, Rubens, & Larson, 1983; Selnes, Niccum, Knopman, & Rubens, 1984; Selnes, Knopman, Niccum, & Rubens, 1985; Knopman et al., 1983; Knopman, Selnes, Niccum, & Rubens, 1984; Mohr et al., 1978; Basso, LeCours, Moraschini, & Vanier, 1985).

The present chapter represents a version of the "revisionist" view, based on clinical and CT observations in a consecutive series of aphasics. The chapter begins with a discussion of methodological issues. An analysis of the anatomic correlates of deficits in five areas of language processing follows: comprehension of single words, comprehension of sentences, confrontation naming, sentence repetition, and speech fluency.

METHODS

Patients with single left-hemisphere ischemic strokes were prospectively and consecutively entered into the study within 1 month of the stroke. Additional inclusion criteria required that the patients be right-handed, native English speakers with a minimum eighth-grade education and score 93 or less on the Western Aphasia Battery 1 month post-stroke. The 54 patients who completed the study ranged in age from 33 to 75 years (mean age 61 years) and included 39 men and 15 women.

DAVID S. KNOPMAN ● Department of Neurology, University of Minnesota, Minneapolis, Minnesota 55455.   OLA A. SELNES ● Department of Neurology, Johns Hopkins University School of Medicine, Baltimore, Maryland 21205.   ALAN B. RUBENS ● Department of Neurology, University of Arizona, Tucson, Arizona 85721.

Computed Tomographic Methods

At 5 months post-stroke, all patients had a CT performed on a high-resolution scanner. The fifth-month scan was used for localization and lesion volume determinations. The scans were analyzed for both lesion localization and volume by neuroradiologists who were blinded to the clinical data. Lesion localization was carried out by parcelling the left hemisphere into 40 regions (Figure 1), each defined by explicit boundaries with the use of a CT atlas (Matsui & Hirano, 1978). The definitions of region boundaries included left hemisphere sulcal markings if visible, left hemisphere ventricles, right hemisphere structures, or circumferential distances from frontal pole to occipital pole. Each region was rated on a 5-point scale, from 0 ("no lesion") to 4 ("total destruction" of the region). Regions that were rated as 2 or higher were considered lesioned. The interrater reliability of the rating system was high. Lesion volume was determined by light-pen tracing of the lesion boundaries on all relevant slices; an estimate of lesion volume in cubic centimeters was then derived.

Interpretation of studies of localization using CT scanning requires several assumptions about the capability of CT to detect and localize brain lesions caused by ischemic stroke. The first assumption is that a cerebral infarction will always appear on CT scans. A second assumption is that within a zone of encephalomalacia on CT, all tissue is nonfunctional. Both of these issues are influenced by the timing of the scan relative to the infarct. Within the first month post-stroke, a lesion may "disappear" because of changing ratios of the water and protein content of the infarct (Becker, Desch, Hacker, & Pencz, 1979). Chronically, small cortical infarcts can be difficult to distinguish from subarachnoid space. In addition, periinfarctional edema may obscure lesion boundaries or give the region of nonviable brain the appearance of being much larger than it will ultimately prove to be. However, even on scans obtained after edema should have resolved, viable tissue may remain in regions that appear infarcted. Furthermore, recent studies with both PET (Metter, Wasterlain, Kuhl, Hanson, & Phelps, 1981) and MRI (DeWitt, Grek, Buonanno, Levine, & Kistler, 1985) scanning show that zones of abnormality appear larger with those techniques than with CT.

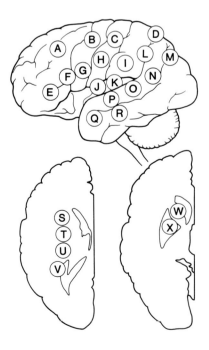

FIGURE 1. Schematic regional parcellation of left hemisphere. Lateral projection of cerebral hemisphere and horizontal sections depicting regions used in CT localization of lesions. To enhance the visual clarity, regions outside of the middle cerebral artery distribution have been omitted. The circles correspond to the following regions: A, superior premotor regions; B, upper precentral gyrus; C, upper postcentral gyrus; D, superior parietal; E, anterior inferior frontal gyrus; F, posterior inferior frontal gyrus; G, lower precentral gyrus; H, lower postcentral gyrus; I, suprasylvian supramarginal gyrus; J, anterior insula; K, posterior insula; L, superior supramarginal gyrus; M, angular gyrus; N, infrasylvian supramarginal gyrus; O, posterior superior temporal gyrus; P, Heschl's gyrus; Q, anterior temporal regions; R, posterior middle and inferior temporal gyri; S to V, subcortical frontoparietal white matter; W, head of caudate; X, putamen–globus pallidus.

A further assumption is that interindividual differences in brain anatomy are minor enough that they will have little impact on lesion localization. There are several reasons to suspect that this assumption is violated. A number of studies have shown that the gyral patterns of the superior temporal lobe and the inferior frontal region are variable among individuals (Rubens, Mahowald, & Hutton, 1976; Whitaker & Selnes, 1976; Chui & Damasio, 1980). Individual differences in functional localization also exist: electrical stimulation studies performed during surgery for intractable epilepsy have shown considerable variability of localization of the regions in which speech arrest or anomia will result (Ojemann, 1983).

To evaluate the extent to which CT can be used for precise localization, we analyzed the relationship between hemiparesis and CT, using scans obtained 5 months post-stroke (Knopman & Rubens, 1986). Hemiparesis was chosen because its basis in stroke patients is, with very few exceptions, in lesions in the corticospinal pathway rostral to its decussation. This pathway, moreover, is readily identified on CT and probably more anatomically and functionally invariant than systems involved in cognition. We found that discrepancies between CT and clinical motor examination were very infrequent but did exist. There were only 3 patients out of 27 with hemiparesis whose corticospinal pathway was not rated as lesioned. In each instance, the lesion was adjacent to the pathway in the corona radiata. On the other hand, there were also 3 patients among the 25 with no paresis who had what appeared to be substantial lesions directly within the corticospinal pathway, again in the corona radiata portion of the pathway. Thus, deviations from expectations occurred in about 12% of patients, but in general, the correspondence between CT and paresis was excellent.

Interpretation of lesion volume estimates was not entirely straightforward. The vast majority of patients in our study had strokes from infarction limited to the territory of the middle cerebral artery, which supplies the perisylvian region in which most language functions appear to be localized. Thus, indices of lesion volume in the present study were referenced to the middle cerebral artery territory; lesion volume estimates in patients with lesions of nonvascular etiologies, e.g., head trauma, may not be comparable (e.g., Ludlow et al., 1986).

Language Methods

Patients were examined with language assessment batteries on a monthly basis. Standardized instruments included the Boston Diagnostic Aphasia Examination (BDAE) (Goodglass & Kaplan, 1972) and the Token Test (DeRenzi & Vignolo, 1962) and were supplemented by several experimental tests. We analyzed linguistic performance from individual subtests that assessed a specific ability: comprehension of single words, confrontation naming, speech fluency, sentence comprehension, and sentence repetition. By analyzing single functions in which performance can be characterized either as a continuous variable or as a categorical variable (i.e., normal or abnormal function), quantitative analysis was possible. Table 1 gives the scoring range for each function

TABLE 1. Performance Categories for Language Tests

Level of severity	Word comprehension (percent correct)	Token test (percent correct)	Confrontation naming (percent correct)	Sentence repetition (percent correct)	Speech fluency rating
Normal	>92	>92	>92	>75	>4.9
Mild deficit	75–92	75–92	75–92	50–75	>2.9, <5
Moderate deficit	50–74	50–74	50–74	<50	<3
Severe deficit	<50	<50	<50	<50	<3

tested that corresponds to ''normal'' or ''impaired'' performance, based on studies in age-matched normal controls.

Interpretation of linguistic performance was complicated by the fact that failure to perform at normal levels may have more than one reason. Error analyses and related function analyses yielded insight into mechanisms of failure, but our study was designed to address the longitudinal nature of recovery from aphasia, so that detailed linguistic data acquisition was not usually feasible. Consequently, interpretation of our findings must be made with the caveat that mechanism of impairment of a particular function cannot usually be specified. However, since impairment of fluency may have precluded naming or repetition, we restricted analysis in those areas to patients who performed above a criterion level on speech fluency.

Our analyses attempted to avoid the difficulties inherent in the approach to aphasic disorders that relied on determination of syndromes. Several recent discussions have questioned the traditional syndrome model on linguistic grounds (Schwartz, 1984; Caramazza, 1984). Problems with using syndrome classification as the primary method of language analysis included (1) the inability to classify a sizable minority of patients, (2) the inefficiency of grouping a small number of patients into a large number of categories, (3) the clinical heterogeneity within some of the syndrome categories at any point in time and especially over time, and (4) difficulties in analysis of changes of syndrome classifications over time. Thus, although we do not dispute the clinical utility of referring to patients by syndrome, we found them to be impractical for analysis of anatomic correlates of aphasia.

Aphasia following ischemic stroke may change dramatically in the months following the initial event. Specification of the timing of the language testing is of critical importance in anatomic localization studies. The homogeneity of patient groups is greatly enhanced when analyses use patients with uniform times post-stroke, at least within the first 4 to 6 months of recovery. We present results for both early (1 month post-stroke) and late (6 months post-stroke) language testing, illustrating the striking differences between the two.

Analytic Methods

The basic issue addressed in the present study is the relationship between language dysfunction and lesions of particular brain structures. The appropriate analytic technique involved determination of the predictive accuracy of one measure (i.e., either linguistic or anatomic) for the other. Predictive accuracy refers to the ratio of the sum of the numbers of true positives and true negatives to the total number of cases. Predictive accuracy captures both the positive associations (i.e., presence of lesion and presence of deficit) and the negative instances (i.e., absence of lesion and normal function). In most other studies, only one part of this analysis was done: determinations of ''true positives.'' Close scrutiny of patients with the smallest lesions and the most severe deficits is a very useful approach and one we use, but it does not give the whole picture. Because ''negative'' instances are as theoretically and practically pertinent as ''positives,'' our analyses included both.

The extent to which a function is localized is a matter of degree, with some functions highly localized, some less discretely localized, and some diffusely localized within the perisylvian core. If motor function is considered the brain function likely to be the most highly localized, along with primary sensory functions, then the degree of localization of lesions causing hemiparesis could be taken as the upper bound on the ability of CT-clinical techniques to predict localization. For hemiparesis, a predictive accuracy for lesions in the corticospinal pathway of 88% was found (Knopman & Rubens, 1986) and would represent that upper bound. The lower bound would be represented by a function in which all regions in question demonstrated about the same level of predictive accuracy.

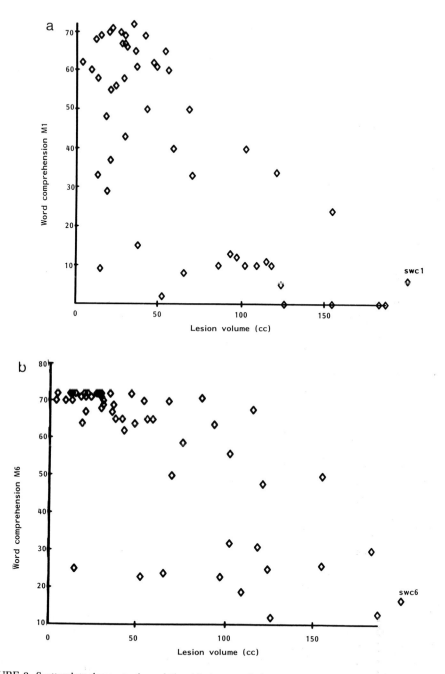

FIGURE 2. Scatterplots demonstrating relationship between lesion volume (in cubic centimeters) and scores on BDAE Word Comprehension test at 1 month (a) and at 6 months (b) post-stroke.

RESULTS

Single-Word Comprehension

Single-word comprehension (Selnes et al., 1984) was assessed with the BDAE word comprehension subtest (Goodglass and Kaplan, 1972). At 1 month post-stroke, deficits in single-word comprehension were common (present in 80% of patients). By 6 months, only 46% of patients still had deficits. The correlation between comprehension of single spoken and single written words was very high ($r = 0.88$ and 0.93, respectively) at months 1 and 6, suggesting that the deficits were supramodal.

The comprehension of single words, either spoken or written, was strongly related ($r = -0.72$) to the volume of the left hemisphere lesion, although there were a number of patients with small lesions and impaired single-word comprehension who later recovered (Figure 2a). The relationship between lesion volume and single-word comprehension was equally strong at 6 months ($r = -0.75$), and there were far fewer patients with small lesions and poor single-word comprehension by that time (Figure 2b). There were more patients with large lesions and normal single-word comprehension by this time.

Lesions in no specific region, including the posterior temporal region, had high predictive accuracy for single-word comprehension deficits at either 1 or 6 months (Figure 3). Lesions in the parietal lobe and underlying white matter showed the highest predictive accuracies, but none were over 80%, and furthermore, most other regions were only slightly lower. This pattern was a result of the strong effect of overall level of perisylvian damage on word comprehension performance. Lesions in the temporal lobe were not uniquely associated with impaired single-word comprehension. In an attempt to control for the effects of lesion volume, we analyzed the data for patients with lesions of less than 60 cc. The results for regional predictive accuracy were not appreciably different.

We identified cutpoints for lesion volume over or under which almost all patients showed similar performance (Table 2). Of the 12 patients (two examples are given in Figure 4) with

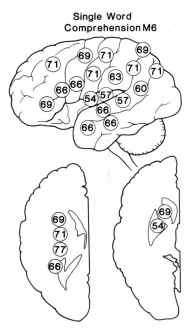

Single Word
Comprehension M6

FIGURE 3. Predictive accuracies (in percent) of regional lesions for presence of deficit on BDAE Word Comprehension Test at 6 months post-stroke. Predictive accuracy refers to the ratio of [(true positives + true negatives)/total number of cases]. For this figure, patients were classified as normal or impaired, and a region's status as not lesioned or lesioned. See Figure 1 for localization of regions.

TABLE 2. Percentage of Patients Scoring in Normal Range for
Single-Word (SW) and Token Test (TT) Compared to
Lesion Volume

Lesion volume (cc)	Month 1		Month 6		N
	SW	TT	SW	TT	
<31	40%	0%	90%	43%	21
31–60	38%	15%	79%	43%	14
61–100	0%	0%	29%	14%	7
>100	0%	0%	8%	0%	12

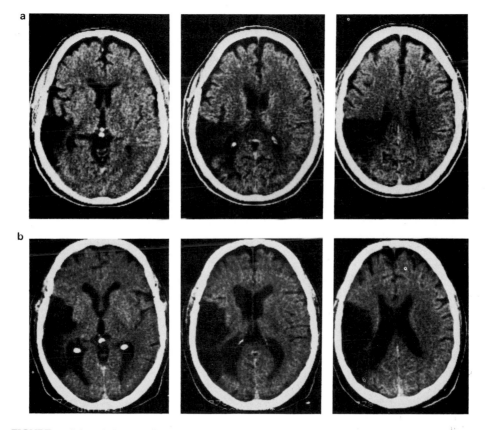

FIGURE 4. Selected CT scans of patients with large lesions. (a) A038. Lesion volume 76 cc. (b) A060. Lesion volume 124 cc. Both patients had moderate (patient a) to severe (patient b) deficits on single-word comprehension, very poor Token Test performance, and very poor naming. Patient a was fluent throughout the poststroke period, whereas patient b was persistently nonfluent. Patient a's temporal–parietal lesion was considerably larger than the similarly configured lesions of the patients shown in Figure 7, which may account for the poorer single-word comprehension scores of the patient a in this figure compared to those in Figure 7. Similarly, patient b's lesion was considerably larger than those of the patient in Figure 13b, which may account for the much worse overall performance of patient b of this figure.

lesions over 100 cc, all but one had severely impaired word comprehension by month 6. The one patient who scored in the normal range had a lesion that was atypical in that there was a substantial anterior cerebral artery distribution component to it, emphasizing the importance of referencing measures of lesion volume to lesions in the middle cerebral artery territory. At the other end of the spectrum of lesion volume, the patients with smaller lesions (<60 cc, $n = 35$) tended to have normal-range single-word comprehension. There were only 5 patients (14%) with small lesions and depressed word comprehension scores (see Figure 13c for one such patient). With white matter lesions, nonfunctional gray matter overlying will not have an abnormal image by CT, whereas functional imaging techniques such as PET may show much greater areas of involvement. Alternatively, a well-situated lesion deep in the white matter underlying the frontoparietal lobes, as suggested by our data, may have disastrous effects on comprehension.

Sentence Comprehension

Comprehension of sentence-length material (Selnes et al., 1983) was assessed using the Token Test (DeRenzi & Vignolo, 1962). At 1 month, only 2 patients scored within the normal range on the Token Test. By 6 months, 30% of patients scored within the normal range. As with single-word comprehension, performance was similar for aural and written presentations.

Deficits in sentence comprehension present at 1 month were correlated with lesion volume ($r = -0.68$) (Table 2, Figure 5a). Patients with lesion volumes greater than 60 cc all had moderate to severe deficits at month 1. Over half of the patients with small lesions also had moderate or severe deficits. There were no regions with high predictive accuracy.

By 6 months, no patients with lesions over 100 cc had normal Token test performance (Table 2, Figure 5b). Although one patient with an inferior temporal–lateral occipital lesion of 76 cc volume had normal-range Token Test performance, no other patient with a lesion over 60 cc had normal Token Test performance. It was only among the subgroup with lesions less than 30 cc that normal or very normal performance was the rule. Overall, the correlation between Token Test scores and lesion volume was high ($r = -0.74$). Thus, normal performance on the Token Test, in contrast to single-word comprehension was much more sensitive to the effects of lesion volume, as one would predict if linguistic difficulty were correlated with quantity of intact perisylvian brain tissue.

Among the patients with lesions less than 60 cc, the predictive accuracies for all regions were low, although the posterior temporal lobe was slightly higher than others (Figure 6). Compared to hemiparesis (and nonfluency and sentence repetition: see below), there were many false positives and false negatives; additional analyses using lower cutpoints for normal Token Test performance increased the number of patients with posterior temporal lesions with "normal" performance even though it decreased the number of patients with impaired Token Test performance who lacked posterior temporal lesions.

In the patients with moderate to severe sentence comprehension deficits and lesions less than 60 cc, the most common finding was complete destruction of the posterior superior temporal region plus the inferior supramarginal gyrus region (two examples are shown in Figure 7a,b). Five (all fluent) of the 8 patients had over 75% destruction of those regions. The 3 without posterior temporal lesions were all nonfluent; based on our conclusions regarding nonfluency and lesion volume (see below), the functional lesions in these patients were substantially larger than the CT lesion. This view was also supported by the low scores on single-word comprehension achieved by these 3. One patient had a lesion of the posterior insula and supramarginal gyrus. The other 2 patients had small subcortical central white matter lesions. Thus, lesions in the posterior temporal lobe were the most critical for producing severe deficits in sentence comprehension, but large lesions that spared the posterior temporal lobe were capable of producing a similar degree of impairment. Moreover, 2 patients scoring in the normal range had substantial posterior

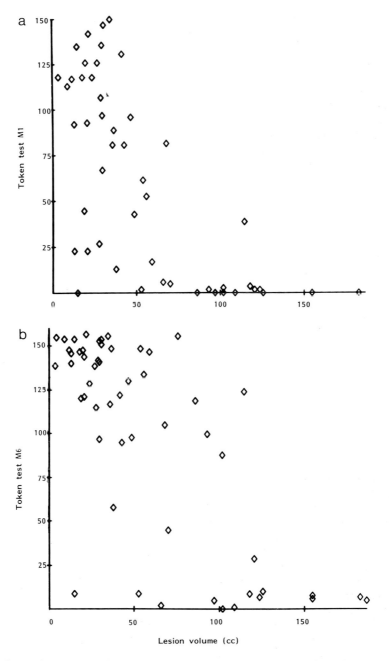

FIGURE 5. Scatterplots demonstrating relationship between lesion volume (in cubic centimeters) and scores on Token Test at 1 month (a) and at 6 months (b) post-stroke.

temporal lesions (Figure 7c), showing that incomplete lesions in the region did not necessarily produce long-lasting deficits in sentence comprehension.

We attempted a more detailed analysis of auditory comprehension for syntactically complex material using an experimental syntax comprehension test that required successful processing of

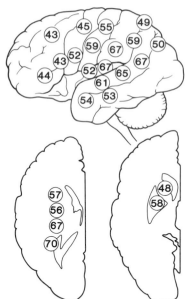

FIGURE 6. Predictive accuracies (in percent) of regional lesions for presence of deficit on Token Test at 6 months post-stroke. Predictive accuracy refers to the ratio of [(true positives + true negatives)/total number of cases]. For this figure, patients were classified as normal or impaired, and a region's status as not lesioned or lesioned. See Figure 1 for localization of regions.

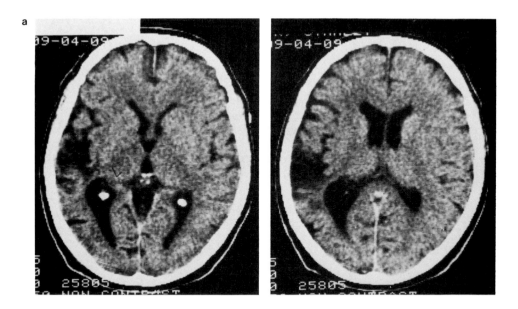

FIGURE 7. Selected CT scans of patients with lesions of the temporal lobe. (a) A007. Overall lesion volume 38 cc. (b) A030. Overall lesion volume 30 cc. (c) A067. Overall lesion volume 18cc. Both patients a and b had very poor single-word comprehension at 1 month but had achieved normal-range scores by 6 months. Both had persistent moderate to severe impairment on the Token Test at month 6 as well as very poor repetition scores and poor naming persistently. On the other hand, patient c recovered single-word, Token Test, and naming performance into the normal range by 6 months. However, this patient did not recover the ability to repeat or to comprehend the syntactically complex sentences.

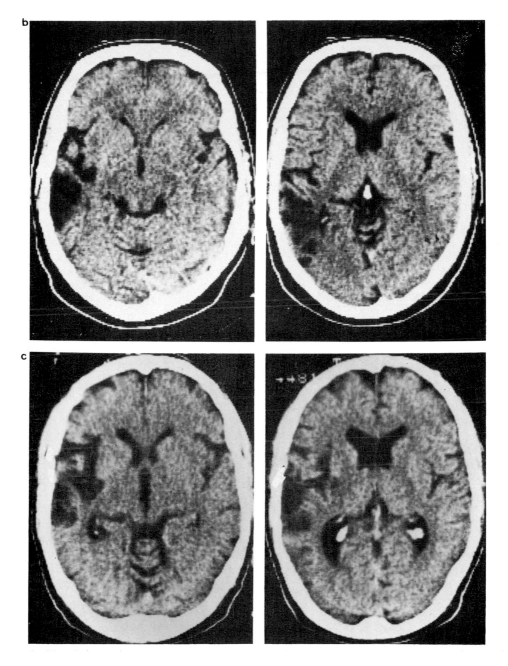

FIGURE 7. (*continued*)

word order. The sentences were constructed using semantically reversible subject–object relations in active, passive, or negative forms. Nonreversible sentences using the same nouns but verbs that allowed only one subject–object relation served as controls for sentence length. Scores on the Token Test were highly correlated with scores on both nonreversible sentences ($r = 0.94$) and

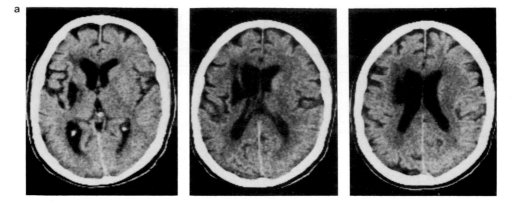

FIGURE 8. Selected CT scans of patients with impaired performance on sentence comprehension involving reversible subject–object relationships and with lesions outside of the temporal lobe. (a) A058. Overall lesion volume 9 cc. This patient's lesion was located in the putamen and subcortical white matter. (b) A031. Overall lesion volume 47 cc. Patient b's lesion was entirely in the parietal lobe. The sagittal and coronal images of this patient's scan were reconstructed from the original overlapping horizontal scans. This capability greatly enhanced the ability to localize the lesion and the Sylvian fissure and, in particular, to conclude that the posterior temporal lobe was not lesioned. Both patients scored in the normal range on the control "nonreversible" sentences but had marked difficulty on the "reversible" sentences. Both had impairment of Token Test scores and sentence repetition in the moderate range at 1 month, and both recovered, although b scored in the mildly impaired range on the Token Test even at 6 months.

reversible sentences ($r = 0.89$). Only those patients who scored in the normal range on the nonreversible sentence comprehension portion of the test were included in the analysis. At 1 month, only 7 patients scored in the normal range on the control nonreversible sentences, and only 1 patient scored in the normal range on the reversible sentences. By 6 months, only 20 patients scored in the normal range on the nonreversible control sentences, and 8 scored in the normal range on the reversible sentences. Anatomic analyses were of necessity confined to the 6-month period.

The analysis of the anatomic correlates of comprehension of syntactically complex material showed that all 20 patients had lesion volumes of 60 cc or less, consistent with the findings on the Token Test. Indeed, 6 of the 8 patients with normal performance on the reversible sentences had lesions of 30 cc or less. None of the patients with normal scores on reversible sentences had Wernicke's area lesions. Of the patients with impairment on reversible sentences, 4 had lesions in the posterior temporal region (Figure 7c). Six patients had lesions in the insula–putamen (Figure 8a), 5 of whom also had extension into the inferior frontal gyrus as well. The other 2 patients had lesions in the supramarginal gyrus (Figure 8b). Thus, there were locations other than the posterior temporal region in which lesions were associated with impairment on reversible sentences. At the same time, among the patients with normal reversible sentence performance, 4 also had subcortical lesions with or without extension into the inferior frontal gyrus, and 2 had suprasylvian supramarginal lesions. We repeated the localization analysis of syntactically-loaded sentence comprehension using lower cutpoints for normal performance on nonreversible sentences and came to the same conclusion, namely, that lesions in no one region had high predictive accuracy for sentence comprehension deficits.

Although the data suggest some role for the posterior temporal–inferior supramarginal region, i.e., Wernicke's area, there was substantial variability. Destruction of this region precluded recovery of sentence comprehension if syntactic demands were high. However, patients with small

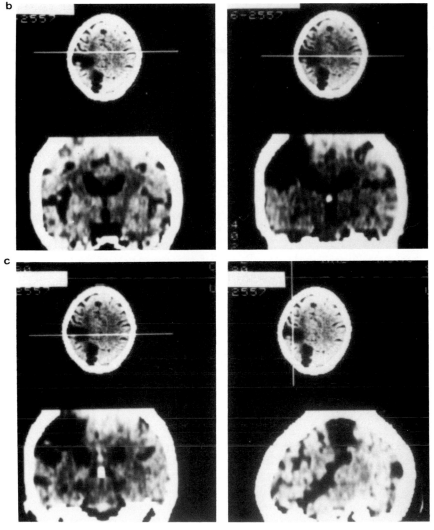

FIGURE 8. (*continued*)

lesions in other perisylvian regions also had syntactic comprehension deficits, and some patients with large perisylvian lesions that spared the posterior temporal region had severe sentence comprehension deficits. Thus, when lesions exceeded 60 cc, sentence comprehension was highly likely to be impaired, in distinction to single-word comprehension, in which the lesion volume threshold for normal performance was much higher. When lesions were less than 60 cc, individual variability prevented any one specific region from having high predictive accuracy for sentence comprehension deficits.

Confrontation Naming

The confrontation naming subtest (Knopman *et al.*, 1984) of the BDAE (Goodglass & Kaplan, 1972) was used. All but 3 patients had deficits on confrontation naming at 1 month. By 6 months, 67% still had deficits. The presence of coexisting nonfluency affected naming of patients

who were nonfluent at 1 month, all had severe naming impairment. The same was also true at 6 months in that only 10% of the persistently nonfluent patients were in the normal range on naming. If the patient were fluent, on the other hand, naming could range from severely impaired to normal, with 49% of fluent patients scoring in the normal range at 6 months.

Confrontation naming showed a strong relationship to lesion volume in that all patients with mild or no impairment had lesions less than 60 cc. For fluent patients alone, the range of lesion volume was restricted, and in that subgroup, there was only a weak relationship to lesion volume at 1 month ($r = -0.42$). There were a number of fluent patients with small lesions who had poor naming at that time (Figure 9a). At 1 month, lesions in any location were associated with deficits in naming. Of the 3 patients scoring in the normal range at 1 month, 2 had small lesions in the supramarginal gyrus, and 1 had a strictly inferior frontal gyrus lesion.

By 6 months, there was a strong relationship between lesion volume and naming score ($r = -0.76$) even within the fluent subgroup in which there was only 1 patient with a lesion over 100 cc (Figure 9b). No fluent patients with lesions over 60 cc had normal or only mild impairment. Of the 2 nonfluent patients who became able to name in the normal range, both had lesions less than 60 cc. All patients with lesions over 60 cc had moderate or severe impairment in naming.

In fluent patients with small lesions (<60 cc), lesions in no one region or regions had high predictive accuracy for anomia (Figure 10). Among patients with moderate to severe naming deficits, most had posterior temporal lesions (see Figure 7c), but there were other patients with similar posterior temporal lesions with normal naming, accounting for the low predictive accuracy of the region as depicted in Figure 10. Lesions in the insula and putamen were also present in some patients with deficits and in some with normal performance (Figure 8a). It appeared that the function of naming can be impaired by any lesion in the perisylvian region of the left hemisphere.

Sentence Repetition

The sentence repetition subtest (Selnes et al., 1985) of the BDAE (Goodglass & Kaplan, 1972) was used. Impaired sentence repetition was highly likely to occur in the presence of coexisting nonfluency, the exception to that rule being patients with transcortical motor aphasia, of which we had 1 example at 1 month only. Our analysis of sentence repetition, therefore, was restricted to fluent patients. At one month post-stroke, 78% of the fluent patients had impaired sentence repetition performance. By 6 months, 51% of fluent patients still had impairment.

At 1 month, sentence repetition deficits did not appear localized to any region (Figure 11). Patients with lesions in various perisylvian regions had impaired repetition. The 7 patients with …mal-range repetition had lesions that spared the posterior temporal and anterior parietal regions . 'though some patients with impaired repetition had similar lesions. All of the patients with normal month ' cpetition had lesion volumes of less than 60 cc, and all but one had lesions less than 40 cc. However, the majority of patients with lesions less than 60 cc had repetition deficits as well.

By 6 months, a clear pattern had emerged. Repetition performance was unrelated to lesion volume. Impaired repetition in fluent aphasics was strongly correlated with lesions in the posterior superior temporal region, extending from Heschl's gyrus back into the infrasylvian supramarginal gyrus region (Figure 7). No patient with a lesion in the posterior temporal region had normal sentence repetition. There were only 3 patients with impaired repetition whose lesions were located out of this region. In the 3 "exception" patients, the lesions were close to the temporal lobe–posterior insular junction. Lesions in other regions, including other parts of the supramarginal gyrus or underlying white matter, had far lower predictive accuracy for repetition deficits than those in the posterior superior temporal region.

Repetition deficits, therefore, were caused by lesions in the immediate vicinity of the primary auditory cortex, i.e., the auditory association area (Wernicke's area). Even incomplete lesions of the region typically resulted in impaired performance. None of our patients had a repetition dis-

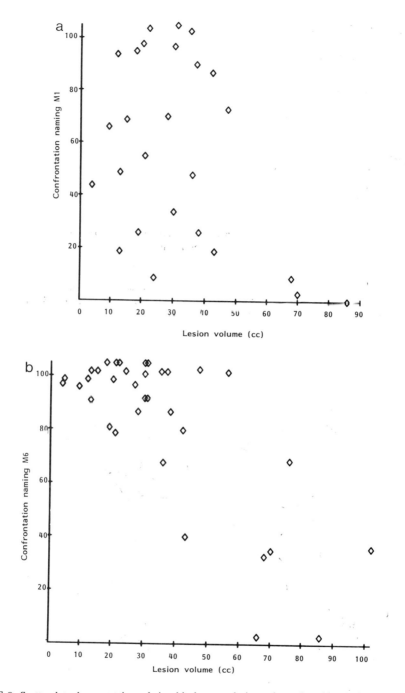

FIGURE 9. Scatterplots demonstrating relationship between lesion volume (in cubic centimeters) and scores on BDAE Confrontation Naming Test at 1 month (a) and at 6 months (b) post-stroke.

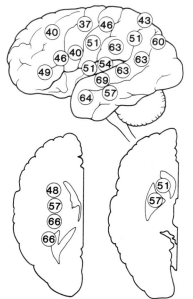

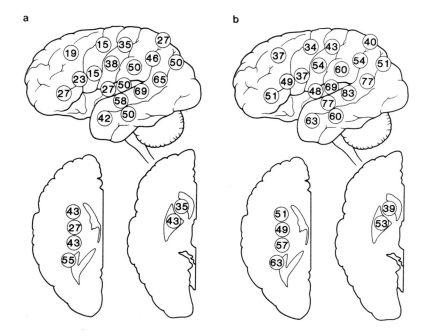

FIGURE 10. Predictive accuracies (in percent) of regional lesions for presence of deficit on BDAE Confrontation Naming Test at 6 months post-stroke. Predictive accuracy refers to the ratio of [(true positives + true negatives)/total number of cases]. For this figure, patients were classified as normal or impaired, and a region's status as not lesioned or lesioned. See Figure 1 for localization of regions.

FIGURE 11. Predictive accuracies (in percent) of regional lesions for presence of deficit on BDAE Sentence Repetition Test at 1 month (a) and 6 months (b) post-stroke. Predictive accuracy refers to the ratio of [(true positives + true negatives)/total number of cases]. For this figure, patients were classified as normal or impaired, and a region's status as not lesioned or lesioned. See Figure 1 for localization of regions.

turbance that resulted from a lesion outside of the posterior superior temporal or posterior temporal–posterior insular junction region. Therefore, the data make a strong case for the attribution of persistent repetition deficits to a gray matter lesion of the posterior temporal lobe.

Nonfluency

Nonfluent speech was operationally defined (Knopman *et al.,* 1983) in terms of reduced number of words per phrase, impaired articulation, altered prosody of speech, and impoverished grammatical style, according to the BDAE criteria (Goodglass & Kaplan, 1972). At 1 month, about half of the patients were nonfluent. By 6 months, there were still 17 nonfluent patients and 4 whose fluency recovered to the intermediate range.

At 1 month, no fluent patients had lesions over 76 cc. At the same time, the majority of the nonfluent patients had lesions over 60 cc, although there were several with smaller lesions. Regional involvement of the rolandic region and underlying white matter was highly predictive of nonfluency, and the absence of lesions in this region was predictive of preserved fluency (Figure 12).

The anatomic relationships were similar at 6 months. Persistent nonfluency was related anatomically to lesions of at least ~50–60 cc in the frontal–parietal operculum adjacent to the Sylvian fissure and centered on the central sulcus, including either frontal or temporal–parietal cortex and underlying white matter (Figures 12 and 13). Lesions that were largely posterior to the central sulcus (for instance, Figure 13b) or lesions largely anterior to the central sulcus were both capable of producing nonfluency if they included some of the lower pre- or postcentral gyrus and the underlying white matter.

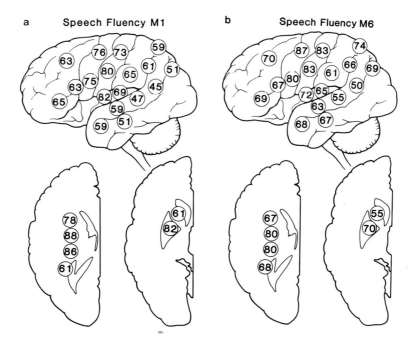

FIGURE 12. Predictive accuracies (in percent) of regional lesions for presence of deficit on BDAE Fluency of Spontaneous Speech at 1 month (a) and at 6 months (b) post-stroke. Predictive accuracy refers to the ratio of [(true positives + true negatives)/total number of cases]. For this figure, patients were classified as fluent or nonfluent, and a region's status as not lesioned or lesioned. See Figure 1 for localization of regions.

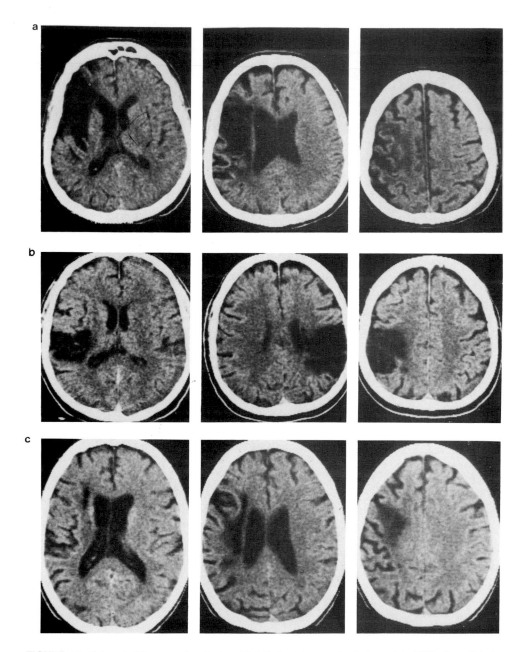

FIGURE 13. Selected CT scans of patients with left frontal–opercular lesions. (a) A009. Overall lesion volume 93 cc. (b) A070. Overall lesion volume 54 cc. (c) A084. Overall lesion volume 15 cc. All three remained persistently nonfluent, but patient b had near-normal-range Token test scores and normal naming. He was impaired on syntax comprehension, however. Despite the lack of a temporal lobe lesion, patient a had poor comprehension, which appears to be a lesion volume effect. Patient c had the smallest lesion that produced persistent nonfluency and very poor single-word and very poor sentence comprehension.

There were no persistently nonfluent patients whose pre- and postcentral gyrus region and underlying white matter were uninvolved. Thus, this region was unique in that a lesion in the region was necessary for the appearance of nonfluency. Inferior frontal gyrus lesions, if less than 60 cc and not involving the pre- or postcentral gyrus, did not produce persistent nonfluency. These findings extend the earlier observations of Mohr *et al.* (1978) and are consistent with observations made in patients with head injuries (Ludlow *et al.,* 1986).

Lesions limited to the lower pre-/postcentral gyrus region, although necessary for producing persistent nonfluency, were not sufficient in all instances. There were 5 patients who had lesions in the region who were not persistently nonfluent. Two of these individuals, who were initially nonfluent, had lesions centered in the opercular perisylvian region that were over 60 cc. The variability in the effects of lesions in the critical region could have resulted from preservation of viable brain tissue in the regions of apparent infarction, allowing for direct left hemisphere connections to brainstem nuclei or for transcallosal output by the left hemisphere. Despite these exceptions, deficits in speech fluency were highly localized to the pre- and postcentral gyrus region of the left hemisphere.

DISCUSSION

Our analyses of the regional anatomic correlates of aphasic deficits have produced several new observations and confirmed older ones. Certain language functions appear to be highly localized, as in the case of sentence repetition in fluent patients and in the case of nonfluent speech. Others are best considered as diffusely represented or nonlocalized, as in the case of single-word comprehension and confrontation naming. This interpretation represents a considerable revision of the concepts of brain–language relationships that existed prior to the CT era. The credibility of our observations depend on the strengths of the methodology: the utilization of a large sample of consecutive patients, the use of linguistic measures that have both clinical and theoretical justification, and the use of CT analyses that allowed precise and reliable lesion localization.

At the extremes of aphasic deficits, our findings are compatible with classical teaching: global aphasics generally have large anterior and posterior lesions, whereas anomic aphasics may have small lesions. Exceptions existed, however, in our series and in the experience of others (Vignolo, Boccardi, & Caverni, 1986). For both naming and single-word comprehension, the strong effects of lesion volume offer a plausible mechanism to account for this pattern. Lesion volume effects could be more important than lesion localization, as demonstrated by the subgroup of patients with severe comprehension deficits whose posterior temporal and inferior supramarginal gyrus regions were spared. The relatively nonspecific pattern of lesions that produced deficits of comprehension of words and sentences was not well appreciated previously.

Our results led us specifically to reject the localization of persistent nonfluency to the foot of the inferior frontal gyrus. Our data show that nonfluency was caused by lesions in the pre-/ postcentral gyrus region that included the underlying white matter. The close association of nonfluency with lesions of the primary motor and sensory cortices suggests that an important element of the deficit is motoric or proprioceptive.

The posterior superior temporal region emerged as critical for sentence repetition in fluent patients and also as an area in which small lesions produced the most severe sentence comprehension deficits. Yet, for sentence comprehension, the effects of lesions in this region varied from patient to patient. For persistent sentence repetition deficits, on the other hand, the predictive accuracy of lesions in the posterior superior temporal lobe was nearly as high as that of the corticospinal pathway for hemiparesis. The strong association between deficits in repetition and lesions very close to the primary auditory cortex indicates that the ability to repeat, in fluent patients, is an important function of the auditory association cortex. Although impaired repetition performance was ubiquitous in the first month post-stroke and possibly compatible with the "arcuate fasciculus'' account, virtually all other linguistic measures were impaired at that time as well. The posterior temporal region has a unique role in producing persistent deficits in repetition.

There was no language measure with perfectly predictable relationships to the CT findings. Some imprecision was necessarily inherent in CT analysis, but an equal or greater source of variability lay in the interindividual differences in the expression of aphasic deficits following stroke. Individual variability was present to some extent in the relationship between all language deficits and brain lesions, and substantial for some. The recognition of numerous exceptions to "classical" teaching is no longer being overlooked; we and others (Selnes, Rubens, Risse, & Levy, 1982; Basso *et al.*, 1985) view this variability as a critical feature of brain–language relationships. If language were uniquely represented in each individual, the aggregate pattern emerging from group studies such as this may fail to capture the essential features of brain–language relationships at the individual level. This is of concern both on practical grounds for the clinician dealing with a particular patient and on theoretical grounds as well. However, we have demonstrated that there are identifiable general trends that can be determined only from group studies.

ACKNOWLEDGMENTS. This work was supported by contract N01-2378 from the National Institutes of Health. Drs. Gail Risse, Nancy Niccum, Douglas Yock, David Larson, and Loren Jordan were of immeasurable assistance during the project.

REFERENCES

Basso, A., Lecours, A. R., Moraschini, S., & Vanier, M. (1985). Anatomoclinical correlations of the aphasias are defined through computerized tomography: Exceptions. *Brain and Language, 26,* 201–229.

Becker, H., Desch, H., Hacker, H., & Pencz, A. (1979). CT fogging effect with ischemic cerebral infarcts. *Neuroradiology, 18,* 185–192.

Caramazza, A. (1984). The logic of neuropsychological research and the problem of patient classification in aphasia. *Brain and Language, 21,* 9–20.

Chui, C. H., & Damasio, A. R. (1980). Human cerebral asymmetries evaluated by computed tomography. *Journal of Neurology, Neurosurgery and Psychiatry, 43,* 873–878.

Damasio, H., & Damasio, A. R. (1980). The anatomical basis of conduction aphasia. *Brain, 103,* 337–350.

DeRenzi, E., & Vignolo, L. A. (1962). The Token Test: A sensitive test to detect receptive disturbances in aphasics. *Brain 85,* 665–678.

DeWitt, L. D., Grek, A. J., Buonanno, F. S., Levine, D. N., & Kistler, J. P. (1985). MRI and the study of aphasia. *Neurology, 35,* 861–865.

Goodglass, H., & Kaplan, E. (1972). *The assessment of aphasia and related disorders.* Philadelphia: Lea & Febiger.

Kertesz, A., Harlock, W., & Coates, R. (1979). Computer tomographic localization, lesion size and prognosis in aphasia and nonverbal impairment. *Brain and Language, 8,* 34–50.

Knopman, D. S., & Rubens A. B. (1986). The validity of computed tomographic scan findings for the localization of cerebral functions. *Archives of Neurology, 43,* 328–332.

Knopman, D. S., Selnes, O. A., Niccum, N., Rubens, A. B., Yock, D., & Larson, D. (1983). A longitudinal study of speech fluency in aphasia: CT correlates of recovery and persistent nonfluency. *Neurology, 33,* 1170–1178.

Knopman, D. S., Selnes, O. A., Niccum, N., & Rubens, A. B. (1984). Recovery of naming in aphasia: Relationship to fluency, comprehension and CT findings. *Neurology, 34,* 1461–1470.

Ludlow, C. L., Rosenberg, J., Fair, C., Buck, D., Schesselman, S., & Salazar, A. (1986). Brain lesions associated with nonfluent aphasia fifteen years following penetrating head injury. *Brain, 109,* 55–80.

Matsui, T., & Hirano, A. (1978). *An atlas of the human brain for computerized tomography.* Tokyo: Igaku-Shoin.

Mazzocchi, F., & Vignolo, L. A. (1979). Localization of lesions in aphasia: Clinical CT-scan correlations in stroke patients. *Cortex, 15,* 627–654.

Metter, E. J., Wasterlain, C. G., Kuhl, D. E., Hanson, W. R., & Phelps, M. E. (1981). FDG positron emission computed tomography in a study of aphasia. *Annals of Neurology, 10,* 173–183.

Mohr, J. P., Pessin, M. S., Finkelstein, S., Funkenstein, H. H., Duncan, G. W., & Davis, K. R. (1978). Broca aphasia: Pathological and clinical aspects. *Neurology, 28,* 311–324.

Naeser, M. A., & Hayward, R. W. (1978). Lesion localization in aphasia with cranial computed tomography and the Boston Diagnostic Aphasia Exam. *Neurology, 28,* 545–551.

Ojemann, G. A. (1983). Brain organization for language from the perspective of electrical stimulation mapping. *The Behavioral and Brain Sciences, 2,* 189–230.

Rubens, A., Mahowald, M., & Hutton, T. (1976). Asymmetry of the lateral (Sylvian) fissures in man. *Neurology, 26,* 620–624.

Schwartz, M. F. (1984). What the classical aphasia categories can't do for us and why. *Brain and Language, 21,* 3–8.

Selnes, O. A., Rubens, A. B., Risse, G. L., & Levy, R. S. (1982). Transient aphasia with persistent apraxia. *Archives of Neurology, 39,* 122–126.

Selnes, O. A., Knopman, D. S., Niccum, N., Rubens, A. B., & Larson, D. (1983). Computed tomographic scan correlates of auditory comprehension deficits in aphasia: A prospective recovery study. *Annals of Neurology, 13,* 558–566.

Selnes, O. A., Niccum, N., Knopman, D. S., & Rubens, A. B. (1984). Recovery of single word comprehension: CT-scan correlates. *Brain and Language, 21,* 72–84.

Selnes, O. A., Knopman, D. S., Niccum, N., & Rubens, A. B. (1985). The critical role of Wernicke's area in sentence repetition. *Annals of Neurology, 17,* 549 557.

Vignolo, L. A., Broccardi, E., & Caverni, L. (1986). Unexpected CT-scan findings in global aphasia. *Cortex, 22,* 55–69.

Whitaker, H. A., & Selnes, O. A. (1976). Anatomic variations in the cortex: Individual differences and the problem of the localization of language functions. *Annals of New York Academy of Sciences, 280,* 844–854.

6

Brain Imaging and Neuropsychological Outcome in Traumatic Brain Injury

RONALD M. RUFF, C. MUNRO CULLUM,
and THOMAS G. LUERSSEN

INTRODUCTION

Historically, the discipline of neuropsychology has been interested in the localization paradigm relating behavioral deficits to focal lesions within the brain. The first generation of modern neuropsychologists (e.g., Benton & Van Allen, 1968; Hecaen & Albert, 1978; Luria, 1966; Milner, 1954; Teuber, 1974; Zangwill, 1969) sought out distinct groups of patients with lesions confined to specific cerebral regions and documented the corresponding behavioral impairments. What may be considered the second generation has relied on similar experimental strategies (i.e., based on double or multiple dissociations) in their efforts to identify the specific cerebral structures involved in various aspects of memory (exemplified by Mishkin, 1978; Pribram & Broadbent, 1970; Squire & Butters, 1985; Weiskrantz, 1968) or language and cognition (e.g., Benson & Geschwind, 1971; Goodglass & Kaplan, 1972).

Within this tradition, patients with traumatic brain injury (TBI) generally have provided limited information with respect to the localization of brain function. Part of the difficulty in using TBI patients for such investigation arises from the fact that craniocerebral trauma tends to result not only in localized damage but also in more widespread and often microscopic damage throughout cortical as well as subcortical regions (Adams, Gennarelli, & Graham, 1982; Bigler, 1987). Moreover, the specific nature of the damage sustained by the CNS from the injury may vary widely between individual cases. This variability arises from a number of factors and may depend on the source or sources of impact (e.g., focal contusion versus coup–contrecoup injury) as well as on the degree of acceleration/deceleration of the cerebrum within the cranial vault. Thus, because of the multifactorial nature of the mechanisms involved in craniocerebral trauma, studies

RONALD M. RUFF • Departments of Psychiatry and Neurosurgery, University of California at San Diego, La Jolla, California 92103. C. MUNRO CULLUM • Department of Psychiatry, University of California at San Diego, La Jolla, California 92103; *present address:* Department of Psychiatry, University of Colorado Health Sciences Center, Denver, Colorado 80262. THOMAS G. LUERSSEN • Department of Neurosurgery, University of California at San Diego, La Jolla, California 92161.

of head-injured patients have tended to fall outside of the neuropsychological tradition of localizing specific brain–behavior patterns.

Studies of TBI patients typically have focused on the severity of the injury in relation to behavioral outcome rather than on the particular regions involved or the specific underlying neuropathology. Accordingly, the description of relatively large samples of patients with TBI tends primarily to address various neurological factors such as length of coma, coma severity (e.g., Glasgow Coma Scale Score; Jennett & Bond, 1975), and duration of posttraumatic amnesia (Levin, Benton, & Grossman, 1982). In addition, various neurological, neurosurgical, and radiological findings pertaining to the presence or absence and evacuation of intracranial hematomas, presence or absence of brainstem signs (e.g., decerebrate rigidity, doll's eyes), midline shift of cerebral structures, as well as the extent of cerebral swelling and level of intracranial pressure have often been a focus of TBI studies.

Unlike much of the neuropsychological literature, which has tended to focus more on localization-of-function issues, studies of TBI often have emphasized recovery of function over time. To elucidate, it is not uncommon to find in the literature a pooling of patients according to severity of injury, with little attention to more specific lesion parameters such as size and location as they may pertain to neurocognitive status. For example, TBI samples may include patients grouped together with lesions involving either or both hemispheres, ranging from anterior to posterior cerebral regions, with varying degrees of subcortical damage, or any combination thereof. Thus, with regard to the issue of localizing specific neurobehavioral functions, the question arises as to whether TBI patients may in fact be too difficult to study because of the heterogeneity of the underlying neuropathological processes.

This chapter addresses this issue by examining the benefits that computerized tomography (CT) and magnetic resonance imaging (MRI) have provided with regard to our understanding of the interrelationships between neurobehavioral deficits and the associated underlying neuropathology in TBI. In addition, the utilization of neuroimaging and neuropsychological techniques is discussed in terms of the acute and long-term treatment phases. Evidence is presented that attempts to demonstrate that neuropathological processes involved in TBI may not be quite as heterogeneous and random as they initially may appear. The need for more specific classifications of underlying neuropathology is proposed in order to raise questions with respect to how neuropsychological functions may be differentially affected based on underlying clusters of neuropathological processes.

NEUROIMAGING IN THE ACUTE AND CHRONIC STAGES OF TBI

Acute Treatment Stages

Neuroimaging, typically in the form of CT scanning, is essential in the early management of craniocerebral injury, in part to aid in the decision of neurosurgical intervention. Although MRI may offer superior resolution to CT, utilization of the MRI technique, in its current form, is limited in terms of the initial evaluation of the TBI patient. For example, head-injured patients sometimes require additional support equipment (e.g., metallic neurosurgical clips, shunts), which may preclude introducing the patient into a strong magnetic field. A second factor is that if the patient is conscious, the greater time needed for the acquisition and quality of the images also requires sustained compliance and may prove problematic in neurologically compromised individuals. Luerssen, Hesselink, Ruff, Healy, and Grote (1986) have argued that despite the ability of MRI to detect relatively small areas of hemorrhage or edema, CT scanning is clearly of sufficient sensitivity to detect those mass lesions that would warrant immediate neurosurgical intervention. Moreover, MRI is not necessarily superior to CT for detecting hemorrhages less than about 48 hr old (Sipponen, Sipponen, & Sivula, 1983). Therefore, because of these potential limitations and the fact that MRI requires more time, technical support, and expense than CT scanning, its present impact appears rather negligible in terms of the acute management of TBI.

Chronic Treatment Stages

The acute treatment stages are generally considered to have been completed once the patient is medically stable. Long-term treatment addresses not only what interventions may facilitate the recovery process but also what supportive care may be needed for the irreversible or chronic difficulties that may be present.

It is primarily for the long-term follow-up and treatment stages that we want to address the potential application and limitations of CT and MRI. Little is yet understood regarding the specific nature and mechanisms involved in the underlying structural changes that may take place during the recovery process. Carefully designed longitudinal studies are called for in order to assess the potential for neurostructural changes over time, which may, in turn, be related to changes in neurobehavioral functioning. Some of the factors that need to be analyzed along these lines include the presence of hemorrhagic contusions, different types of edema, changes in ventricular size, and localized versus diffuse atrophic processes.

Hematoma

It has been estimated that between 15% and 35% of adult patients with severe closed-head injury experience hematoma (Clifton et al., 1980; Cullum & Bigler, 1985; Galbraith & Smith, 1976). Both CT and MRI can detect the location and general size of all surgically significant hematomas, but for smaller hematomas, MRI is clearly better for detecting areas of hemorrhage. This is particularly true in regions not visualized well with CT, i.e., the anterior temporal lobes and the structures of the posterior fossa. As mentioned, MRI is limited in the acute phases until the hemoglobin in the region is converted to methemoglobin, a process that occurs over 48 to 72 hr (Sipponen et al., 1983). Fresh hemorrhage produces a signal very similar to that of brain tissue. After conversion to methemoglobin, the T_1 relaxation time is shorter, and this results in bright signal on both T_1 and T_2-weighted images. As hematomas age further, hemosiderin appears, which has a shorter T_2 relaxation time than either methemoglobin or brain tissue and therefore appears as a dark, low-signal area on the images. Hematomas as a result of TBI often occur concomitantly with other sequelae such as ventricular dilation, focal contusion, and occasionally infarction (see Figure 1).

Edema

Swellings that occur in the parenchyma are important variables that change over time, can be followed serially by neuroimaging, and can be detected with both CT and MRI. However, areas of cerebral edema tend to appear as a secondary response to hemorrhagic contusions and intracranial hemorrhage and are more readily detected by MRI. Since MRI uses the hydrogen proton as the element studied most commonly for imaging purposes, it is extremely sensitive for detecting and localizing normal and abnormal areas of brain water (see Figure 2). Cerebral edema falls into the latter category. Areas of brain water can appear different on MRI, presumably because of differences in chemical composition, and these areas of fluid are seen best in T_2-weighted images. Depending on the injury, edemas can be characterized as to time of occurrence, location, magnitude, and type (e.g., vasogenic, cytotoxic, and ischemic edema; Brant-Zawadzki, Pereira, & Weinstein, 1986; Tornheim, 1985).

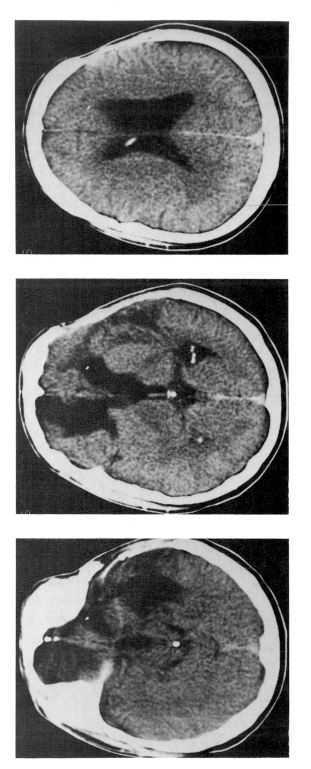

FIGURE 1. Computed tomographic scan in a patient 4 years after severe traumatic brain injury. This scan demonstrates many of the component aspects of residual cerebral damage as a result of TBI. The scan on the left demonstrates residual contusion with necrotic tissue at the base of the brain in the frontal region bilaterally as well as extension of an infarct into the mesial aspect of the left frontal region. This infarct was the result of massive cerebral edema that in turn compressed the anterior cerebral artery. There are scattered density changes throughout the frontal and temporal regions, again associated with patchy areas of necrotic tissue secondary to contusion. There is compensatory ventricular enlargement that is particularly evident in the right lateral ventricle. There is also a skull defect in the right lateral frontal region. Behaviorally, this patient developed prominent frontal lobe symptoms as a result of the bilateral extensive frontal damage.

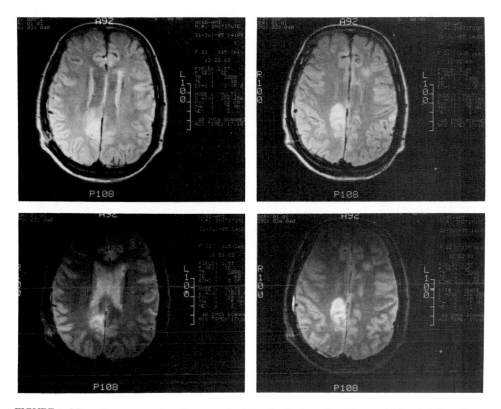

FIGURE 2. Magnetic resonance imaging scans depicting focal edema in a 21-year-old female TBI patient.

Ventricular Enlargement

One of the common changes in gross brain morphology following TBI is the delayed development of ventricular enlargement (see Figures 3 and 4), and this consistently has been found in past studies using pneumoencephalography (Boller, Albert, Lemay, & Kertesz, 1972; Hawkins, Lloyd, Fletcher, & Hanlca, 1976) and CT (Kishore *et al.*, 1978). Using ventricle–brain percentage ratios, Levin, Meyers, Grossman, and Sarwar (1981) observed dilation of the lateral ventricles in 72% of patients with severe TBI at least 30 days post-trauma. Similarly, using computer-assisted volumetric estimates of ventricular size, Cullum and Bigler (1986) recently reported ventricular enlargement in 77% of their sample of 48 patients with moderate to severe closed-head injuries. Although the precise mechanism for ventricular dilation following TBI is less than clear, it has been suggested that the enlargement results from a reduction in the bulk of cerebral white matter, thus allowing the ventricles to enlarge as in an *ex-vacuo* process (Levin *et al.*, 1981; Han *et al.*, 1984). Regional ventricular dilation also may be observed in the proximity of cerebral lesions (see Figure 5).

Whereas signs of hydrocephalus can, of course, be readily visualized by CT as well as MRI and followed over time, MRI allows for a much clearer distinction between regions of cerebrospinal fluid and periventricular white matter (i.e., by comparing T_1- and T_2-weighted images). This may be particularly useful in cases of apparent periventricular CSF leakage and ventricular distension or gliosis (e.g., in frontal or posterior horn regions; see Figure 6). In addition, the ability of MRI to visualize the ventricular system more clearly in the coronal as well as horizontal plane may allow for even more accurate assessments of ventricular size.

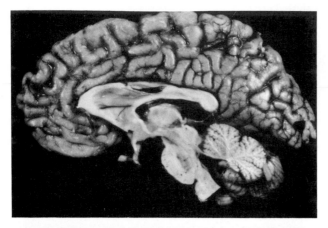
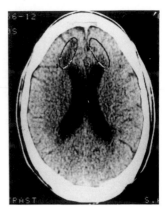
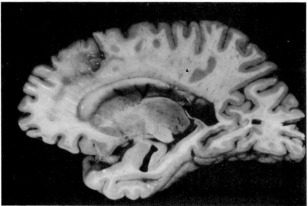
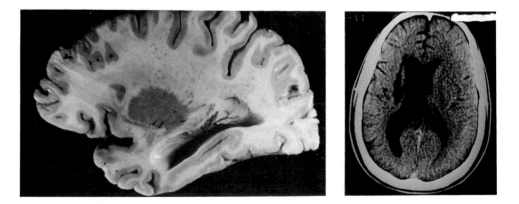

FIGURE 3. The sagittal views on the left are from a postmortem study in a patient who had previously sustained a severe traumatic brain injury. Note the ventricular dilation, particularly the prominence of the temporal horn and the thinning of the corpus callosum. In the bottom sagittal section, note the internal frontal contusion with gliotic changes and diminished density in and around the site of that lesion. On the upper right, note the CT scan and the frontal hypodense lesions that are circled. These are analogous to the necrotic changes that were present in the postmortem examination (not the same patient). The bottom CT scan on the right demonstrates a contusion along the axis of the internal capsule. This white matter shearing lesion left this cavitation and likewise resulted in complete contralateral hemiplegia. These CT scans demonstrate the classic configuration of ventricular dilation associated with necrotic tissue loss and contusions (from Bigler, 1987).

Atrophy

Cortical atrophy is another alteration in brain morphology that may result from cerebral trauma (Levin *et al.*, 1982; also see Figure 7). Among patients with a history of severe TBI and loss of consciousness, van Dongen and Braakman (1980) noted cortical atrophy in approximately 86% of patients on CT evaluation between 1 and 2 years post-trauma. Levin *et al.* (1981) subsequently reported mild sulcal widening in approximately one-fourth of their sample of patients with TBI. Using computer-assisted volumetric estimates of cortical atrophy, Cullum and Bigler (1986) found significantly greater indices of atrophy among TBI patients as a group.

The precise mechanisms involved in the development of cortical atrophy are not thoroughly understood, although factors such as direct contusing effects and potential hypoxia of neurons within the cortical mantle may be of key importance. With the enhanced resolution of MRI, a more precise delineation of the extent of cortical atrophy may be facilitated and subject to quantification, which may yield greater insights into the atrophic process.

Smaller areas of focal density changes also can be better detected with MRI, especially in the deep white matter and brainstem (Luerssen *et al.*, 1986; Levin, Kalisky, *et al.*, 1985). Degeneration of cerebral white matter is of particular interest in TBI (see Figure 3), and histological studies suggest that with loss of consciousness on impact, diffuse axonal injuries are common (Strich, 1956; Adams, Graham, Murray, & Scott, 1982). It has been suggested that TBI results in a degeneration of white matter, which in turn contributes to the neurobehavioral sequelae of what is commonly referred to as "diffuse closed-head injury" (Levin *et al.*, 1981, 1982). In a more recent paper based on MRI data, however, Levin and coinvestigators (Levin, Handel, Goldman, Eisenberg, & Guinto, 1985) suggested that a more precise interpretation may be that multifocal lesions are involved (predominantly in the cerebral white matter), which may correspond with the neurobehavioral sequelae of diffuse brain injury. Prior to MRI, clear visualization of areas of degenerated white matter was difficult and not uniformly obtained with typical CT resolution. However, MRI is significantly more sensitive in terms of fine resolution and, for example, has a successful track record of disclosing lesions of cerebral white matter in multiple sclerosis that may have otherwise been underestimated or undetected by CT (Bydder *et al.*, 1982; Crooks *et al.*, 1982).

Whereas studies using CT techniques have demonstrated their utility in the visualization and measurement of structural cerebral parameters, it seems reasonable to suggest that MRI may facilitate clearer delineation of many of the factors involved in TBI (i.e., hemorrhagic contusions, different types of edema, changes in ventricular size, and localized versus diffuse atrophy). Although CT scanning proved an excellent tool for imaging the rather gross structural status of the cerebrum (i.e., larger cavitations, ventricle size, and atrophic regions), areas of smaller cavitation, especially in the deep white matter, are better delineated with MRI.

PATTERNS OF NEUROPATHOLOGY

To advance our understanding both clinically and from a research perspective, it seems critical that patterns of underlying neuropathology of individual patients be more clearly delineated. One of the most challenging aspects is to come to a better and more detailed understanding of the processes involved in what are referred to as "focal lesions" versus "diffuse brain damage." The term "focal" describes a circumscribed area of damage or a space-occupying lesion. Given its multifactorial nature, TBI can result in discrete lesions that can be classified as cortical contusions or ischemic lesions, which are superimposed on more diffuse damage. Secondary effects of diffuse edema may produce another type of brain lesion associated with tentorial herniation, which may include hemorrhage and infarction in the brainstem or parahippocampal gyrus region (Harenstein, Chamberlin, & Conomy, 1967). Thus, what constitutes "focal" or "diffuse" damage may actually vary neuropathologically across individual cases.

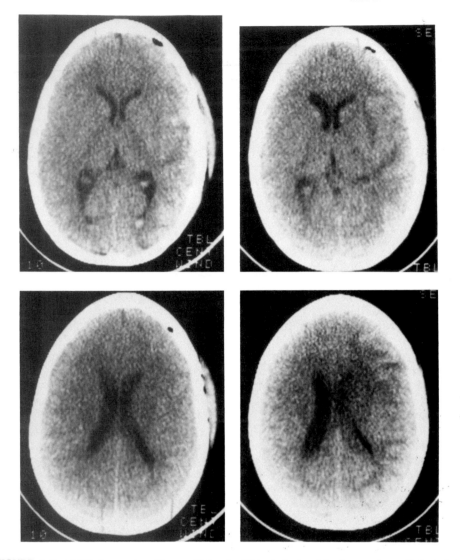

FIGURE 4. Serial CT changes in a 12-year-old female child who had sustained severe traumatic brain injury. The two CT images on the left depict a relatively normal scan on admission to the hospital, although it is noted that there is a pocket of air in the right frontal region; there were also some patches of intracerebral hemorrhaging. The right-column images were obtained 7 days post-injury and demonstrate right cerebral hemisphere edema with partial collapse of the ventricular system and right-to-left shift. The two CT scans on the next page are corresponding images taken 2 months post-injury. These images indicate ventriculomegaly, particularly of the right lateral body of the ventricle, associated with diffuse tissue loss of the right hemisphere, which was secondary to edema effects as well as cortical contusion (from Bigler, 1987).

Cerebral contusions or "bruises" occur over the surface of the brain and typically are most severe at the crests of the gyri. These contusions, however, are not only confined to the cortex but may affect subcortical structures and white matter. In addition, over time these contusions may result in the formation of scar tissue (Adams, Graham, & Gennarelli, 1982). Within this

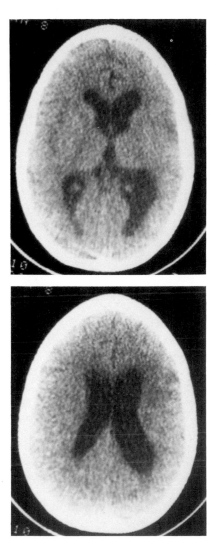

FIGURE 4. (*continued*)

context, the point that needs to be made is that these cerebral contusions are not necessarily to be considered as random. That is, focal contusions generally occur in the vicinity of the impact site, and, in addition, there appears to be a somewhat predictable pattern that is somewhat independent of the source of the impact (See Figures 3 and 8). Cerebral contusions commonly appear to be bilateral but may be asymmetric and involve the frontopolar, orbitofrontal, anterotemporal, and laterotemporal surfaces (Clifton *et al.*, 1980; Ommaya & Gennarelli, 1974; see also Figures 9 and 10). Thus, regardless of the original site of impact, other factors such as the heightened vulnerability of various cortical regions with respect to the internal topography of the skull must be considered. Still other factors include the neuropathological effects of multiple impacts and the rate of acceleration/deceleration of the cerebrum as it contacts the bony skull (Bigler, 1987).

In addition to cerebral contusions, TBI may result in ischemic vascular lesions, which can

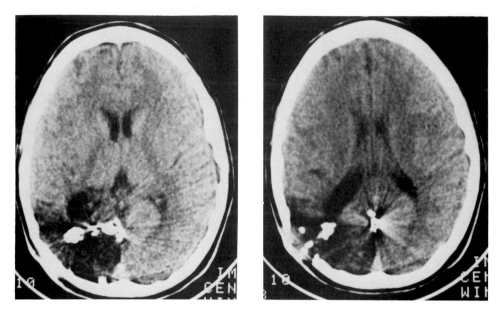

FIGURE 5. Regional ventricular dilation of the posterior horn of the left lateral ventricle in a patient who sustained a gunshot wound to the left parietal–occipital area. Note that there are still bone and metal fragments in the brain as well as a skull defect. The patient was 13 years of age at the time of this CT scan, and the scan was obtained some 2 years post-injury. The patient had dyslexia and spelling dyspraxia as well as a complete homonymous hemianopia of the right visual field (from Bigler, 1988).

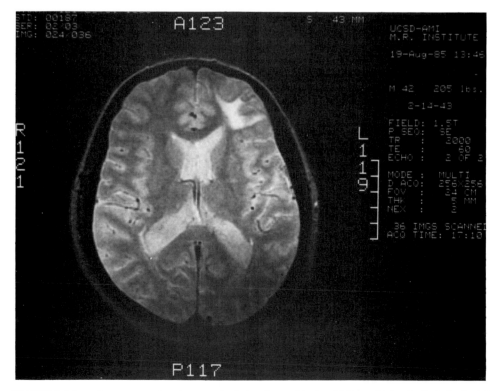

FIGURE 6. Magnetic resonance imaging scan depicting the ventricular dilation with focal gliosis in the left frontal region.

be restricted to the territory of a specific vessel. More specific to TBI, tentorial herniation can result in the occlusion of the posterior cerebral artery, which in turn can result in secondary damage in the mediotemporal and/or thalamic regions. Focal brain damage can also occur secondary to raised intracranial pressure, which can be attributed to the development of mass lesions such as intracranial hematomas or cerebral edema. Furthermore, this may result in pressure necrosis in regions such as one or both of the parahippocampal gyri (Signoret, 1985). Moreover, this may also include damage to the brainstem or result in ischemic changes in the territory of the posterior cerebral artery (Adams, Graham, & Gennarelli, 1982).

With respect to diffuse brain damage, the two major forms are diffuse axonal injury (DAI) and hypoxic damage. As discussed earlier, shearing lesions (most likely caused by acceleration/deceleration effects) cause axonal damage, which results in degeneration, specifically in the regions of the parasagittal and rostral midbrain areas. Diffuse axonal injury also has been associated with damage to the corpus callosum and the dorsal lateral quadrant of the rostral brainstem in the area of the superior cerebral peduncles (Adams, Graham, Murray, et al., 1982). It appears, therefore, that secondary to TBI, the diffuse axonal injuries may not be as random as might initially be assumed. It is interesting to note that an association has been suggested that would correlate the severity of posttraumatic coma with the severity of DAI (Gennarelli et al., 1982).

In summary, two general patterns of global neuropathology can be observed following TBI: the first involves focal damage sustained on direct impact, which generally results in damage to the poles of the frontal and temporal lobes (see Figure 9). The second, DAI, refers to more widespread axonal damage in the white matter, which may particularly involve the dorsolateral quadrants of the midbrain as well as the corpus callosum (see Figure 3).

NEUROIMAGING CORRELATES OF NEUROPSYCHOLOGICAL FUNCTIONING

Ventricular Enlargement

To assist in the exploration of the possible relationships between ventricular dilation and neurocognitive status, a variety of methods for quantifying ventricular size have been proposed (e.g., see Bird, 1982; Damasio et al., 1983). Some of these techniques involve simple linear measurements, ratios, or planimetric calculations of the area of the ventricles from a single CT slice (Levin et al., 1981; Meese, Kluge, Grumme, & Hopfenmulle, 1980; Synek & Reuben, 1976). Such methods may not result in accurate size estimates, however, because of their unidimensional nature. Other techniques for obtaining measures of ventricle size involve computer-derived volumetric (i.e., three-dimensional) estimates, which may be accomplished by manually tracing the structures in question and then feeding these data into a computer for algorithmic analysis (e.g., Cullum & Bigler, 1986; Turkheimer, Yeo, & Bigler, 1983), or by specialized CT analysis programs such as that of Jernigan and co-workers (Jernigan, Zatz, & Naeser, 1979) that utilize raw digital data directly from the scanner.

Clinically, ventricular enlargement in TBI, even despite differences in measurement techniques, generally has been found to be associated with residual cognitive deficit (Gudeman et al., 1981; Kishore et al., 1978; Levin et al., 1981; Timming, Orrison, & Mikula, 1982). Deficits in terms of new learning, attention, and memory abilities in particular have frequently been noted, yet more specific relationships with ventricular size have also been reported. Cullum and Bigler (1986) reported moderate associations between volumetric ventricle–brain size estimates and measures of new-learning or "fluid" cognitive abilities, with smaller correlations associated with better-ingrained, "crystallized" cognitive abilities. In addition, various measures of verbal and nonverbal abilities (e.g., verbal learning and constructional tasks) demonstrated some tendency to

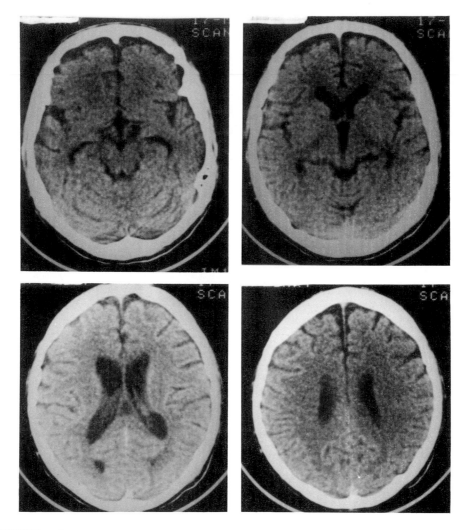

FIGURE 7. Computed tomographic scan from an 18-year-old male patient who had sustained a severe traumatic brain injury 1 year previously. Note the generalized and diffuse nature of cerebral atrophy with corresponding ventricular dilation. Also note the prominence of the Sylvian fissure as well as the interhemispheric fissure and the temporal horns of the lateral ventricular system. The patient had a corresponding traumatic dementia associated with these generalized atrophic changes (from Bigler, 1987).

be more strongly associated with left- and right-hemispheric ventricular size, respectively, thereby suggesting that regional ventricular dilation may be somewhat differentially associated with various neuropsychological abilities. It was suggested that better-established, "crystallized" verbal functions such as those assessed by WAIS Verbal IQ tasks appear to be less affected by craniocerebral trauma and ventricular dilation, whereas the more "fluid" cognitive measures appeared to be more sensitive to the gross morphological and pathophysiological changes induced by TBI.

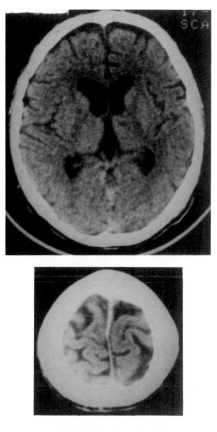

FIGURE 7. (*continued*)

Atrophy

There exists a relative dearth of studies attempting to assess the degree of atrophy present in patients with TBI, although similar investigations with dementia patients have been more numerous. Despite overall mixed results with respect to correlating measures of cortical atrophy with cognitive status in patients with dementia or degenerative disease (e.g., Earnest, Heaton, Wilkinson, & Manke, 1979; Wu, Schenkenberg, Wing, & Osborn, 1981), a number of these studies have demonstrated some significant relationships (Bigler, Hubler, Cullum, & Turkheimer, 1985; Damasio *et al.*, 1983; de Leon & George, 1983; Huckman, Fox, & Topel, 1975; Kaszniak, Garron, Fox, Bergen, & Huckman, 1979).

In the Cullum and Bigler (1986) study of TBI patients, volumetric estimates of atrophic tissue demonstrated significant correlations with a number of neuropsychological functions, most consistently in terms of memory deficits. When volumetric estimates of atrophy were examined along right–left and anterior–posterior dimensions, the greatest amount of atrophic tissue was noted in the frontal regions, and some specific relationships with neuropsychological measures were observed, with the strongest associations involving left frontal atrophy measures. The finding of greater atrophy in frontal brain regions is not surprising, since the areas most vulnerable to damage in head injury are the frontal and temporal areas because of their situation within the skull (Adams, Mitchell, Graham, & Doyle, 1977; also see Figure 3). It was suggested that the

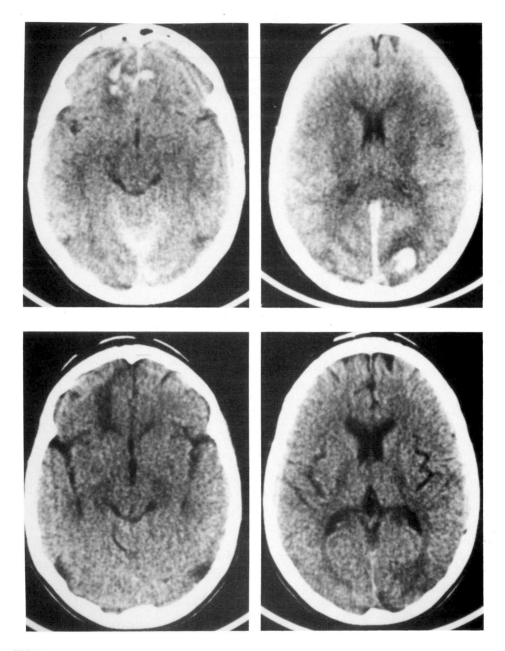

FIGURE 8. Coup–contrecoup effects of traumatic brain injury as visualized by CT imaging. The top row demonstrates an area of contusion in the left frontal region and the right occipital region. These scans were taken on the day of admission to the hospital. The bottom row shows corresponding cuts analogous to the images just above but obtained 6 weeks later. Note the infarction that has taken place in the left frontal region corresponding to the area of focal contusion and the contrecoup lesion in the right occipital area (from Bigler, 1987).

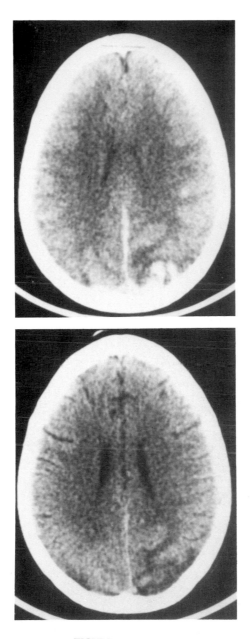

FIGURE 8. (*continued*)

loss of functional tissue in the anterior region of the dominant hemisphere may be particularly sensitive in TBI with respect to the abilities (e.g., attention, concentration, and complex problem solving) required on certain neuropsychological tasks.

Whereas cortical atrophy and ventricular dilation both represent atrophic processes, they nevertheless appear to be somewhat independent (Arai *et al.*, 1983). Cullum and Bigler (1986), for example, reported a correlation of 0.50 between computer-derived volumetric estimates of

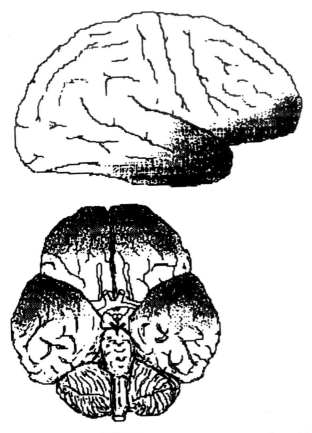

FIGURE 9. Depiction of areas most often subject to damage in TBI.

ventricle size and cortical atrophy. More specifically, examining the relationship between cortical atrophy and ventricular dilation in a sample of TBI patients, Massman, Bigler, Cullum, and Naugle (1986) found that focal damage effects were most apparent in measures of cortical atrophy, whereas generalized effects tended to be more associated with ventricular dilation. Ventricle size in TBI was also found to be more strongly associated with neuropsychological status than cortical atrophy, whereas the reverse was true in a group of dementia patients.

Hematoma

Studies of TBI patients with hematoma evaluated neuropsychologically and neuroradiologically are few and generally have focused on the different types of hematoma (i.e., epi- or subdural) that occur and their long-term prognostic significance. In a study of matched closed-head-injury patients with and without a history of hematoma, Cullum and Bigler (1985) found particular impairments in attention and memory abilities among those with a history of hematoma. In addition, greater estimates of cortical atrophy also were associated with the prior presence of hematoma. It was suggested that hematoma subsequent to head injury may be associated with additional pathological effects on gross brain morphology and memory functions.

Thus, whereas the number of studies of TBI patients attempting to relate neuropsychological functions with underlying neuropathology has been relatively small, the results have been at least somewhat encouraging. In addition to the need to explore further these and more specific issues

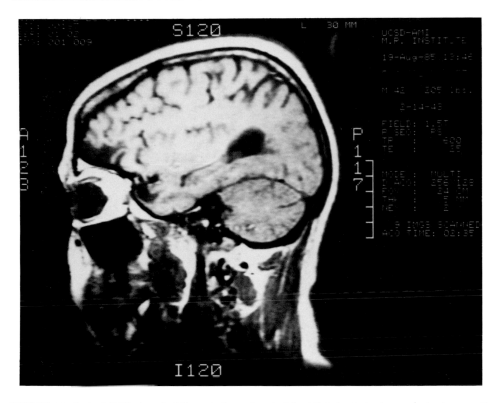

FIGURE 10. Sagittal MRI view depicting anterior and ventral frontal and anterior temporal atrophy as well as ventricular dilation in a 42-year-old TBI patient.

regarding brain–behavior relationships in TBI by using modern neuropsychological and neuro-imaging techniques, it may be possible at some point to be better able to predict outcome with the combined use of various indices of neuropsychological and brain morphology status.

Magnetic Resonance Imaging Studies

There are only a handful of published reports that have sought to correlate neuropsychological test performance with specific MRI data in TBI, and these primarily involve case studies. Levin, Handel, *et al.* (1985) presented both CT and MRI findings on a young woman who sustained a severe diffuse brain injury, with follow-up data over 5 years documenting the patient's recovery. It was tentatively postulated that the MRI and neuropsychological results were compatible with multifocal lesions predominantly in the cerebral white matter. It was further suggested that a refinement of the more global term of "diffuse brain damage" may be in order.

In a subsequent study, Levin and colleagues (Levin, Kalisky, *et al.*, 1985a) reported on the findings of 4 head-injured patients, again correlating neuropsychological data with MRI findings. They noted that the initial follow-up CT scans provided no direct evidence of cerebral white matter injury, although MRI documented increased density in the cerebral white matter in all 4 patients. The MRI results also led the authors to conclude that the dilation of the lateral ventricles was caused secondary to the degeneration of periventricular white matter. From a neurobehavioral perspective, their data suggested that damage to mesial frontal and temporal white matter in particular corresponded with a reduced ability to generate verbal and figural material. Moreover,

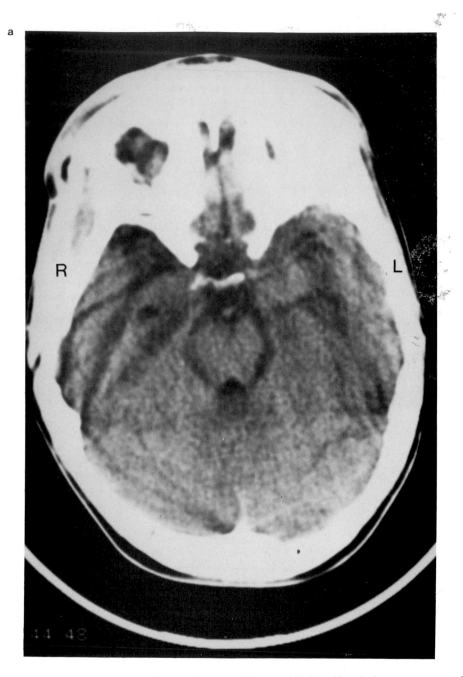

FIGURE 11. This CT scan (a) was interpreted to be within normal limits, although there was an area of questionable density change in the temporal area. However, the MRI scan on the facing page (b) clearly demonstrates damage in the right temporal region, which was not detected in the CT scan.

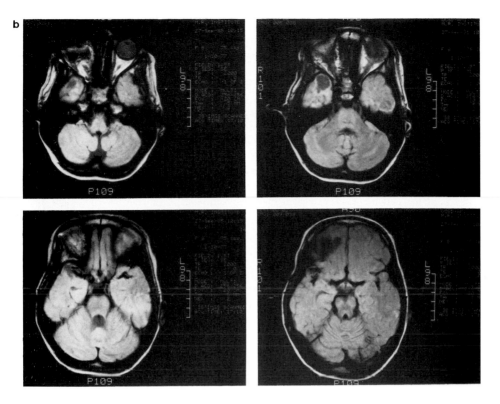

FIGURE 11. (*continued*)

3 of the 4 patients suffered impairment of long-term memory processing on both verbal and visuospatial tests. They noted, however, that only 1 of the 4 cases evidenced focal increased temporal lobe intensity on MRI. From these results, it appeared that the correlations between aspects of memory and temporal lobe dysfunction were less impressive than the aforementioned findings pertaining to frontal lobe functions.

In a recent study by Luerssen *et al.* (1986), five TBI cases were examined neuropsychologically and with MRI. In one of the cases (a 12-year-old child), the CT scan obtained 2 months post-injury was essentially normal. However, MRI demonstrated clear evidence of structural injury to the left temporal region. In concert with this finding, the pattern of neuropsychological performance indicated primary memory deficits for material processed in the verbal rather than the visuospatial mode. To illustrate, cases such as that in Figure 11 similarly have demonstrated that whereas no abnormalities may be evident on CT, various neuropathological findings may become manifest on MRI.

As studies of head injured patients have progressed, some general statements regarding the overall utility of MRI can be made. For severely head-injured patients who show persistent neurological or neurobehavioral disturbance, the MRI studies will almost always demonstrate abnormalities of signal intensity in the white matter or may reflect general cortical volume loss. For moderately head-injured patients, abnormalities of signal intensity have been observed, especially in the T_2-weighted pulse sequences early after injuries in patients whose CT scans were normal. These signal abnormalities, which may be caused by localized regions of traumatic edema, seem to resolve with time. Typically, these patients show neurological or neurocognitive improvement (Levin *et al.*, 1987). Magnetic resonance imaging studies of mildly head-injured patients also may show similar signal abnormalities, but in our experience this finding has been extremely variable. As with moderately injured patients, the likelihood of detecting abnormalities with MRI tends to be higher if the patient demonstrates neurological or neuropsychological abnormalities. Moreover, the time of the evaluation may play a role. For example, MRI studies of patients suffering cerebral concussion with persistent subjective complaints did not generally detect significant abnormalities (Levin *et al.*, 1987). It should be noted that these studies were conducted long after any acute changes such as edema or mild focal hemorrhage were present and thus do not represent acute injury effects.

In summary, current studies have indicated that evaluations of brain injury with MRI are more likely to show abnormalities in the early stages following the injury and are more likely to be positive if the patients show persistent objective neurological or neuropsychological disturbance. Finally, as with CT scan studies of TBI, abnormalities detected by MRI will be more frequent and of greater magnitude as the severity of injury increases.

DIRECTION OF FUTURE STUDIES

Advances in neuroimaging techniques over the last two to three decades have been remarkable and have allowed increasingly precise *in-vivo* visualizations of the cerebrum. At the present time, MRI represents the most sophisticated window to the neuroanatomic substrate of cerebral injury and therefore provides a unique opportunity to address more carefully the interrelationships between structural cerebral status and the behavior of the patient. It seems essential that, with time, more sophisticated classification schemas be developed for TBI to refine the extant global classifications of "mild," "moderate," and "severe" head injury. This, in turn, may serve to enhance our ability to make both more accurate diagnoses and prognostic statements. These more sophisticated classifications will need to include a combination of neuroimaging, neurophysiological, and neurobehavioral measures, and these combined classifications will need to be longitudinally evaluated and correlated with multiple outcome measures.

In order to achieve a more comprehensive assessment of TBI, the focus will need to be expanded beyond the issues addressed in this chapter on neuroimaging and neuropsychology. That is, CT and MRI techniques depict the underlying structural integrity of the brain and represent relatively static measures. In contrast, neuropsychological techniques and neurophysiological measures such as positron emission tomography, brain electrical activity mapping, and event-related potentials capture brain functioning over time and therefore in a more dynamic fashion. This is a particularly important point to be noted in those cases with no detectable abnormalities on MRI but clear neurobehavioral disturbance. Both static and particularly dynamic measures will need to be orchestrated to obtain more comprehensive evaluations and to allow for a greater understanding of the highly complex interrelationships between brain and behavior following TBI.

With respect to neurobehavioral measures, the focus traditionally has been on cognitive performance. In comparison, the affective status and subjective complaints of patients and their family members have received far less attention (see Chapter 10). In the future, studies will need to bridge the gap between, on the one hand, cognitive performance measures and, on the other hand, the patient's actual and perceived quality of life. It is of fundamental importance to understand to what extent TBI has changed the psychosocial, emotional, as well as neuropsychological and mental domains of a person's life. No doubt, the challenge will be to assess systematically and quantify the various interrelationships among these domains. Finally, more well-controlled outcome studies regarding the rehabilitation of TBI patients may be facilitated when more sophisticated and comprehensive neuroimaging and neurobehavioral assessment and classification paradigms are developed.

REFERENCES

Adams, J. H., Mitchell, D. E., Graham, D. I., & Doyle, D. (1977). Diffuse brain damage of immediate impact type. *Brain, 100,* 489–502.

Adams, J. H., Gennarelli, T. A., & Graham, D. I. (1982). Brain damage in non-missile head injury: Observations in man and subhuman primates. In W. T. Smith & J. B. Cavanagh (Eds.), *Recent advances in neuropathology.* London: Churchill Livingston.

Adams, J. H., Graham, D. I., Murray, L. S., & Scott, G. (1982). Diffuse axonal injury due to nonmissile head injury in humans: An analysis of 45 cases. *Annals of Neurology, 12,* 557–563.

Arai, H., Kobayashi, K., Ikeda, K., Nagao, Y., Ogchara, R., & Kosaka, K. A. (1983). A computed tomography study of Alzheimer's disease. *Journal of Neurology, 229,* 69–77.

Benson, D. F., & Geschwind, N. (1971). Aphasia and related cortical disturbances. In A. B. Baker & L. H. Baker (Eds.). *Clinical neurology.* New York: Harper & Row.

Benton, A. L., & Van Allen, M. W. (1968). Aspects of neuropsychological assessment in patients with cerebral disease. In C. M. Gaitz (Ed.), *Aging and the brain.* New York: John Wiley & Sons.

Bigler, E. D. (1987). Neuropathology of acquired cerebral trauma. *Journal of Learning Disabilities, 20,* 458–473.

Bigler, E. D. (1988). *Diagnostic clinical neuropsychology.* Austin, TX: University of Texas Press.

Bigler, E. D., Hubler, D. W., Cullum, C. M., & Turkheimer, E. (1985). Intellectual and memory impairment in dementia: CT volume correlations. *Journal of Nervous and Mental Disease, 173,* 347–352.

Bird, J. M. (1982). Computerized tomography, atrophy and dementia: A review. *Progress in Neurobiology, 4,* 91 115.

Boller, F. C., Albert, M. L., Lemay, M., & Kertesz, A. (1972). Enlargement of the Sylvian aqueduct: A sequel of head injuries. *Journal of Neurology, Neurosurgery and Psychiatry, 35,* 463–467.

Brant-Zawadski, M., Pereira, E., & Weinstein, P. (1986). MR imaging of acute experimental ischemia in cats. *American Journal of Neuroradiology, 7,* 7–12.

Bydder, G. M., Steiner, R. E., Young, I. R., Hall, A. S., Thomas, D. J., Marshall, T., Pallis, C. A., & Legg, N. J. (1982). Clinical NMR imaging of the brain: 140 cases. *American Journal of Roentgenology, 139,* 215–236.

Clifton, G. L., Grossman, R. G., Makala, M. E., Miner, M. E., Handel, S., & Sadhu, V. (1980). Neurological course and correlated computerized tomography findings after severe closed head injury. *Journal of Neurosurgery, 52,* 611–624.

Crooks, L. E., Mills, C. M., Davis, P. L., Brant-Zawadzki, M., Hoenninger, J., Arakawa, M., Watts, J., & Kaufman, L. (1982). Nuclear magnetic resonance. Visualization of cerebral and vascular abnormalities by NMR imaging. The effects of imaging parameters on contrast. *Radiology, 144,* 843–852.

Cullum, C. M., & Bigler, E. D. (1985). Late effects of hematoma on brain morphology and memory in closed head injury. *International Journal of Neuroscience, 28,* 279–283.

Cullum, C. M., & Bigler, E. D. (1986). Ventricle size, cortical atrophy and the relationship with neuropsychological status in closed head injury: A quantitative analysis. *Journal of Clinical and Experimental Neuropsychology, 8,* 437–452.

Damasio, H., Eslinger, P., Damasio, A. R., Rizzo, M., Huang, H. K., & Demeter, S. (1983). Quantitative computer tomographic analysis in the diagnosis of dementia. *Archives of Neurology, 40,* 715–719.

de Leon, M. J., & George, A. E. (1983). Computed tomography in aging and senile dementia of the Alzheimer's type. In R. Mayeux & W. G. Rosen (Eds.), *The dementias,* pp. 103–122. New York: Raven Press.

Earnest, M. P., Heaton, R. K., Wilkinson, W. E., & Manke, W. F. (1979). Cortical atrophy, ventricular enlargement and intellectual impairment in the aged. *Neurology, 29,* 1138–1143.

Galbraith, S., & Smith, J. (1976). Acute traumatic intracranial hematoma without skull fracture. *Lancet, 1,* 501–503.

Gennarelli, T. A., Thibault, L. E., Adams, J. H., Graham, D. I., Thompson, C. J., & Marcincin, R. P. (1982). Diffuse axonal injury and traumatic coma in the primate. *Annals of Neurology, 12,* 564–574.

Goodglass, H., & Kaplan, E. (1972). *The assessment of aphasia and related disorders.* Philadelphia: Lea & Feibiger.

Gudeman, S. K., Kishore, R. S., Becker, D. P., Lipper, M. H., Girevendulis, A. K., Jeffries, B. F., & Butterworth, J. F. (1981). Computed tomography in the evaluation of incidence and significance of posttraumatic hydrocephalus. *Neuroradiology, 141,* 397–402.

Han, J. S., Kaufman, B., Alfidi, R. J., Yeung, H. N., Benson, J. E., Haaga, J. R., El Yousef, S. J., Clampitt, M. E., Bonstelle, C. T., & Huss, R. (1984). Head trauma evaluated by magnetic resonance and computed tomography: A comparison. *Radiology, 150,* 71–77.

Harenstein, S., Chamberlin, W., & Conomy, J. (1967). Infarction of the fusiderm and calcarne regions: Agitated delirium and hemianopia. *Trans American Neurological Association, 92,* 85–89.

Hawkins, T. D., Lloyd, A. D., Fletcher, G. I. C., & Hanka, R. (1976). Ventricle size following head injury: A clinico-radiological study. *Clinical Radiology, 27,* 279–289.

Hecaen, H., & Albert, M. (1978). *Human neuropsychology.* New York: John Wiley & Sons.

Huckman, M. S., Fox, J., & Topel, J. (1975). The validity of criteria for the evaluation of cerebral atrophy by computed tomography. *Neuroradiology, 116,* 85–92.

Jennett, B., & Bond, M. (1975). Assessment of outcome after severe brain damage: A practical scale. *Lancet, 1,* 480–487.

Jernigan, T. L., Zatz, L. M., & Naeser, M. A. (1979). Semiautomated methods for quantitating CSF volume on cranial computed tomography. *Radiology, 132,* 463–466.

Kaszniak, A. N., Garron, D. C., Fox, J. H., Bergen, D., & Huckman, M. (1979). Cerebral atrophy, EEG slowing, age, education and cognitive functioning in suspected dementia. *Neurology, 29,* 1273–1279.

Kishore, P. R., Lipper, M. H., Miller, J. D., Girevendulis, A. K., Becker, D. P., & Fine, F. S. (1978). Posttraumatic hydrocephalus in patients with severe head injury. *Neuroradiology, 16,* 261–265.

Levin, H. S., Meyers, C. A., Grossman, R. G., & Sarwar, M. (1981). Ventricular enlargement after closed head injury. *Archives of Neurology, 38,* 623–629.

Levin, H. S., Benton, A. L., & Grossman, R. G. (1982). *Neurobehavioral consequences of closed head injury.* New York: Oxford University Press.

Levin, H. S., Kalisky, Z., Handel, S. F., Goldman, A. M., Eisenberg, H. M., Morrison, D., & Von Laufen, A. (1985). Magnetic resonance imaging in relation to the sequelae and rehabilitation of diffuse closed head injury: Preliminary findings. *Seminars in Neurology, 5,* 221–231.

Levin, H. S., Handel, S. F., Goldman, A. M., Eisenberg, H. M., & Guinto, F. C. (1985). Magnetic resonance imaging after "diffuse" nonmissile head injury: A neurobehavioral study. *Archives of Neurology, 42,* 963–968.

Levin, H. S., Amparo, E., Eisenberg, H. M., Williams, O. H., High, W. H., McArdle, C. B., & Weiner, R. C. (1987). Magnetic resonance imaging and computerized tomography in relation to the neurobehavioral sequelae of mild and moderate head injuries. *Journal of Neurosurgery, 66,* 706–713.

Luerssen, T. G., Hesselink, J. R., Ruff, R. M., Healy, M. E., & Grote, C. A. (1986). Magnetic resonance imaging of craniocerebral injury. *Concepts in Pediatric Neurosurgery, 8,* 190–208.

Luria, A. R. (1966). *Higher cortical functions in man.* New York: Basic Books.

Massman, P. H., Bigler, E. D., Cullum, C. M., & Naugle, R. I. (1986). The relationship between cortical atrophy and ventricular volume. *International Journal of Neuroscience, 30,* 87–99.

Meese, W., Kluge, W., Grumme, T., & Hopfenmulle, T. (1980). CT evaluation of the CSF spaces of healthy persons. *Neuroradiology, 19,* 131–136.

Milner, B. (1954). The intellectual function of the temporal lobes. *Psychological Bulletin, 15,* 42–62.

Mishkin, M. (1978). Memory in monkeys severely impaired by combined but not separate removal of the hippocampus. *Nature, 273,* 297–298.

Ommaya, A. K., & Gennarelli, T. A. (1974). Cerebral concussion and traumatic unconsciousness. *Brain, 97,* 633–654.

Pribram, K. H., & Broadbent, D. E. (Eds.). (1970). *Biology of memory.* New York: Academic Press.

Signoret, J. L. (1985). Memory and amnesias. In M. M. Mesulam (Ed.), *Principles of Behavioral Neurology* (pp. 169–192). Philadelphia: F. A. Davis.

Sipponen, J. T., Sipponen, R. E., & Sivula, A. (1983). Nuclear magnetic resonance (NMR) imaging of intracerebral hemorrhage in the acute and resolving phases. *Journal of Computer Assisted Tomography, 7,* 954–959.

Squire, L., & Butters, N. (Eds.). (1984). Neuropsychology of memory. New York: Guilford Press.

Strich, S. J. (1956). Diffuse degeneration of the cerebral white matter in severe dementia following head injury. *Journal of Neurology, Neurosurgery and Psychiatry, 19,* 163–185.

Synek, V., & Reuben, J. R. (1976). The ventricular–brain ratio using planimetric measurement of EMI scans. *British Journal of Radiology, 49,* 233–237.

Teuber, H. L. (1974). Recovery of function after lesions of the central nervous system: History and prospects. *Neurosciences Research Program Bulletin, 12,* 132–145.

Timming, R., Orrison, W. W., & Mikula, J. A. (1982). Computerized tomography and rehabilitation outcome after severe head trauma. *Archives of Physical Medicine and Rehabilitation, 63,* 154–159.

Tornheim, P. A. (1985). Traumatic edema in head injury. In D. P. Becker & J. T. Povlishock (Eds.), *NIH central nervous system trauma status report* (pp. 431–442). Bethesda: NINCDS.

Turkheimer, E., Yeo, R., & Bigler, E. D. (1983). Digital planimetry in APLSF. *Behavior Research Methods and Instrumentation, 15,* 471–473.

van Dongen, K. J., & Braakman, R. (1980). Late computed tomography in survivors of severe head injury. *Neurosurgery, 7,* 14–22.

Weiskrantz, L. (Ed.). (1968). *Analysis of behavior change.* New York: Harper & Row.

Wu, S., Schenkenberg, T., Wing, S. D., & Osborn, A. G. (1981). Cognitive correlates of diffuse cerebral atrophy determined by computed tomography. *Neurology, 31,* 1180–1184.

Zangwill, O. L. (1969). Neuropsychological models of memory. In G. Talland & N. Waugh (Eds.), *The pathology of memory* (pp. 161–166). New York: Academic Press.

7

Brain Imaging and Neuropsychological Identification of Dementia of the Alzheimer's Type

RICHARD I. NAUGLE and ERIN D. BIGLER

INTRODUCTION

This chapter reviews the neuropathology and neuropsychological concomitants of dementia of the Alzheimer's type (DAT) and the nature of the degenerative changes as visualized on computerized tomography (CT) scanning and magnetic resonance imaging (MRI). Efforts to quantify CT scan results formally and objectively and to determine the nature of the relationship between the morphological changes and neuropsychological deficit patterns are reviewed. The frequent practice of dichotomizing DAT patients into presenile and senile groups is presented as well as more recent attempts to uncover distinct clusters or subgroups among the heterogeneous population of affected patients.

It should be noted that, unless otherwise stated, the patients included in the studies discussed herein typically were diagnosed as suffering from dementia of the Alzheimer's type on the basis of clinical data (rather than postmortem histological studies); consequently, they are referred to as "DAT" patients rather than "Alzheimer's disease" patients. As will be discussed, a considerable number of patients thus diagnosed in fact suffer from disorders in addition to or other than Alzheimer's disease that produce similar clinical presentations.

CLINICAL CHARACTERISTICS OF DAT

The degenerative changes of DAT are evident clinically as the deterioration of motor, sensory, language, memory, and higher cognitive functions. The nature and rate of decline vary within each of these areas as well as across patients. The clinical presentation of a given DAT patient, then, is dependent in part on one's stage in the course of the disease process and in part on one's premorbid levels of ability. It is of paramount importance that the clinician be familiar with the neuropsychological changes resulting from the disease process in order to identify or rule

---

RICHARD I. NAUGLE ● Section of Neuropsychology, Department of Psychiatry, Cleveland Clinic Foundation, Cleveland, Ohio 44106. ERIN D. BIGLER ● Austin Neurological Clinic and Department of Psychology, University of Texas at Austin, Austin, Texas 78705.

out other etiologies that might be responsible for apparent changes. (For a comprehensive description of the stages that are typical of the disease process, see Strub & Black, 1981). Consider the cases shown in Tables 1 and 2 and Figure 1a,b, which allow for a comparison of the neuropsychological and CT data of two individuals referred for evaluation of possible dementia.

TABLE 1. Neuropsychological Data on a DAT
Patient (Case 1a)[a]

Age	61	
Education	High school	
Motor		
	Dom	Ndom
Finger oscillation	40	31
Grip strength (kg)	42	32
Sensory		
	Right	Left
Visual extinctions[b]	0	0
Auditory extinctions[b]	2	2
Tactile		
Extinctions[b]	2	2
Graphesthesia	4 errs	2 errs
Finger gnosis	2 errs	3 errs
Trailmaking form A	349	
Trailmaking form B	Cannot complete	

Speech and language
  Aphasia Screening Test: Significant for constructional dyspraxia, spelling dyspraxia, dyscalculia on subtraction and multiplication, mild dysnomia

Memory	
Memory quotient	62
Information	3
Orientation	1
Mental control	0
Immediate verbal	2
20-min delayed verbal	0 (with prompting)
Immediate visual	1
20-min delayed visual	0
Associate learning	6
Rey–Osterrieth	Cannot complete

Intellectual functioning	
WAIS	
Information	7
Digit span	5
Vocabulary	9
Arithmetic	5
Comprehension	4
Similarities	4

TABLE 1. (Continued)

Verbal IQ	80
Picture completion	3
Picture arrangemt	2
Block design	2
Object assembly	4
Digit symbol	1
Performance IQ	66
Full-scale IQ	73
Ravens CPM	4/36 correct

[a] These neuropsychological data were collected from a 61-year-old custodial supervisor at a public elementary school who presented with declining job performance and "forgetfulness." He had an exemplary work history for 30 years, but over the past year, his performance began to decline, and his most recent work evaluation recommended that he be removed from his supervisory position. This prompted a medical evaluation, which in turn led to the neurological and neuropsychological studies that were consistent with DAT. Note the significant decline in intellectual abilities from presumed premorbid level of function, particularly the performance scores in a man who was described premorbidly as having excellent motor skills. Memory studies demonstrate marked dysfunction. Figures 1b and 1c demonstrate the visual memory deficits as well as constructional apraxia. Compare this patient's performance with the normal patient's performance in Table 2.

[b] Errors on double simultaneous stimulation.

TABLE 2. Neuropsychological Data on a Normal, Depressed Patient (Case 1b)[a]

Age	64	
Education	Law degree	
Motor		
	Dom	Ndom
Finger oscillation	52	47
Grip strength (kg)	44	39
Sensory		
	Right	Left
Visual extinctions[b]	0	0
Auditory extinctions[b]	0	0
Tactile		
Extinctions[b]	0	0
Graphesthesia	0 errs	0 errs
Finger gnosis	0 errs	0 errs
Trailmaking form A	34 s	
Trailmaking form B	59 s	
Speech and language		
Aphasia Screening Test: within normal limits		
Memory		
Memory quotient	>143	
Information	6	
Orientation	5	
Mental control	9	
Immediate verbal	20	
20-min delayed verbal	19	

(continued)

TABLE 2. (Continued)

Immediate visual	14
20-min delayed visual	14
Associate learning	19
Intellectual functioning	
WAIS	
Information	17
Digit span	14
Vocabulary	16
Arithmetic	13
Comprehension	15
Similarities	14
Verbal IQ	135
Picture completion	9
Picture arrangement	11
Black design	11
Object assembly	9
Digit symbol	10
Performance IQ	117
Full-scale IQ	132
Raven CPM	35/36 correct

[a]These data were collected from a 64-year-old attorney who presented with what he regarded as a significant memory loss. However, his neuropsychological studies were all within normal limits with above-average to superior ability noted on all intellectual measures. This was diagnosed as a pseudodementia with cognitive efficiency reduced because of depression.
[b]Errors on double simultaneous stimulation.

INCIDENCE AND PREVALENCE

The prevalence of severe dementia in the United States has been estimated at 1.3 million cases, of which 50–60% are the result of DAT (Terry & Katzman, 1983). Clearly, it is essentially an adult disorder, with most cases beginning past the fifth decade of life (Bigler, 1984; Joynt, 1981). Rocca, Amaducci, and Schoenberg (1986) cite the prevalence ratio for patients aged 65 and over at 1.9 to 5.8%. One writer estimates that only 4% of DAT cases are individuals under the age of 40 (Cook, Ward, & Austin, 1979). As with dementia in general, the incidence and prevalence of the disorder increase with age. Rocca et al. (1986) report the annual incidence among those aged 40 to 60 to be 2.4 cases per 100,000; the incidence increases to 127 cases per 100,000 for those over the age of 60. Katzman (1976) more generously estimates that 10% of adults over the age of 65 are afflicted with DAT. With the increased longevity of the population, then, the prevalence of the disease is very likely going to increase dramatically (Rocca et al., 1986). The prevalence ratios are comparable for both blacks and whites, and the dementia is regarded to be more common among females than males (Rocca et al., 1986; Schoenberg, Anderson, & Haerer, 1985).

It should be noted that because of the lack of pathognomonic indicators of the disease on EEG, CT, and laboratory studies, the diagnosis of DAT is based on behavior in the absence of other readily identifiable causes for a given patient's symptoms. Diagnosis is not certain, then, and the error rate is considerable. Of those dementia patients who come to autopsy, 20–50% are diagnosed as having some disorder other than Alzheimer's disease (Fuld, 1983; McKhann et al., 1984; Terry & Katzman, 1983).

The etiology of the disease is still unknown. It has been associated with familial history of Down's syndrome (Heyman et al., 1983, 1984; Wisniewski, Wisniewski, & Wen, 1985), premorbid head trauma (Heyman et al., 1984; Mortimer, French, Hutton, & Schuman, 1985), serum aluminum (Crapper, Quittrat, Krishnau, Dalton, & De Bon, 1980; Perd & Brody, 1980), thyroid disease (Heyman et al., 1983, 1984), familial history of dementia (Amaducci et al., 1986; Heyman et al., 1983), and advanced age of the mother at the patient's birth (Amaducci et al., 1986).

NEUROPATHOLOGY AND MORPHOLOGICAL BRAIN CHANGES OF DAT

Although the etiology of the disease is still uncertain, the macroscopic and microscopic changes in the brain have been known since 1906, when the disease was first identified by Alois Alzheimer (McMenemey, 1970). The first changes (among the cholinergic subsystem) are the degeneration of presynaptic terminals and dendritic loss. Following this degeneration, the commonly cited finding— that of the neurofibrillary tangles of Alzheimer—develops and contributes to further neuronal degeneration by obstructing axoplasmic flow (Strub & Black, 1981). Continued degeneration of those cells results in the "senile plaque of Marinesco," the third finding of DAT. The plaques and tangles are distributed throughout the neocortex, paleocortex, and deep in the gray matter through the pontine tegmentum; the basal ganglia, thalamus, and substantia nigra are typically spared (Terry & Katzman, 1983).

As gray matter diminishes as a consequence of the disease process (Hubbary & Anderson, 1981), morphological changes occur. Cortical atrophy, which is commonly associated with DAT, is visible to some extent on CT and MRI scans in the form of prominent sulci and enlarged lateral and/or third ventricles. In fact, in their study of 39 DAT patients and 36 healthy controls, Creasey, Schwartz, Frederickson, Haxby, & Rapoport (1986) documented a relationship between ventricular space and loss of gray matter. For a detailed review of the biological substrates of DAT, the interested reader is referred to the recent work of Scheibel and Wechsler (1986).

Although the presence of cortical atrophy in the patient presenting with symptoms of dementia is considered to be one of the diagnostic signs associated with DAT (see, for example, Cutler et al., 1984; Damasio et al., 1983), the mere presence of atrophy on CT imaging is not pathognomonic of dementia or DAT. The identification of cortical atrophy on computerized axial tomography scanning is associated with a wide variety of neuropathological conditions including progressive degenerative diseases, cerebral trauma, and secondary effects of cerebrovascular disease and anoxia (see Figure 2).

Cortical atrophy and ventricular enlargement are also often found in "normal" elderly subjects devoid of cognitive compromise (Barron, Jacobs, & Kinkee, 1976; Brinkman, Sarwar, Levin, & Morris, 1981; Earnest, Heaton, Wilkinson, & Manke, 1979; Gonzalez, Lanhert, & Nathan, 1978; Jacoby & Levy, 1980; Schwartz et al., 1985). Conversely, some researchers have reported cases of patients with dementia who showed no neuroradiological evidence of atrophy (Bigler, Hubler, Cullum, & Turkheimer, 1985; Jacobsen & Farmer, 1979). As pointed out above, the gross pathological alterations shown on radiographic depictions of the brain do not necessarily indicate that the micropathological changes associated with DAT are present (Huckman, Fox, & Ramsey, 1977; Neary et al., 1986). It is only in cases in which the patient shows both the

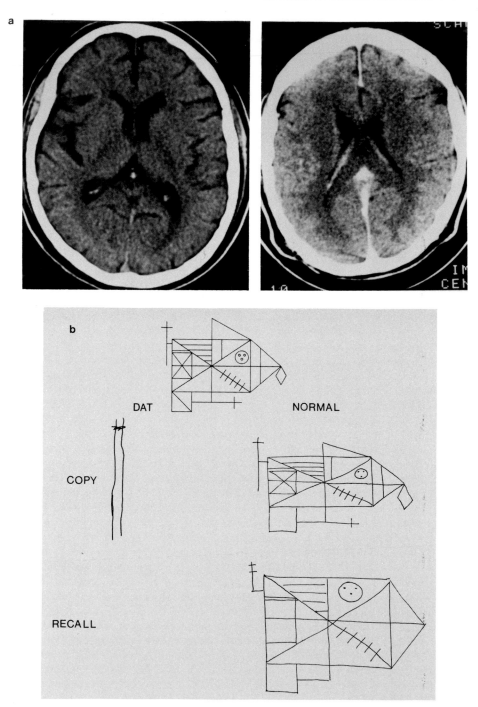

FIGURE 1. (a) The CT scan on the left demonstrates the classic findings of cortical atrophy seen with DAT. Note the prominence of the sulcal clefts and in particular the left Sylvian fissure. This DAT patient's neuro-psychological data are presented in Table 1. Compare this with the CT scan on the right, which is from a somewhat older man who was evaluated for DAT but was found to have pseudodementia secondary to depres-

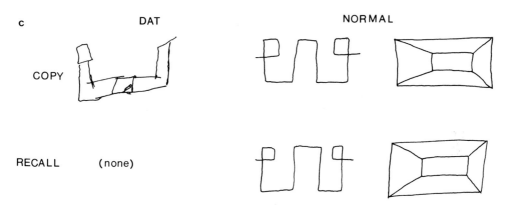

sion. This normal patient's neuropsychological test results are presented in Table 2. (b) Rey–Osterrieth Complex Figure Design drawings from the DAT patient (left) and the pseudodementia patient (right) whose CT scans are presented in a. Note the severe constructional apraxia in the DAT patient, whereas the normal patient has no difficulty in copying or recalling the figure. The original stimulus figure for the Rey–Osterrieth design is presented at top center. (c) Visual memory subtest performance by the DAT patient (right) and the normal patient (right). Again note the significant constructional apraxia by the DAT patient in his attempt to perform the immediate-recall segment of the visual memory subtest. Compare the DAT patient's performance with the intact patient's performance, which is an exact replica of the actual stimulus figure, both for immediate recall (top view, labeled copy) and delayed recall (bottom view, labeled recall).

cognitive and emotional signs of dementia and the cortical atrophy on the CT scan that DAT can be assumed to be associated with the atrophy.

It has been suggested that atrophy as depicted on CT images is of limited clinical utility in part because of the inadequate means of assessing it (Damasio et al., 1983). Initial rating scales (see Bird, 1982, for review) were unsuccessful in demonstrating significant relationships between CT-documented atrophy and dementia. However, these rating scales used either visual inspection, simple linear measurements, or radiographic rating scales. Such negative findings certainly could be questioned on the basis of the simplicity of the assessment technique, but despite the use of more specific measurement techniques (such as Evans' ratio, summed sulcal width measurements, ventricle–brain ratios; see Figure 3), significant relationships were not found (Fox, Kazniak, & Huckman, 1979; Ramani, Loewenson, & Gold, 1979; Roberts, Caird, Grossart, & Steven, 1976).

Note that the measures depicted in Figure 3 are taken from the horizontal cuts of the CT scan. To date, there has been no research using volumetric data from MRI scans to study atrophic brain changes as a consequence of DAT; the sagittal and coronal planes that are visible on the MRI scan would likely prove to be valuable means of cross validating atrophic brain assessments taken from the horizontal plane. The potential usefulness of such measures is obvious from the MRI scans shown in Figure 4.

Damasio et al. (1983) contend that earlier studies relating CT and dementia are flawed in part because of their failure to take into account the wide variation of brain and skull size among subjects. Using tracings from CT scans, they measured the areas subtended by the third ventricle, frontal horns, the bodies of the lateral ventricles, and the interhemispheric fissure and then determined the ratio between each of those areas and the entire brain at the level that each measure was taken. They found that demented patients (matched with controls on the basis of age and gender) were characterized by significantly greater ratios on all measures. Demented patients showed a 73% increase of the third ventricle, 71% increase of the bodies of the lateral ventricles,

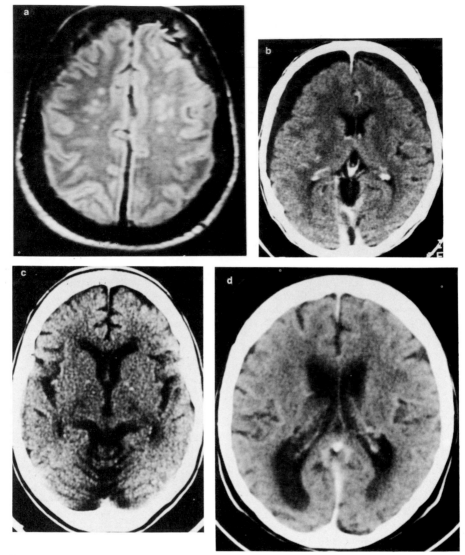

FIGURE 2. These scans demonstrate various causes other than DAT that may produce dementia-type syndromes. These other syndromes typically present very distinguishing features via CT or MRI neuroimaging techniques that differentiate them from DAT. (a) This is a case of multiinfarct dementia. The MRI scan clearly shows multiple lucency changes throughout the central white matter consistent with multiple infarcts. (b) This CT scan reveals bilateral subdural hygromas in a patient who presented with memory loss and confusion. Once the hygromas were evacuated, the patient's dementia resolved. (c) This is a CT scan from a patient with histologically verified Jakob–Creutzfeldt disease. Although this scan shows prominent cortical atrophy that is similar to that seen in DAT, the prominent cerebellar atrophy differentiates it from what would typically be seen in DAT, where cerebellar atrophy is not present. (d) The CT scan reveals marked frontal and temporal lobe atrophy in a patient believed to be suffering from Pick's disease. Note that the atrophy is restricted primarily to the frontal pole and anterior temporal region. (e) This CT scan demonstrates enlarged ventricles secondary to degenerative changes associated with alcohol abuse. This patient has an alcohol-induced dementia. The patient's initial neuropsychological evaluation revealed a verbal IQ of 69, performance IQ of less than 48, and an MQ of 61. These deficits have remained stable over the past 3 years with no further deterioration noted. In alcohol dementia patients, it appears that abstinence results in arrest of degenerative changes other than those that would be expected with age. (f) Two MRI slices that reveal large occipital lobe infarcts in a patient who was thought to have Alzheimer's disease but had an aphasic syndrome secondary to cerebral infarction.

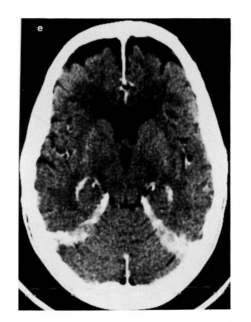

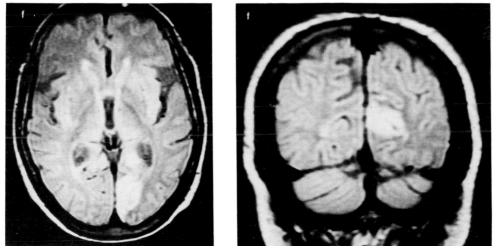

FIGURE 2. (*continued*)

a 47% increase in the interhemispheric fissure, and a 45% increase in the frontal horns. After weighting the size of the bodies of the lateral ventricles and the interhemispheric fissure, Damasio *et al.* were able to predict correctly the group classification—demented or control—of 84% of their subjects through the use of a stepwise linear discriminant analysis.

Albert, Naesar, Levine, and Garvey (1984a) applied a number of semiautomated computerized and linear measurements to the CT scans of 8 mildly to moderately impaired male presenile DAT patients and 10 controls matched on the basis of age. Linear measures of the lateral and third ventricles completed by the experimenters correctly predicted group membership in 78.9% of the subjects. However, they found that the semiautomated computerized measure of fluid vol-

FIGURE 3. Several planimetic estimates of cortical atrophy are presented
here, as follows: A', bifrontal span of
the lateral ventricles; A, brain width
correction for measurement A'; B,
transverse diameter of the left frontal
horn; C, oblique diameter of the right
frontal horn; D, caudate diameter; E,
width of the third ventricle; F, width of
the Sylvian fissure; G, distance between
the third ventricle and the Sylvian fissure; H, width of the cortex at the wid-

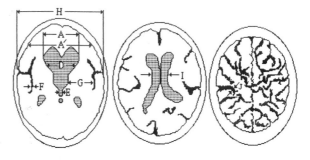

est lateral cut; I, minimal width of the ventricular bodies; J, width of an enlarged sulcus. Other measures
include Evans ratio or the anterior horn index (A/H), the bicaudate index (D/H), and the Huckman number
(A + D). Adapted from Bird (1982, p. 94) and DeLeon and George (1983, p. 108).

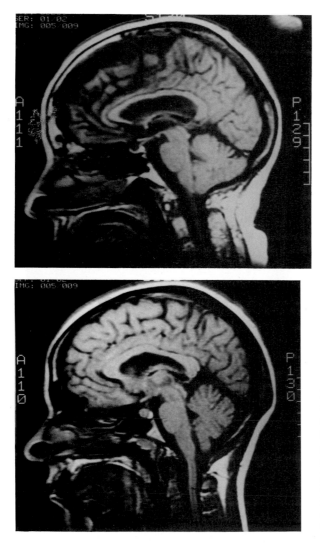

FIGURE 4. The sagittal MRI view
on the top depicts severe frontal
atrophy (as compared with the normal MRI on the bottom) in a DAT
patient. To date, studies regarding
atrophy have used the horizontal
planes taken from CT scans; as is
evident from these sagittal views, a
clear depiction of cortical atrophy is
available from the sagittal plane as
well.

ume taken from the CT slice depicting the maximum width of the lateral ventricles correctly predicted group membership in nearly 90% of the subjects. The addition of other computerized measures of cerebral spinal fluid (CSF) area and CT density did not enhance this prediction rate.

In the case of senile dementia, the authors found very different results (Albert *et al.*, 1984b). Studying 13 patients—11 women and 2 men—with senile dementia of the Alzheimer type and 18 male, age-matched controls, the authors found that CT density numbers most efficiently predicted the group membership of 77.42% of the subjects. When the fluid volume on the CT slice at the level of the maximum width of the third ventricle was added to the prediction equation, the discrimination power increased to 93.55%. The authors concluded that perhaps there is a disease-related change that is compounded by age-related changes, thereby resulting in the different macroscopic markers of the two forms of the disease. Combining the pre- and senile patient groups resulted in a marked decline in the accuracy of these prediction rates. It should be noted that the sex differences between their samples is a serious confounding variable and potentially limits the generalizability of their results.

Sulcal width has most often been quantified by measuring the four largest sulci on the two or three uppermost CT slices (Gonzalez, Lanhert, & Nathan, 1978; Huckman *et al.*, 1975). Measurement inaccuracies with such small distances (the average sulcal width is 3.4 mm) is a problem though, as the CT scanner measures to the nearest 5 mm (Bird, 1982). Furthermore, the sulcal widths pictured in those uppermost cuts are greatly affected by both the position of the patient's head and the level of the cuts (Bird, 1982). The resolution in some of the upper slices suffers as a result of the inclusion of "noise" in the image caused by the cranial bones (McCullough *et al.*, 1976; Wolpert, 1977); this "masking" of cortical structures by the bone makes any measure of the higher cuts less reliable.

QUANTIFYING CEREBRAL MEASUREMENTS

The precise measurement of lesion size and location is critical in establishing brain/behavior relationships (cf. Gazzaniga & Blakemore, 1975). Accordingly, it was assumed that by increasing the precision of measurement of neuropathological findings on the CT image, similar precision could be given to human brain/behavior relationships. By adapting partial volume estimation techniques (Turkheimer, Yeo, & Bigler, 1983), the size of these three-dimensional structures can now be more accurately estimated, as is described below.

In the normal brain, the outer mantle of the cerebral cortex meets the inner surface of the skull, being separated only by the width of the meninges and surrounding vasculature (i.e., arachnoid and subarachnoid space). Likewise, in the normal brain, the gyral pattern is such that the amount of space accounted for by the sulcal cleft is negligible. Consequently, the actual volume accounted for by the meninges, sulcal space, and vasculature in the normal brain is minimal; the amount of original brain volume can be estimated by measuring the inner circumference of the skull cavity. By subtracting the accumulative size of the sulcal width from the estimate of the original brain size, an estimation of the "shrinkage" can be obtained (see Figure 5). This actually permits both an estimate of brain volume and an estimate of the space or volume left by the atrophy (i.e., the space between the outer surface of the brain and the inner skull surface of the cranium). To assure that head size does not contribute to atrophy estimates, larger skull volume is controlled through the use of ratio scores between brain size and skull volume (see Bigler, 1988).

Positioning of the brain and partial volume effects also must be controlled. Since the CT sections are not continuous but are separated by 5 to 10 mm, the volume has to be estimated between each slice. To ensure uniformity among subjects, a consistent starting point must be

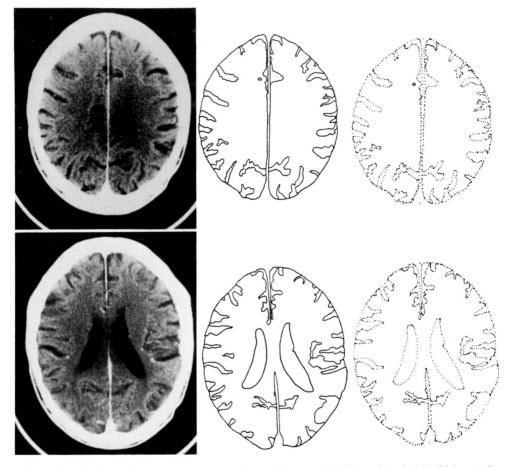

FIGURE 5. This illustration demonstrates the tracing technique used in the studies cited by Bigler *et al.* (1985) and Naugle *et al.* (1985). The CT scan (on the left) was first manually traced (middle figure) with particular attention given to tracing the details of all visible sulci and the ventricular system. These tracings were then digitized using a Summagraphics Bit Pad by moving a cursor manually around the perimeter of the traced structures. The bit pad recorded the X/Y coordinates of the cursor every 0.2 s. A computer-generated representation derived from the X/Y plot coordinates is presented at the far right. This data provided a computer reconstruction of the brain and ventricular system from which ventricular volume, brain volume, and cranial volume could be computed.

defined. In our studies, eight sequential CT slices (separated by 10 mm) were used, starting at the base of the temporal lobes where the inferior horn of the lateral ventricular system can first be visualized and then proceeding dorsally. This procedure excludes some cerebral tissue from analysis; the omitted tissue includes the more ventral aspects of the temporal and, in some cases, frontal areas and the most dorsal aspect of the posterior frontal and parietal areas. Also, the brainstem and most of the cerebellum are excluded. Despite its omission of these areas, this standardized approach quantifies approximately two-thirds of brain volume; the critical areas of interest in degenerative disease are fully included (i.e., third and lateral ventricles, Sylvian fissure, and frontal interhemispheric fissure).

Ventricular volumetric measures are more straightforward. By using this volumetric measurement, aspects of the ventricular system can be traced and volume estimates established. On current-generation CT and MRI scanners, the distinct contrast between cerebral spinal fluid and cerebral tissue renders the ventricular system easily visualizable and, consequently, accurately traced. These tracings have been shown to have high interrater reliability (Turkheimer *et al.*, 1984). One remaining problem with this technique, however, deals with the border zone between ventricle and brain surface on the face of the ventricle. The CT picture elements (i.e., pixels) surrounding the ventricles are partially volumed—that is, they are neither all brain nor all ventricle but are part of both. Thus, using a simple tracing technique may slightly over- or underestimate actual volume size. Albert *et al.* (1984b) have shown that this problem can be minimized by using an automated program based on CT density number configuration. Unfortunately, few labs store such data directly from the digitation of original CT scans; typically, laboratories retain the x-ray film as a "hard" copy. Thus, the advantage of the tracing technique is that it can be applied to the CT film image kept on hand, even though there may be some sacrifice with respect to accuracy.

Utilizing the above technique for estimating brain size and ventricular volume, Bigler *et al.* (1985) studied 42 patients with putative Alzheimer's disease. The mean values of the cranial, cerebral, and ventricular measurements are presented in Table 3. The reader is reminded that the total estimate of brain volume of these subjects—858.6 cc—is only a partial estimate. (The mean estimate for total brain is 1333 cc according to Blinkov & Glezer, 1968.) It bears repeating that this technique does account for approximately two-thirds of the cerebral volume and includes the critical structures of importance to the documentation of atrophy.

The individual brain volumes were corrected for head size by taking ratio scores of brain volume to skull volume. Precise cutoff scores for what is a "normal" or "pathological" degree of atrophy are not available for several reasons. The major impediment to such a distinction is that a "normal" degree of atrophy (i.e., that which occurs solely as a result of aging), as assessed by CT scan analysis, may overlap considerably with atrophy associated with DAT (Schwartz *et al.*, 1985). Rather than focusing on "cutoff" scores of normal versus abnormal, then, this analysis centered on examining the correlational relationships between the degree of cortical atrophy and cognitive functioning.

In this sample, the mean ventricular volume was 65.56 cc. Although there is considerable

TABLE 3. Mean Cranial, Brain, and Ventricular
Volume

	Mean (cc)	SD (cc)
Cranial volume	910.97	87.41
Right cranial volume	455.56	47.45
Left cranial volume	455.42	47.51
Brain volume	858.60	83.98
Right brain volume	432.44	42.05
Left brain volume	426.16	44.26
Total ventricular volume	65.66	27.95
Right ventricular volume	32.03	12.90
Left ventricular volume	33.42	15.67
Lateral ventricular volume	61.27	27.25
Third ventricular volume	3.22	1.47
Fourth ventricular volume	1.29	2.40

variability in what is considered to be normal ventricle size, Blinkov and Glezer (1968) have suggested 39 cc as an estimate of the upper limit of normal volume. The mean ventricular volume in this sample of DAT patients is approximately 60% greater than that upper limit of normal. The relationship between these various brain volumetric analyses and neuropsychological performance is discussed in the following sections.

NEUROPSYCHOLOGICAL FINDINGS

The presence of dementia implies that there has been some detectable decline in cognitive functioning from presumed premorbid levels (Albert, 1984; Friedland, Budinger, Brant-Zawadski, & Jagust, 1984; Wells, 1984). In terms of measured memory functioning, the Wechsler Memory Scale (WMS) (Wechsler, 1945) is perhaps the most widely researched standardized memory test. Based on the results of the WMS, a memory quotient (MQ) can be derived that has a mean and standard deviation identical to those of the Wechsler Adult Intelligence Scale (WAIS) IQ. This permits direct comparison of a given patient's MQ and IQ; because they are highly correlated premorbidly, and because several verbal IQ measures are initially resistant to change in DAT (Bigler, 1984), the degree of superiority of IQ relative to the MQ provides an estimate of memory decline. With respect to subtest analysis, the Logical Memory, Visual Memory, and Paired Associate subtests have been shown to be particularly impaired in patients suffering from dementing illnesses (Albert, 1984; Bigler, 1984; Storandt, Botwinick, Danziger, Berg, & Hughes, 1984).

To assess intellectual deterioration, the standard neuropsychological assessment has included the WAIS (Wechsler, 1955). There has been a plethora of WAIS studies (see Russell, 1979) examining the effect of organic brain disorders on intellectual functioning. The concensus from these studies has been that verbal subtests on the WAIS (which tend to be regarded as more overlearned and are therefore considered to be "crystallized" cognitive functions) are typically less affected by generalized cerebral dysfunction (Cullum, Steinman, & Bigler, 1984). Performance subtests rely on more visual–spatial and visual–motor abilities and tend to tap new learning skills. These are regarded as more "fluid" cognitive processes, which are more vulnerable to cognitive dysfunction. Hence, the WAIS verbal IQ scores of a demented population tend to be more intact than performance IQ scores. A performance IQ that is inferior to verbal IQ by one standard deviation or more may be a more discriminating measure of intellectual deterioration (Bigler, 1984; Brinkman & Braun, 1984; Fuld, 1983).

In addition to verbal IQ versus performance IQ comparisons, examination of performance on select subtests is useful in determining the extent of intellectual decline. Subtests that "hold up" with age, i.e., are less affected by cerebral damage/dysfunction, provide an estimate of premorbid ability levels (see Wechsler, 1958). Such "hold" subtests are Information, Vocabulary, Comprehension, Object Assembly, and Picture Completion. The "don't hold" subtests consist of Block Design, Digit Symbol Picture Arrangement, Similarities, Arithmetic, and Digit Span. These latter measures tend to be more impaired in patients with organic cerebral disorders, irrespective of whether the damage or dysfunction is diffuse, lateralized, or focal (Bigler, 1984; Russell, 1979). Comparison of the "hold" and "don't hold" subtests provides another, albeit rough, estimate of cognitive decline.

Eslinger and Benton (1983) found that dementia patients showed considerably more decline with regard to visuoperceptual performance than did age-matched controls. Whereas steady, moderate deterioration was apparent with increasing age in normal subjects, the visuoperceptual deficits of demented patients were far more severe. So striking was this discrepancy that Eslinger and Benton suggested the use of such measures in the differential diagnosis of normal aging versus pathological decline.

This is not to say that language function is necessarily spared; in their comparison of 30

DAT patients and 70 normal controls, Cummings, Benson, Hill, and Read (1985) reported that the DAT patients frequently presented in a manner comparable to transcortical sensory aphasic patients. Severity of the dementia correlated with the degree of language compromise. Semple, Smith, and Swash (1982) noted a dissociation between DAT patients' performance on the Token Test (which requires that the patient listen and respond to verbal commands of gradually increasing difficulty) and the Reporter's Test (which requires that the patient describe actions based on commands taken from the Token Test). Four of their sample of 9 patients fell below the fifth percentile on the former, whereas 8 of the 9 performed below that level on the latter, suggesting greater expressive than receptive language difficulties. In a more comprehensive study of language capacity, Emery and Emery (1983) investigated DAT patients' language patterning with regard to phonology, morphology, syntax, and semantics. Of these four categories, the DAT patients were least impaired on tasks requiring repetition. It was concluded that repetition was performed phonetically; i.e., the sounds were mimicked without processing of their meaning. Linguistic performance dropped as tasks increasingly required the integration of sound and meaning. Decoding of meaning was more defective as sentences became more abstract, logical, and complex.

Although it typically occurs later in the course of the disease process, language impairments have been reported as initial symptoms in some DAT patients (Horenstein, 1971; Kirshner, Webb, Kelly, & Wells, 1984; Mesulam, 1982; Wechsler, 1977). These are reviewed in greater detail below.

From the psychometric standpoint, the presence of dementia was originally thought to entail a rather uniform decline in neuropsychological functioning. However, this notion has been abandoned in favor of the heterogeneity or diversity of declining functioning, which is most readily apparent in the early stages of DAT. For example, Rosen, Mohs, and Davis (1984) have demonstrated four cognitive dimensions (memory, word recognition, language functions, and motor/constructional praxic functions) and two noncognitive dimensions (mood state and behavioral disorders) that differentiate DAT from non-DAT patients. Such findings suggest that the neuropsychological assessment of patients with dementing illness must tap a variety of cognitive and behavioral dimensions.

Accordingly, Storandt et al. (1984) were able to classify correctly 98% of patients with senile dementia of the Alzheimer's type using only four neuropsychological measures. The Mental Control and Logical Memory subtests of the WMS, Trailmaking Form A (Reitan & Davison, 1974), and word fluency using the letters S and P were used to arrive at the differential classification. Discriminant function analysis demonstrated significant impairment on the Logical Memory, Trailmaking Form A, and word fluency subtests in the DAT patients, with intact level of performance on Mental Control for the normals (thus, the Mental Control results functioned as a suppressor variable—that is, it positively correlated with normalcy but not with dementia). This work suggests that examination of neuropsychological function over a range of tasks that tap new learning and memory (Logical Memory subtest), verbal fluency (which requires expressive and receptive language, immediate recall, as well as recall from long-term storage), and sequential reasoning (the verbal sequencing of information on the WMS Mental Control subtest and visual motor sequencing on the Trailmaking Form A) may show cognitive deficits in dementia not only to be limited to but to be highly discriminatory with respect to the disorder (see also Eslinger, Damasio, Benton, & Van Allen, 1985).

NEUROPSYCHOLOGICAL AND CT SCAN INTERRELATIONSHIPS

Researchers (Albert et al., 1984a,b; Creasey et al., 1986; Eslinger, Damasio, Graff-Radford, & Damasio, 1984; Turkheimer et al., 1984) have long speculated that a relationship between cortical atrophy and neuropsychological deficits exists but have been unable to identify consistently significant findings. As previously stated, this was believed to be related to measurement error. That is, the problem of contradictory findings was considered to result at least in part from the many different means of measuring this complexly shaped three-dimensional space using one- and two-dimensional representations. Each of the linear and planimetric measurements that have been devised to estimate ventricular volume has a different sensitivity to the detection of atrophy (Fox, Topel, & Huckman, 1975; Fox, Ramsey, Huckman, & Proske, 1976; Gonzalez et al., 1978; Haug, 1977). (See Figure 3 for a review of these measurements.) With increased sophistication of measuring techniques, it was thought that a discernible relationship between cortical atrophy and neuropsychological deficits could be demonstrated. The evolution of studies in this area over the past decade has been considerable.

Despite the fact that many researchers have been unsuccessful in their search for a significant relationship between cortical atrophy and cognitive impairment (e.g., Claviera, Moseley, & Stevenson, 1977; Earnest et al., 1979; Ford & Winter, 1981; Gado & Hughes, 1978; Hughes & Gado, 1981; Kazniak et al., 1978), others have reported positive findings (e.g., Albert et al., 1984a; Brinkman et al., 1981; de Leon et al., 1979; Jacoby & Levy, 1980; Roberts & Caird, 1976; Roberts et al., 1976). Still others have found mixed results (Bigler et al., 1985; Eslinger et al., 1984). Attempts to determine those methodological characteristics that differentiate studies resulting in positive findings from those with negative results suggest that the latter used subjects who were less carefully assessed and classified (Bird, 1982) or used groups that were more heterogeneous with regard to age and/or diagnosis (de Leon & George, 1983).

De Leon et al. (1980) evaluated a variety of these measures by applying them to a 43-patient sample and attempting to relate them to cognitive changes. None of the seven linear/planimetric measures used was as efficient as a subjective rank ordering of scans. The authors attributed the superiority of the rank-ordering method to the ability of the observer to evaluate several pathological features of the dilated ventricles rather than reduce the atrophy pictured on the CT scan slice to a single linear measure.

Others have studied morphological change by investigating the correlates of ventricular enlargement and cortical atrophy separately. Marguez (1983) had a neuroradiologist rank CT scans in terms of cortical atrophy and ventricular dilation. These ratings were then correlated with a number of neuropsychological measures. She found that ventricular enlargement was more highly correlated with aphasic symptoms, agnosia, and verbal memory deficits, whereas cortical atrophy ratings correlated more highly with visual–spatial reasoning and haptic memory functioning. Albert et al. (1984a) found that four neuropsychological measures—verbal memory, verbal abstraction, verbal fluency, and constructional ability—correlated highly with both linear and semiautomated computerized measures of the third and lateral ventricles as depicted on CT.

Littman, Berg, and Novelly (1984) administered a limited neuropsychological battery to their sample of 56 patients aged 50 or older. Their measures (chosen to provide maximum information in minimum time) were four subtests of the Wechsler Adult Intelligence Scale (Vocabulary, Similarities, Block Design, and Picture Arrangement), delayed (30 min) verbal and figural recall of Russell's Revised Wechsler Memory Scale, Finger Oscillation Test, and the Purdue Pegboard (using the 30-s administration). They concluded that sulcal enlargement and ventricular dilation represent independent processes with separate behavioral correlates. Specifically, ventricular dilation appeared to be associated with impaired verbal memory. Although the small sample size and limited neuropsychological battery used in this study limit the extent to which these results can be interpreted, the need to separate cortical atrophy and ventricular volume measures for research purposes is made clear by this and similar studies (e.g., Albert & Stafford, 1986; de Leon et al., 1980).

In an effort to evaluate comprehensively the possible relationship between CT findings and neuropsychological deficits, Bigler et al. (1985) have examined the relationship between neuropsychological performance on the Wechsler Adult Intelligence Scale and the Wechsler Memory Scale and volumetric CT findings in the group of 42 patients previously described who met the criteria for probable DAT. As reported earlier, CT volume estimates of ventricular volume in this group of DAT patients revealed dilation to nearly twice the size of what is considered to be normal. This very significant ventricular enlargement did not correlate significantly with neuro-psychological performance, however. Thus, although ventricular dilation appears to be a highly discriminating factor for the presence of DAT in patients who meet the behavioral and neuro-psychological criteria for the disease (see Damasio et al., 1983), there appears to be little rela-tionship between ventricular dilation and WAIS or WMS performance.

The marked variability of ventricular size and the type and degree of attendant cognitive impairments that is responsible for the lack of such a relationship is also convincingly demon-strated by the cases that are presented in Figure 6a. In this illustration, two patients with nearly identical performance on the WMS are presented. Despite their similarity with regard to psycho-metric performance, the two patients show striking differences in the degree of ventricular en-largement. Such variability precludes significant correlations between ventricular size and cogni-tive impairment.

The same dilemma exists with regard to pericerebral atrophy. There is a considerable range among DAT patients with regard to the extent of atrophic brain changes. This variability is made apparent in Figure 6b. Despite that variability, the cortical atrophy measures were found to cor-relate significantly with a number of intelligence and memory measures (Bigler et al., 1985). The results of this analysis are presented in Table 4. Although verbal IQ measures were not signifi-cantly correlated with the degree of atrophy, performance IQ measures were. This relationship was apparent regardless of which hemisphere was more affected. As previously indicated, the performance subtests and the overall performance IQ are considered to tap new learning and consequently are more sensitive to the presence of diffuse organic dysfunction. This contention was corroborated by these studies. Furthermore, these findings suggest that there is a significant relationship between increasing cortical atrophy and impaired manipulospatial or visual–motor functioning as assessed by the WAIS performance subtests.

To examine further the relationship between the degree of atrophy and neuropsychological performance, an atrophy index (ATVOL) was used to dichotomize the patients into two groups with respect to the degree of atrophy. The results of this analysis are presented in Table 5. Review of these findings indicates that the group characterized by greater atrophy did indeed exhibit greater impairment on most measures. The subtests that were not significantly different likely were not because of "floor" effects; that is, both groups did very poorly on these particular measures (i.e., Digit Span, Logical Memory, and Visual Memory subtests). Those subtests tap new learning and short-term memory functioning and have likewise been demonstrated to be consistently impaired in patients with DAT (see Storandt et al., 1984). Massman, Bigler, Cullum, and Naugle (1986) further examined the relationship between high and low atrophy index scores in dementia and closed-head injury patients and confirmed this relationship between degree of cortical atrophy and neuropsychological impairment in DAT.

It is of interest to note in this analysis between groups of higher and lower atrophy that there was not a significant difference between WMS MQ scores. Although memory deficits often are considered to be the hallmarks of DAT, and the mean MQ values were indicative of overall memory disturbance, significant differences between groups with higher and lower atrophy in terms of MQ results were negligible. The only significant WMS difference between these groups was on the Associate Learning subtest. It is of interest to compare these findings with those of Storandt et al. (1984; discussed previously), in which only the Logical Memory subtest of the WMS was found to be among the neuropsychological measures in their discriminant analysis that differentiated DAT patients from unaffected subjects. However, between DAT groups with high and low atrophy, Logical Memory was not found to relate systematically to morphological brain change but was rather uniformly affected independent of the degree of atrophy. It is also interest-ing that there were not more significant findings relating changes in brain morphology and neuro-psychological performance in patients with DAT. These results suggest there may be some rela-

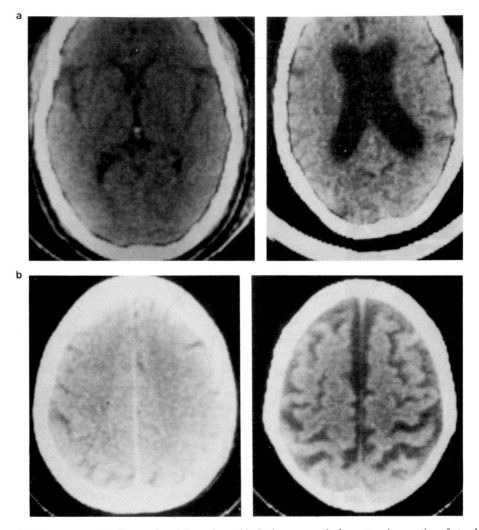

FIGURE 6. (a) The CT-film results of the patient with the largest ventricular system irrespective of atrophy (top right) as well as the patient (top left) with the smallest ventricular volume in the study of 42 patients with DAT. The patient depicted in the top right scan was a 74-year-old female. Her WAIS and WMS results were as follows: verbal IQ 74, performance IQ 67, full-scale IQ 69, MQ 72. She had a high school education and was a homemaker. The patient depicted in the top left scan was a 53-year-old male. His WAIS and WMS results were: verbal IQ 105, performance IQ 91, full-scale IQ 99, MQ 70. It is of interest to note in these scans that despite significant differences in ventricular size, there is no significant difference in MQ. Accordingly, it is not surprising that there is little relationship between MQ and ventricular volume in DAT. (From Bigler *et al.*, 1985.) (b) The CT-film results of the patient with the greatest degree of pericerebral atrophy (bottom right scan) irrespective of ventricular volume and the patient (bottom left scan) with the least degree of cortical atrophy in the study of 42 patients with DAT. The patient depicted in the lower right scan was a 78-year-old female. Her WAIS and WMS scores were: verbal IQ 97, performance IQ 73, full-scale IQ 85, MQ 87. She was a retired school teacher and had a college education. The patient depicted in the bottom left scan was a 74-year-old female whose WAIS and WMS scores were: verbal IQ 93, performance IQ 82, full-scale IQ 88, MQ 68. She had a high school education and was a homemaker. These results reveal the considerable range in the extent of pericerebral atrophy in this group of DAT patients. (From Bigler *et al.*, 1985.)

TABLE 4. Correlation Matrix between ATVOL[a] and Intellectual and Memory Indices

Neuropsychological measure	ATVOL	ATVOL-right	ATVOL-left
WAIS verbal IQ	-0.15	-0.17	-0.12
	(0.16)	(0.14)	(0.22)
WAIS performance IQ	-0.47	-0.45	-0.41
	(0.001)	(0.002)	(0.005)
WAIS full-scale IQ	-0.38	-0.36	-0.34
	(0.008)	(0.01)	(0.02)
WMS MQ	-0.24	-0.25	-0.21
	(0.07)	(0.06)	(0.10)

[a] ATVOL, atrophy volume index = (cranial volume − brain volume)/cranial volume.

tionship between cognitive deficits and morphological brain change, but the relationship is clearly not a linear one.

A major complicating factor regarding this lack of linearity is the gap between the more molar measure that the CT image represents and the microscopic neuronal changes characteristic of DAT (see Scheibel & Wechsler, 1986). In light of the microscopic nature of this pathology, it is not surprising that the molar measures obtained from CT imaging would show only a few significant relationships.

Another factor accounting for the lack of relationship between WAIS or WMS performance

TABLE 5. Neuropsychological Performance Comparisons between High- and Low-ATVOL[a] Groups

Neuropsychological measure	High ATVOL		Low ATVOL		t-test significance (P<)
	Mean	SD	Mean	SD	
WAIS verbal IQ	90.5	11.8	100.4	17.2	0.04
WAIS performance IQ	80.1	10.9	93.2	18.6	0.01
WAIS full-scale IQ	85.3	10.6	98.6	16.0	0.01
Information	6.9	2.1	9.5	3.8	0.01
Comprehension	7.1	2.5	9.7	4.1	0.02
Arithmetic	6.4	2.4	7.4	3.9	0.32
Similarities	6.5	3.5	8.7	4.1	0.07
Digit span	6.9	3.2	7.0	2.8	0.88
Vocabulary	8.3	3.5	9.7	3.3	0.20
Picture completion	4.1	2.2	7.2	3.5	0.01
Block design	3.6	2.8	5.7	4.3	0.07
Picture arrangement	3.9	2.2	5.7	3.0	0.04
Object assembly	3.9	2.8	6.1	3.7	0.05
Digit symbol	3.0	2.6	4.0	3.2	0.28
WMS MQ	78.5	15.1	84.4	19.9	0.31
Logical memory	4.2	3.1	5.6	3.4	0.18
Digits	8.6	2.6	8.9	1.3	0.62
Visual memory	2.8	2.7	3.6	3.7	0.41
Associate learning	6.2	3.6	9.3	4.9	0.03

[a] ATVOL, atrophy volume index = (cranial volume − brain volume)/cranial volume.

and CT findings in DAT is that these measures may not be accurately assessing the nature of dysfunction in DAT. For example, Eslinger *et al.* (1984) have shown that more robust correlations between ventricular enlargement and memory disturbance in DAT can be obtained using more complex short-term measures than the WMS.

Last, de Leon and George (1983) point out that the heterogeneous nature of subject samples very likely accounts for some researchers' failure to uncover a relationship between CT measures and cognitive dysfunction. They contend that, with greater care to select patients who are more homogeneous with regard to age and diagnosis, clear and replicable relationships will be uncovered (see also Eslinger *et al.*, 1984).

PRESENILE VERSUS SENILE DEMENTIA

Many regard patients suffering from dementia of the Alzheimer's type (DAT) as a dichotomous population. Generally, age of onset of the disorder has been the basis for categorizing patients: those in whom the disease process begins before the age of 65 are designated as having presenile dementia, and those whose disease process begins at the age of 65 or later are referred to as having senile dementia. Recent speculation concerning the heterogeneity of DAT has supported the notion that these two forms may be different with regard to the nature and degree of neuropsychological deficit, the rate of progression of the disease and longevity, the findings on computerized axial tomography (CT), and/or various pathophysiological factors (Albert *et al.*, 1984a,b; Barclay, Zemcov, Blass, & McDowell, 1985; Chui, Teng, Henderson, & Moy, 1985; Koss, Ober, Friedland, & Jagust, 1985; Loring & Largen, 1985; McDuff & Sumi, 1985; Seltzer & Sherwin, 1983). Other researchers have dismissed these findings, preferring to regard the disorder as a more unitary entity (Appel, 1981; Kazniak *et al.*, 1978; Seguerra, 1984; Terry, 1978).

Seltzer and Sherwin (1983) found that among their total sample of 65 patients, those cases with a presenile onset of dementia were characterized by language disturbance, a considerably shorter survival time, and greater incidence of left handedness or sinistrality. The authors hypothesized a heightened selective vulnerability of the left hemisphere in early-onset cases of DAT. Unfortunately, they offered little morphological data to support this possibility. Chui *et al.* (1985) found that language dysfunction was more prevalent and more severe among their presenile patients relative to their senile patients. Likewise, Filley, Kelly, and Heaton (1986) reported that, relative to their 18 senile dementia patients, their 23 presenile patients presented with more language impairment in the form of significantly lower verbal IQs on the WAIS and Reitan–Indiana Aphasia Screening Test and a tendency toward decreased auditory comprehension on the Boston Diagnostic Aphasia Examination. Conversely, the older patients had more impairment of visuoconstructional ability. Incidence of sinistrality, gender ratios, duration of the illness, and years of education were comparable for the two groups. The authors concluded that, in addition to the left hemisphere vulnerability of presenile dementia, the disease process beginning in the senium primarily affects the right hemisphere, thereby further diminishing nonverbal functioning, which tends to decline among unaffected elderly individuals (Hochandel & Kaplan, 1984).

Also in agreement with the Seltzer and Sherwin study, Barclay *et al.* (1985) found significantly shorter survival times in DAT patients with onset before the age of 65, suggesting a more malignant course of the disease with an earlier age of onset. Using an abbreviated neuropsychological battery administered to a sample of 13 presenile and 24 senile onset patients, Loring and Largen (1985) found that the earlier-onset cases showed a greater degree of impairment of sustained concentration and mental tracking than those individuals in whom the disease process began at a later age. This impairment was most evident in Digits Backward, an embedded figures task, and graphomotor speed. The authors concluded that the presenile form of dementia typically has an accelerated course.

Koss *et al.* (1985) reported a more complex age-dependent decline of function: senile dementia patients showed greater declines than their younger counterparts, particularly in the areas of language-related tasks. The pattern of neuropsychological decline in presenile patients, however, was related to the relative patterns of cerebral hemispheric hypometabolism as indicated by positron emission tomography using 18-fluorodeoxyglucose. Those patients with greater right hemispheric hypometabolism than left showed greater declines on visual–spatial tasks than those patients with the opposite pattern of hypometabolism.

In their comparisons of ventricular volume and density numbers of CT scans of presenile and senile patients with healthy age-matched controls, Albert et al. (1984a,b) found differences between effective predictor variables for each of the two groups versus their age-matched control groups (discussed above). The authors concluded that, because the two groups were optimally discriminated from controls through the use of different brain morphological measures, presenile and senile forms of dementia may be morphologically distinct entities.

In their study of subcortical structures in Alzheimer's disease, McDuff and Sumi (1985) found neuritic plaques and neurofibrillary tangles in the thalamus, hypothalamus, and mamillary bodies of 25, 22, and 17 patients, respectively, in their sample of 28 patients. Moreover, they reported that the severity of those lesions appeared to be greater in those patients with an early onset of the disorder irrespective of the duration of the disease. It was again concluded that presenile dementia is the neuropathologically more severe form.

Despite findings suggesting a fundamental distinction between dementia arising in the senium and presenium, other eminent authors (Appel, 1981; Kaznlak et al., 1978, Seguerra, 1981; Terry, 1978) have regarded the presenile and senile forms of the disorder as qualitatively comparable and have dismissed the need to distinguish between the two either clinically or neuropathologically. One of the most adamant among these has been Seguerra (1984), who dismissed the conclusion of Seltzer and Sherwin (1983, reviewed above) that early cases of Alzheimer's disease are more likely to show focal involvement. He argued that the left-handed patients showing language compromise naturally are more likely to show that pattern of impairment as a consequence of diffuse neuropathology because their language ability is more likely to be bilaterally represented. Others (e.g., Appel, 1981; Kazniak et al., 1978) state simply that the brain morphological changes or neuropsychological changes are identical in the two groups and the distinction between the groups on the basis of age is nothing more than arbitrary.

In an attempt to explain this discrepancy, Naugle, Cullum, Bigler, and Massman (1986) compared their sample of 41 presenile and 97 senile patients (which failed to replicate any neuropsychological or handedness differences between the two groups) to those of others' samples. They concluded that it may be the case that those samples of presenile patients who do not present as more impaired than their older counterparts may have been evaluated at an earlier stage in the course of their disease than those of the other samples. That is, when it is manifest, the greater divergence between the two groups may be in part a reflection of the more advanced disease among the younger patients. In support of this contention is the fact that the mean full scale IQ of their sample of presenile subjects was 90.9; Loring and Largen (1985)—who found considerably more impairment in their presenile subjects relative to their senile subjects—reported an average full scale IQ of 79.4 for their presenile dementia subjects, thereby suggesting a more advanced disease in their presenile group. Naugle et al. (1986) hypothesized that those individuals who contract the disease at a younger age are simply less able to withstand its effects. In those patients, the effects of the dementia are more severe, and the disorder tends to run a faster course. Those who are sufficiently healthy to reach the older age range are less vulnerable to the disease process. It was hypothesized that this accounts for the finding that the disease runs a more gradual course in those older patients (Barclay et al., 1985; Loring & Largen, 1985; Seltzer & Sherwin, 1983). This hypothesized relationship between age and course among presenile and senile patients is presented visually in Figure 7.

It is also interesting to note that the gender composition of the two groups differed considerably, with the presenile group being comprised primarily of males (27 : 14) and the senile group consisting primarily of females (60 : 37). This is likely the result of important social, biological, diagnostic, and sampling factors. For example, it may be the case that females, because they tend

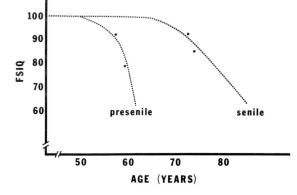

FIGURE 7. Results of Loring and Largen (1985) and Naugle *et al.* (1985). Squares represent data points from the former study, whereas dots represent data points from the latter. (Note the reversed positions of the data points of the two studies.) It is hypothesized that evaluating presenile patients earlier in the course of the disease will reveal a different result relative to their senile counterparts than when those subjects are evaluated at a later stage of the disease. Research in which the presenile patients are seen later in the course of the disease will more likely find those patients to be more severely impaired than the senile-onset patients.

to have longer life spans, outnumber males in the 65-and-older age group; their prevalence in the senile group is simply a reflection of their greater longevity. It is also interesting to note that the proportion of females in their senile group—60 of a total of 97 subjects, or approximately 62%—corresponds closely to that reported by Barclay *et al.* (1985), who found 93 females out of a total of 143 subjects (65%) in their senile dementia group.

The presenile group was predominantly (nearly 65%) male. Perhaps this reflected the greater likelihood that males at that age (mean = 57.05) will be employed, will more likely show deficits in the course of their jobs, and, consequently, may tend to be referred for formal evaluation at an earlier age than females. If this is indeed the case, it may be that presenile samples that are, on the average, older will be less likely to demonstrate such a disproportion. In support of this notion, the sample studied by Barclay *et al.* (1985) (which included 22 males among the 56 presenile subjects, i.e., 39%) averaged 63.3 years of age. Clearly, the stage within the course of the disease process and various sample characteristics such as patients' genders, age, and employment status are important considerations in speculating about differences among samples of presenile and senile dementia patients.

Ventricular volume differences between the groups were largely accounted for by aging effects alone. Gross brain atrophy as measured from CT was not significantly different between the groups even when the effect of age was not covaried in the analysis. Rather than showing less extensive cortical atrophy on CT, as might be expected, the younger subjects presented with measured atrophy levels comparable to those of the older patients.

To investigate the issue of handedness and its relationship to severity of neuropsychological compromise, Naugle, Bigler, Cullum, and Massman (1987) matched 7 left-handed DAT patients to 7 right-handed patients on the basis of gender, age, years of education, and stage of the disease and compared the two groups with regard to a composite measure of neuropsychological functioning. The sinistral subjects were, on the average, more impaired than the dextral group; the extent of the difference between them approached but did not reach statistical significance, however. Thus, although definitive empirical confirmation was lacking, there was a trend in the data that suggested some support for the notion that sinistral DAT patients are more neuropsychologically impaired than dextral patients.

SUBGROUPS OF DAT PATIENTS

As alluded to earlier, there is considerable heterogeneity in the presenting symptomatology of DAT. Mayeux, Stern, and Spanton (1985) concluded after a 4-year investigation of 121 DAT patients that four subgroups were discernible: "benign," with minimal progression; "myoclonic," with marked intellectual decline and frequent mutism following an onset at a relatively early age; "extrapyramidal," with severe intellectual and functional decline and frequently evidence of psychosis; and "typical," with a gradual, progressive deterioration of intellectual and functional ability but without other distinguishing characteristics. Unfortunately, there was no attempt to correlate these deficit patterns with brain pathology. Botwinick, Storandt, and Berg (1986), in their report of a 4-year study of senile-onset DAT patients, also suggested that distinct subroups may exist. Of particular interest in their study of 18 patients and 30 controls, is a subsample of 5 DAT patients whose neuropsychological compromise did not progress at a rate comparable to the other 13.

As in the speculation of a relationship between cerebral atrophy and cognitive impairment in DAT, it has been hypothesized that a relationship exists between presenting symptomatology and morphological brain change in DAT and other degenerative diseases (Brouwers, Cox, Martin, Chase, & Fedio, 1984; Direnfeld et al., 1984; Fuld, 1983). For example, it has been speculated that more lateralized or focal atrophy may indeed have a direct bearing on a given patient's presenting symptomatology. Mesulam (1982) (see also Wechsler, Verity, Rosenschein, Fried, & Scheibel, 1982; Bigler, 1984) argues through the use of six case illustrations that prominent language disturbance in the early stages of a dementing illness suggests a greater lateralization of atrophy to the dominant hemisphere, and, in some cases, such findings may be more suggestive of Pick's disease than DAT (see also Morris, Cole, Banker, & Wright, 1984; Munoz-Garcia & Ludwin, 1984). Gustafson, Hagberg, and Ingvar (1978) reported that demented patients with nonfluent dysphasic deficits in their study had focal cerebral blood flow abnormalities in the frontal lobe of the dominant hemisphere (see also Duara et al., 1984; Foster et al., 1984; MacInnes et al., 1984; Kitagawa, Meyer, Tachibana, Mortel, & Rogers, 1984).

Wechsler (1977) described the case of a 67-year-old male whose presenting deficits were impaired repetition and comprehension and an aphasia involving paraphasic distortions. Memory loss, personality changes, and a generalized dementia developed over the ensuing 2 years. CT scanning revealed particularly prominent dilation of the left Sylvian fissure superimposed on a diffuse cortical atrophy. Kirshner et al. (1984) reported on 6 patients who initially presented with aphasic symptoms that resembled those of focal left hemisphere lesions. In all, no focal brain lesion could be demonstrated by extensive clinical and laboratory evaluations. Crystal, Horoupian, Katzman, and Jatkowitz (1982) have reported a case of histologically verified Alzheimer's disease that presented initially with focal right parietal symptomatology (in the form of sensory perceptual/visual–spatial deficits) in the absence of significant language disturbance or other cognitive impairment. Based on these studies, it appears that on an individual basis, focal atrophic processes may result in focal symptomatology.

To study this possibility further, Naugle, Cullum, Bigler, and Massman (1985) examined the relationship between cognitive impairment and CT volumetric findings in dementia: 138 patients having a primary dementing disorder were identified. All patients received a detailed neuropsychological examination including the following measures: Wechsler Adult Intelligence Scale (WAIS), Wechsler Memory Scale (WMS), Reitan–Klöve Sensory Perceptual Examination, Reitan–Indiana Aphasia Screening Test, Lateral Dominance Examination, Trailmaking Forms A and B, Finger Oscillation Test, and Strength of Grip (see Reitan & Davison, 1974, for a description of those measures). The CT scan results were analyzed using volumetric measures as previously outlined.

To eliminate errors from arbitrary or subjective classification of patients, the neuropsychological data were cluster analyzed. This statistical procedure permitted the "clustering" of patients into discrete, relatively homogeneous groups based on psychometric data. In the course of doing this, a variety of cluster solutions are computer generated, and a "best" cluster solution is established (Spath, 1980). In their sample, a five-cluster solution was found to differentiate clearly the groups without producing any redundancy in classification. This five-cluster solution (see Figure 8) was best characterized in terms of verbal versus visuospatial functioning. The following clusters were uncovered: cluster 1 ("low-functioning subgroup"; $N=42$) displayed generalized deficits on all measures and differed from the others with respect to severity of impairment. Subjects

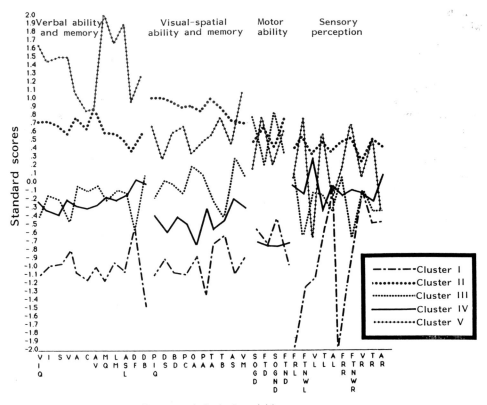

FIGURE 8. Neuropsychological profiles of the five clusters. All values are given in z-score form. Of particular relevance are the Verbal Ability and Memory and the Visual–Spatial and Memory graph lines. Cluster 1
was named the "low-functioning" subgroup; clusters 2 through 5 were each named for their relative neuropsychological strengths as follows: cluster 2, intact visual–spatial subgroup; cluster 3, intermediate/intact
visual–spatial subgroup; cluster 4, intermediate/intact verbal subgroup; and cluster 5, intact verbal subgroup.
Variables abbreviated along the abscissa are VIQ, verbal IQ; I, information; S, similarities; V, vocabulary;
A, arithmetic; C, comprehension; AV, aphasia screening test, verbal score; MQ, memory quotient; LM,
logical memory; ASL, associate learning; DF, digits forward; DB, digits backwards; PIQ, performance IQ;
DS, digit span; BD, block design; PC, picture completion; OA, object assembly; PA, picture arrangement;
TA, trailmaking test form A; TB, trailmaking test form B; AS, aphasia screening test, spatial score; VM,
visual memory; SOGD, strength of grip, dominant hand; FTD, Finger oscillation test, dominant hand; SOGND,
Strength of grip, nondominant hand; FTND, finger oscillation test, nondominant hand; FRL, finger recognition
on the left hand; FTNWL, fingertip number writing on the left hand; VL, visual extinctions in the left visual
field; TL, tactile extinctions on the left body side; AL, auditory extinctions of the left ear; FRR, finger
recognition on the right hand; FTNWR, fingertip number writing on the right hand; VR, visual extinctions in
the right visual field; TR, tactile extinctions on the right hand; and AR, auditory extinctions of the right ear.

of cluster 2 ("intact visual–spatial subgroup"; $N = 39$) were found to be relatively intact on
visuospatial tasks but had relatively lowered functioning on verbal language measures. Cluster 3
("intermediate/intact visual–spatial subgroup"; $N = 21$) and cluster 4 ("intermediate/intact verbal
subgroup"; $N = 42$) were, as their names imply, considered to be intermediate subgroups that
were virtually indistinguishable in terms of verbal and memory ability but who did exhibit signif-

TABLE 6. Demographic Breakdown of the Five DAT Clusters

Cluster	Total	Females	Males	Age	Education
1. Low-functioning cluster	25	18	7	70.84	11.83
2. Intact visual–spatial cluster	39	15	24	63.36	14.54
3. Intermediate/intact visual–spatial cluster	21	5	16	66.33	11.00
4. Intermediate/intact verbal cluster	42	30	12	74.74	11.67
5. Intact verbal cluster	11	7	4	69.36	13.00
Total	138	75	63	69.11	11.84

icant differences on visual–spatial functioning with the latter group having more prominent deficits. (It should be noted that these two groups made more verbal/language-related errors than group 2.) Subjects within cluster 5 ("intact verbal subgroup"; $N = 11$) were characterized by verbal language and memory ability that was superior to all of the other clusters but performed considerably more poorly on visual–spatial tasks. Demographic data regarding the gender composition and mean age and years of formal education for each of the five clusters are presented in Table 6.

Although the groups were studied with regard to their performance on a number of motor tasks, none of the motor findings differentiated among the groups. This is consistent with recent findings that have suggested little relationship between motor deficits and presentation or progression of DAT (Koller, Wilson, Glatt, & Fox, 1984). Likewise, sensory perceptual deficit patterns were also of minimal heuristic value.

Across the five subgroups, subjects did differ significantly with regard to age. The patients in clusters 2 and 3 were younger than those of the other groups. Both had visual–spatial scores consistently higher than verbal–memory scores. Combining these two groups yields a mean age of 64.4. Combining clusters 4 and 5, which had the opposite combination—visual–spatial inferior to verbal–memory ability—resulted in a mean age of 73.6. This discrepancy between the subgroups with regard to age is consistent with the observation of Seltzer and Sherwin (1983) that a greater prevalence of language dysfunction is evident among those patients with a presenile onset of the disease.

The clusters also differed with respect to gender composition. Despite the greater number of females in the total sample, males outnumbered females in clusters 2 and 3 (intact visual–spatial groups relative to verbal functioning) by a factor of 2 : 1. Combining clusters 4 and 5 (i.e., those with intact verbal–memory ability relative to visual–spatial ability) resulted in an equally disproportionate ratio in the opposite direction: 37 females to 16 males. Several avenues of research have suggested that females may show fewer focal deficits with lateralized injury than do males (see review by Corballis, 1983; Springer & Deutsch, 1981) and may show greater facility with language-based functions, whereas males may have greater facility with spatial abilities (Yeo, Turkheimer, & Bigler, 1984; however, see Bornstein & Matarazzo, 1984). This pattern now appears to be evident in the course of cognitive declines as a consequence of DAT.

It is noted from Table 6 that the two "intact" subgroups with the highest level of functioning (that is, clusters 2 and 5) are characterized by more years of formal education. Two interpretations are possible. First, it is possible that those patients having a higher level of premorbid ability pursued more years of education and that that higher level of ability premorbidly offered a "buffer" against the ravages of the disease. A second possibility is that those patients with lower levels of education were employed in less demanding positions in which the subtle, early effects of the disease were not made apparent; detection of those effects was possible only after the course of the degeneration had progressed. On the other hand, clusters 2 and 5 are more likely to include individuals who are employed in higher-level positions, which are more taxing intellectually and,

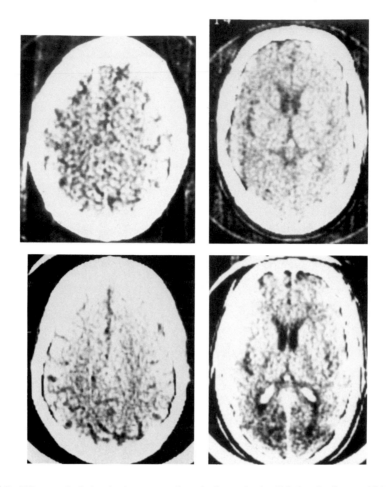

FIGURE 9. CT scans depicting in the course of cerebral atrophy in Alzheimer's disease. This patient was first diagnosed in 1977 (top scans) as having DAT. The CT scan findings showed some prominence of the Sylvian fissures bilaterally and slight prominence of the anterior aspect of the frontal interhemispheric fissure. Otherwise, the scan was within normal limits for this 55-year-old college-educated engineer. In that year, his WAIS and WMS results indicated the following: Verbal IQ 93, Performance IQ 87, Full-Scale IQ 90, MQ 63. Eighteen months later, there were some changes noted on CT examination (bottom scans), namely, in the presence of sulcal enlargement. His WAIS and WMS at that time indicated the following scores: Verbal IQ 85, Performance IQ 72, Full-Scale IQ 79, MQ 57. He was seen again in 1981, but no CT was performed that year. The following WAIS and WMS results were obtained: verbal IQ 59, performance IQ 62, full-scale IQ 57, MQ 48; he could perform no aspect of the WMS. Prior to this writing, he was seen again (in 1984), and there was a complete absence of any higher mental functions, and the patient presented with a generalized paresis. CT scanning (facing page) depicts advanced cortical atrophy, widespread sulcal enlargement, and very prominent bilateral Sylvian fissures. It is of interest to note that there is relatively little change in the size of the ventricular system from 1978 to 1984.

consequently, demonstrate cognitive deficits more quickly after they have begun. Clearly, these results suggest that there is a relationship between neuropsychological impairment profiles and premorbid abilities as they are inferred from patients' genders, years of education, and ages of onset of the disease.

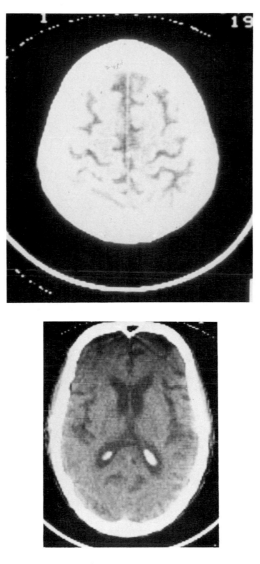

FIGURE 9. (*continued*)

If lateralized atrophy affects symptomatology in a systematic fashion, it would be predicted that CT results would show that cluster 1 would have generalized atrophy, cluster 2 lateralized left hemisphere atrophy, and cluster 5 lateralized right hemisphere involvement. Similarly, based on presenting symptomatology as outlined in Table 5, a relationship between such symptoms and CT findings could be predicted. The results of volumetric analysis, however, showed no such relationship. Likewise, ventricular size volume measurements consistently failed to be statistically significantly related to cognitive deficits. Accordingly, there was no significant morphological feature that corresponded to cluster grouping.

Recently, Raz *et al.* (1988) have also examined this problem by using asymmetry indices of both brain and intellectual performance in DAT patients. Cortical atrophy was assessed for each hemisphere using the method described above, and a cognitive asymmetry estimate was computed using VIQ–PIQ difference scores from the WAIS. Thus, a lower Verbal IQ relative to Performance IQ was considered to represent greater verbal deficits, and the opposite pattern (greater Verbal IQ relative to Performance IQ) was an indication of impaired visuomotor ability. Their study demonstrated some modest correlations between left hemispheric atrophy and verbal IQ. Consequently, their research suggests that there might be some unique groupings of DAT patients in whom lateralized pathology may have a relationship to pattern of cognitive impairment.

In conclusion, although case studies have occasionally demonstrated that focal symptomatology is related to focal atrophy in dementia, studies using a cluster analysis of neuropsychological psychometric data suggest that this is not uniformly the case; broad conclusions cannot be made from those case studies. Along these lines, Kirshner *et al.* (1984) reviewed 6 cases of patients with clear dementia without focal or lateralized CT findings but who had language disturbance as the most prominent initial symptom. Even though they had what appeared to be focal presenting symptomatology (dysphasia), this was not associated with lateralized findings on CT-scan analysis. However, on 1-year follow-up it was discovered that one of those patients did develop subsequent left perisylvian atrophy. It may be that the nature of the progressive deterioration with dementing illness is such that there is a critical time period for the expression of symptoms but that the onset of such symptoms may not correspond to immediately observable CT changes. Microstructural degenerative changes clearly would occur before more gross atrophic changes could be visualized. Obviously, for sufficient cortical degeneration to occur to be detected on CT scanning, it would take weeks to months to accrue; within this time frame, it may be that that atrophy does not correspond to presenting symptoms. Thus, there may be little relationship among initial symptoms, CT analysis, and subsequent cognitive symptomology because of the imprecise relationship between atrophy and neuropsychological status (Bigler, 1988).

PREDICTION OF DETERIORATION

Initial studies using CT scan ratings to predict progression of DAT have been disappointing (Kazniak *et al.*, 1978), whereas neuropsychological studies have been quite promising. Barclay *et al.* (1985) reported that although Hachinski ischemic scores and signs of vascular disease had no correlation with survival in their sample of 199 DAT patients, Haycox behavioral scores (which are indicative of severe behavioral impairment) were significantly correlated with survival at 500 days following diagnosis of the disease on the basis of clinical data. To date, the most comprehensive study has been that of Berg *et al.* (1984), who found that two neuropsychological measures—the Digit Symbol subtest of the WAIS and an aphasia battery—correctly predicted the stage of dementia 1 year later in 95% of their patients studied. Based on their findings, the greater the neuropsychological impairment on these measures, the greater was the severity of dementia and rate of progression within a year. The CT scan and electrophysiological measures (EEG and evoked potential) were of no predictive value. The case study detailed in Figure 9 illustrates these various points.

Using the more detailed procedures for volumetric analysis outlined above, we examined the deterioration indices in DAT and found them to correlate only rarely—with Performance IQ measures—and only modestly with the degree of cortical atrophy (Bigler *et al.*, 1985). It appears that cognitive measures alone, particularly those based on complex short-term memory, manipulo-spatial abilities, and certain verbal-language functions are better predictors of deterioration in DAT than are a variety of CT measures.

In their study of factors that impact on survival, Barclay *et al.* (1985) found that gender and age at onset of the disease process were significantly associated with mortality. In general, males had a significantly shorter duration of survival at 500 and 1000 days post-diagnosis. Patients under the age of 65 tended to expire earlier in the course of the disease than those affected at ages beyond 65 years.

CONCLUSION

The investigation of the neuropathological and associated neuropsychological sequelae of DAT continues to progress. As long as the histological changes of the disease remain beyond the resolving power of brain-imaging techniques, however, the relationship between the results of those brain-imaging procedures and behavioral manifestations will necessarily be modest. Perhaps improved imaging resolution that allows more precise quantification among limited cortical and subcortical regions will reveal a clearer relationship between neuropathological and neuropsychological characteristics.

It may be that the molar measurements of "cortical atrophy" by CT are going to be too far removed from critical sites of specific degeneration that may account for the major aspects of cognitive impairment in DAT. Improved resolution of MRI will permit a clearer differentiation among DAT, vascular dementia, and combinations of the two (Hershey, Modic, Greenough, & Jaffe, 1987). Also, multiquantification methods using horizontal, coronal, and sagittal planes may provide a more accurate three-dimensional estimation of cerebral atrophy. This better differentiation may permit a more careful analysis of neuroimaging variables in relationship to neuropsychological impairment. It is anticipated that, as increasing proportions of variance are reduced (by subgrouping DAT patients according to premorbid characteristics, demographic variables, and/or screening non-AD patients from study) and brain imaging techniques become more sophisticated to the extent that artifactual data will be negligible, these relationships will become more striking and reliable.

REFERENCES

Albert, M. (1984). Assessment of cognitive function in the elderly. *Psychosomatics, 25*, 310–317.

Albert, M., & Stafford, J. L. (1986). CT scan and neuropsychological relationships in aging and dementia. In G. Goldstein & R. E. Tarter (Eds.), *Advances in clinical neuropsychology* (pp. 31–53). New York: Plenum Press.

Albert, M., Naesar, M. A., Levine, H. L., & Garvey, A. J. (1984a). Ventricular size in patients with presenile dementia of the Alzheimer's Type. *Archives of Neurology, 41*, 1258–1263.

Albert, M., Naeser, M. A., Levine, H. L., & Garvey, A. J. (1984b). CT density numbers in patients with senile dementia of the Alzheimer's type. *Archives of Neurology, 41*, 1264–1269.

Amaducci, L. A., Fratiglioni, L., Rocca, W. A., Fieschi, C., Livrea, P., Pedone, D., Bracco, L., Lippi, A., Gandolfo, C., Bino, G., Prencipe, M., Bonatti, M. L., Girotti, F., Carella, F., Tavolato, B., Feria, S., Lenzi, G. L., Carolei, A., Gambi, A., Grigoletto, F., & Schoenberg, B. S. (1986). Risk factors for clinically diagnosed Alzheimer's disease: A case-control study of an Italian population. *Neurology, 36*, 922–931.

Appel, S. H. (1981). A unifying hypothesis for the cause of amyotrophic lateral sclerosis, parkinsonism and Alzheimer's disease. *Annals of Neurology, 10*, 499–505.

Barclay, L. L., Zemcov, A., Blass, J. P., & McDowell, F. H. (1985). Factors associated with duration of survival in Alzheimer's disease. *Biological Psychiatry, 20*, 86–93.

Barron, S. A., Jacobs, L., & Kinkee, W. L. (1976). Changes in size of normal lateral ventricles during aging, determined by computerized tomography. *Neurology, 26*, 1011–1013.

Berg, L., Danziger, W. L., Storandt, M., Cohen, L. A., Gado, M., Hughes, C. P., Knesevich, J. W., & Botwinick, J. (1984). Predictive features in mild senile dementia of the Alzheimer type. *Neurology, 34*, 563–569.

Bigler, E. D. (1984). *Diagnostic clinical neuropsychology.* Austin: University of Texas Press.

Bigler, E. D. (1988). Neuropsychological and CT identification in dementia. In H. A. Whittaker (Ed.), *Neuropsychological studies of non-focal brain damage: Dementia and trauma*, pp. 61–85. New York: Springer-Verlag.

Bigler, E. D., Hubler, D. W., Cullum, C. M., & Turkheimer, E. (1985). Intellectual and memory impairment: CT volume correlations. *Journal of Nervous and Mental Disease, 173*, 347–352.

Bird, J. M. (1982). Computerized tomography, atrophy and dementia: A review. *Progress in Neurobiology, 19*, 91–115.

Blinkov, S. M., & Glezer, I. I. (1968). *The human brain in figures and tables.* New York: Basic Books.

Bornstein, R. A., & Matarazzo, J. D. (1984). Relationship of sex and the effects of unilateral lesions on the Wechsler intelligence scales. *Journal of Nervous and Mental Disease, 172*, 707–710.

Botwinick, J., Storandt, M., & Berg, L. (1986). A longitudinal, behavioral study of senile dementia of the Alzheimer type. *Archives of Neurology, 43*, 1124–1127.

Brinkman, S. D., & Braun, P. (1984). Classification of dementia patients by a WAIS profile related to central cholinergic deficiencies. *Journal of Clinical Neuropsychology, 6*, 393–400.

Brinkman, S. D., Sarwar, M., Levin, H. S., & Morris, H. H. (1981). Quantitative indexes of computed tomography in dementia and normal aging. *Radiology, 138*, 89–92.

Brouwers, P., Cox, C., Martin, A., Chase, T., & Fedio, P. (1984). Differential perceptual–spatial impairment in Huntington's and Alzheimer's dementias. *Archives of Neurology, 41*, 1073–1076.

Chui, H. C., Teng, E. L., Henderson, V. W., & Cook, A. C. (1985). Clinical subtypes of dementia of the Alzheimer type. *Neurology, 35*, 1544–1550.

Claviera, L. E., Moseley, L. F., & Stevenson, J. F. (1977). The clinical significance of "cerebral atrophy" as shown on CAT. In G. H. DuBoulay & L. F. Moseley (Eds.), *Computed axial tomography in clinical practice.* Berlin: Springer.

Cook, R. H., Ward, B. E., & Austin, J. H. (1979). Studies in aging of the brain: IV. *Neurology, 29*, 1402.

Corballis, M. C. (1983). *Human laterality.* New York: Academic Press.

Crapper, D. R., Quittrat, S., Krishnau, S. S., Dalton, A. J., & DeBon, U. (1980). Intranuclear aluminum content in Alzheimer's disease, dialysis encephalopathy and experimental aluminum encephalopathy. *Acta Neuropathologia, 50*, 19–24.

Creasey, H., Schwartz, M., Frederickson, H., Haxby, J. V., & Rapoport, S. I. (1986). Quantitative computed tomography in dementia of the Alzheimer type. *Neurology, 36*, 1563–1568.

Crystal, H. A., Horoupian, D. S., Katzman, R., & Jotkowitz, S. (1982). Biopsy-proved Alzheimer disease presenting as a right parietal lobe syndrome. *Annals of Neurology, 12*, 186–188.

Cullum, C. M., Steinman, D. R., & Bigler, E. D. (1984). Relationship between "fluid" and "crystallized" cognitive function using category and WAIS test scores. *Clinical Neuropsychology, 6*, 172–174.

Cummings, J. L., Benson, F., Hill, M. A., & Read, S. (1985). Aphasia in dementia of the Alzheimer type. *Neurology, 35*, 394–397.

Cutler, N. R., Duara, R., Creasy, H., Grady, C. L., Haxby, J. V., Schapiro, M. B., & Rapaport, S. I. (1984). Brain imaging: Aging and dementia. *Annals of Internal Medicine, 101*, 355–369.

Damasio, H., Eslinger, P., Damasio, A. R., Rizzo, M., Huang, H. K., & Demeter, S. (1983). Quantitative computer tomographic analysis in the diagnosis of dementia. *Archives of Neurology, 40*, 715–719.

de Leon, M. J., & George A. E. (1983). Computed tomography in aging and senile dementia of the Alzheimer's type. In R. Mayeux & W. G. Rosen (Eds.), *The dementias* (pp. 103–122). New York: Raven Press.

de Leon, M. J., Ferris, S. H., Blau, I., George, A.E., Reisberg, B., Kricheff, I. I., & Gershon, S. (1979). Correlations between CT changes and behavioral deficits in senile dementia. *Lancet, 20*, 859–860.

de Leon, M. J., Ferris, S. H., George, A. E., Reisberg, B., Kricheff, I. I., & Gershon, S. (1980). Computed tomography in senile dementia of the Alzheimer's type. *Neurobiology of Aging, 1*, 69–70.

Direnfeld, L. K., Albert, M. L., Volice, L., Langlais, P. J., Marquis, J., & Kaplan, E. (1984). Parkinson's disease: The possible relationship of laterality to dementia and neurochemical findings. *Archives of Neurology, 41*, 935–941.

Duara, R., Grady, C., Haxby, J., Ingvar, D., Sokoloff, L., Margolin, R. A., Manning, R. G., Cutler, N. R., & Rapaport, S. I. (1984). Human brain glucose utilization and cognitive function in relation to age. *Annals of Neurology, 16*, 702–713.

Earnest, M. P., Heaton, R. K., Wilkinson, W. E., & Manke, W. R. (1979). Cortical atrophy, ventricular enlargement and intellectual impairment in the aged. *Neurology, 29*, 1138–1143.

Emery, O. B., & Emery, P. E. (1983). Language in senile dementia of the Alzheimer type. *Psychiatric Journal of the University of Ottawa, 8*, 169–178.

Eslinger, P. J., Benton, A. L. (1983). Visuoperceptual performances in aging and dementia: Clinical and theoretical implications. *Journal of Clinical Neuropsychology, 5*, 213–220.

Eslinger, P. J., Damasio, H., Graff-Radford, N., & Damasio, A. R. (1984). Examining the relationship between computed tomography and neuropsychological measures in normal and demented elderly. *Journal of Neurology, Neurosurgery and Psychiatry, 47*, 1319–1325.

Eslinger, P. J., Damasio, A. R., Benton, A. L., & Van Allen, M. (1985). Neuropsychologic detection of abnormal mental decline in older persons. *Journal of the American Medical Association, 253,* 670–674.

Filley, C. M., Kelly, J., & Heaton, R. K., (1986). Neuropsychologic features of early- and late-onset Alzheimer's disease. *Archives of Neurology, 43,* 574–576.

Ford, C. V., & Winter, J. (1981). Computed axial tomograms and dementia in elderly patients. *Gerontology, 36,* 164–169.

Foster, N. L., Chase, T. N., Mansi, L., Brooks, R., Fedio, P., Patronas, N. J., & DiChiro, G. (1984). Cortical abnormalities in Alzheimer's disease. *Annals of Neurology, 16,* 649–654.

Fox, J. H., Topel, J. L., & Huckman, M. S. (1975). Use of computed tomography in senile dementia. *Journal of Neurology, Neurosurgery and Psychiatry, 38,* 948–953.

Fox, J. H., Ramsey, R. G., Huckman, M. S., & Proske, A. E. (1976). Cerebral ventricular enlargement: Chronic alcoholics examined by computerized tomography. *Journal of the American Medical Association, 236,* 365–368.

Fox, J. H., Kaszniak, A. W., & Huckman, M. (1979). Computerized tomographic screening not very helpful in dementia—nor in craniopharyngioma. *New England Journal of Medicine, 300,* 437.

Friedland, R. P., Budinger, T. E., Brant-Zawadzki, M., & Jagust, W. J. (1984). The diagnosis of Alzheimer-type dementia: A preliminary comparison of positron emission tomography and proton magnetic resonance. *Journal of the American Medical Association, 252,* 2750–2752.

Fuld, P. A. (1983). Psychometric differentiation of the dementias: An overview. In B. Reisberg (Ed.), *Alzheimer's disease: The standard reference* (pp. 201–210). New York: Free Press.

Fuld, P. A. (1984). Test profile of cholinergic dysfunction of Alzheimer-type dementia. *Journal of Clinical Neuropsychology, 6,* 380–392.

Gado, M., & Hughes, C. (1978). Cerebral atrophy and aging. *Journal of Computer Assisted Tomography, 2,* 520–522.

Gazzaniga, M. S., & Blakemore, C. (1975). *Handbook of Psychobiology.* New York: Academic Press.

Gonzalez, C. F., Lanhert, R. L., & Nathan, R. J. (1978). The CT scan appearance of the brain in the normal elderly population: A correlative study. *Neuroradiology, 16,* 120–122.

Gustafson, L., Hagberg, B., & Ingvar, D. (1978). Speech disturbances in presenile dementia related to local cerebral blood flow abnormalities in the dominant hemisphere. *Brain and Language, 5,* 103–118.

Haug, G. (1977). Age and sex dependence of the size of normal ventricles on computed tomography. *Neuroradiology, 14,* 201–204.

Hershey, L. A., Modic, M. T., Greenough, G., & Jaffe, D. F. (1987). Magnetic resonance imaging in vascular dementia. *Neurology, 37,* 29–36.

Heyman, A., Wilkinson, W. E., Hurwitz, B. J., Schmechel, D., Sigmon, A. H., Weinberg, T., Helms, M. J., & Swift, M. (1983). Alzheimer's disease: Genetic aspects and associated clinical disorders. *Annals of Neurology, 14,* 507–515.

Heyman, A., Wilkinson, W. E., Stafford, J. A., Helms, M. J., Sigmon, A. H., & Weinberg, T. (1984). Alzheimer's disease: A study of epidemiological aspects. *Annals of Neurology, 15,* 335–341.

Hochandel, G., & Kaplan, E. (1984). Neuropsychology of normal aging. In M. L. Albert (Ed.), *Clinical neurology of aging.* New York: Oxford University Press.

Horenstein, S. (1971). Amnestic, agnostic, apractic and aphasic features in dementing illness. In C. E. Wells (Ed.), *Dementia.* Philadelphia: F. A. Davis.

Hubbard, B. M. & Anderson, J. M. (1981). A qualitative study of cerebral atrophy in old age and senile dementia. *Journal of Neurological Science, 50,* 135–145.

Huckman, M. S., Fox, J., & Topel, J. (1975). The validity of criteria for the evaluation of cerebral atrophy by computed tomography. *Radiology, 116,* 85–92.

Huckman, M. S., Fox, J. H., & Ramsey, R. G. (1977). Computed tomography in the diagnosis of degenerative disease of the brain. *Seminars in Roentgenology, 1,* 63–75.

Hughes, C. P., & Gado, M. (1981). Computed tomography and aging of the brain. *Radiology, 139,* 391–396.

Jacobson, P. L., & Farmer, T. W. (1979). The ''hypernormal'' CT scan in dementia: Bilateral isodense subdural haematomas. *Neurology, 29,* 1522–1524.

Jacoby, R., & Levy, R. (1980). CT scanning and the investigation of dementia: A review. *Journal of the Royal Society of Medicine, 73,* 366–369.

Joynt, R. J. (1981). Neurology of aging. *Seminars in Neurology, 1,* 1–59.

Katzman, R. (1976). The prevalence and malignancy of Alzheimer's disease. *Archives of Neurology, 33,* 217.

Kazniak, A. W., Fox, J., Gandell, D. L., Garron, D. C., Huckman, M. S., & Ramsey, R. G. (1978). Predictors of mortality in presenile and senile dementia. *Annals of Neurology, 3,* 246–252.

Kirshner, H. S., Webb, W. G., Kelly, M. P., & Wells, C. E. (1984). Language disturbance: An initial symptom of cortical degenerations and dementia. *Archives of Neurology, 41,* 491–496.

Kitagawa, Y., Meyer, J. S., Tachibana, H., Mortel, K. F., & Rogers, R. L. (1984). CT–CBF correlations of cognitive deficits in multi-infarct dementia. *Stroke, 15,* 1000–1009.

Koller, W. C., Wilson, R. S., Glatt, S. L., & Fox, J. H. (1984). Motor signs are infrequent in dementia of the Alzheimer type. *Annals of Neurology, 16,* 514–516.

Koss, E., Ober, B. A., Friedland, R. P., & Jagust, W. J. (1985). Cognitive deterioration in Alzheimer-type dementia is related to age and cerebral metabolic asymmetries. Presented at the Thirteenth Annual Meeting of the International Neuropsychological Society, San Diego.

Littman, E. G., Berg, G., & Novelly, R. A. (1984). Relationship of sulcal or ventricular enlargement to behavior without reference to dementia. Presented at the Twelfth Annual Meeting of the International Neuropsychology Society, Houston.

Loring, D. W., & Largen, J. W. (1985). Neuropsychological patterns of presenile and senile dementia of the Alzheimer type. *Neuropsychologia, 23,* 351–357.

MacInnes, W. D., Golden, C. J., Gillen, R. W., Sawicki, R. F., Quaife, M. Uhi, H. S. M., & Greenhouse, A. J. (1984). Aging, regional cerebral blood flow, and neuropsychological functioning. *Journal of the American Geriatric Society, 32,* 712–718.

Marguez, J. A. (1983). Computerized tomography and neuropsychological tests in dementia. *Clinical Gerontologist, 2,* 13–22.

Massman, P. J. (1986). The relationship between cortical atrophy and ventricular volume. *International Journal of Neuroscience, 30,* 87–99.

Massman, P. J., Bigler, E. D., Cullum, C. M., & Naugle, R. I. (1986). The relationship between cortical atrophy and ventricular volume in Alzheimer's Disease and closed head injury. *International Journal of Neuroscience, 30,* 87–99.

Mayeux, R., Stern, Y., & Spanton, S. (1985). Heterogeneity in dementia of the Alzheimer type: Evidence of subgroups. *Neurology, 35,* 453–461.

McCullough, E. C., Payne, J. T., Baker, H. L., Hattery, R. R., Sheedy, P. F., Stephens, D. H., & Gedgaudus, E. (1976). Performance evaluation and quality assurance of computed tomography scanners, with illustrations from the EMI, ACTA and delta scanners. *Radiology, 120,* 173–188.

McDuff, T., & Sumi, S. M. (1985). Subcortical degeneration in Alzheimer's disease. *Neurology, 35,* 123–126.

McKhann, G., Drachman, D., Folstein, M., Katzman, R., Prize, D., & Stadlan, E. M. (1984). Clinical diagnosis of Alzheimer's disease: Report of the National Institute of Neurological and Communicative Disorders and Stroke–Alzheimer's Disease and Related Disorders Association work group under the auspices of the Department of Health and Human Services Task Force on Alzheimer's Disease. *Neurology, 34,* 939–944.

McMenemey, W. H. (1970). Alois Alzheimer and his disease. In G. E. W. Wolstenholme & M. O'Connor (Eds.), *Alzheimer's disease and related conditions* (pp. 5–9). London: Churchill.

Mesulam, M. M. (1982). Slowly progressive aphasia without generalized dementia. *Annals of Neurology, 11,* 592–598.

Morris, J. C., Cole, M., Banker, B. Q., & Wright, D. (1984). Hereditary dysphasic dementia and the Pick–Alzheimer spectrum. *Annals of Neurology, 16,* 455–466.

Mortimer, J. A., French, L. R., Hutton, J. T., & Schuman, L. M. (1985). Head injury as a risk factor for Alzheimer's disease. *Neurology, 35,* 264–267.

Munoz-Garcia, D., & Ludwin, S. K. (1984). Classic and generalized variants of Pick's disease: A clinicopathological, ultrastructural, and immunocytochemical comparative study. *Annals of Neurology, 16,* 467–480.

Naugle, R. I., Cullum, C. M., Bigler, E. D., & Massman, P. J. (1985). Neuropsychological and CT volume characteristics of empirically-derived subgroups. *Journal of Nervous and Mental Disease, 10,* 596–604.

Naugle, R. I., Cullum, C. M., Bigler, E. D., & Massman, P. J. (1986). Neuropsychological characteristics and atrophic brain changes in senile and presenile dementia. *Archives of Clinical Neuropsychology, 1,* 219–230.

Naugle, R. I., Bigler, E. D. Cullum, C. M., & Massman, P. J. (1987). Handedness and dementia. *Perceptual and Motor Skills, 65,* 207–210.

Neary, D., Snowden, J. S., Mann, D. M. A., Bowen, D. M., Sims, N. R., Northen, B., Yates, P. O., & Davison, A. N. (1986). Alzheimer's disease: A correlative study. *Journal of Neurology, Neurosurgery and Psychiatry, 49,* 229–237.

Perd, D. P., & Brody, A. R. (1980). Alzheimer's disease: X-ray spectrometric evidence of aluminum accumulation in neurofibrillary tangle-bearing neurons. *Science, 208,* 297–299.

Ramani, S. V., Loewenson, R. B., & Gold, L. (1979). Computerized tomographic scanning and the diagnosis of dementia. *New England Journal of Medicine, 300,* 1336–1337.

Raz, N., Raz, S., Yeo, R. A., Turkheimer, E., Bigler, E. D., & Cullum, C. M. (1987). Relationship between cognitive and morphological asymmetry in dementia of the Alzheimer type: A CT scan study. *International Journal of Neuroscience. 35,* 235–243.

Reitan, R. M., & Davison, L. A. (1974). *Clinical neuropsychology: Current status and applications.* Washington: Winston.

Roberts, M. A., & Caird, F. I. (1976). Computerized tomography and intellectual impairment in the elderly. *Journal of Neurology, Neurosurgery and Psychiatry, 39,* 986–989.

Roberts, M. A., Caird, F. L., Grossart, K. W., & Steven, J. L. (1976). Computerized tomography and the diagnosis of cerebral atrophy. *Journal of Neurology, Neurosurgery and Psychiatry, 39,* 909–915.

Rocca, W. A., Amaduci, L. A., & Schoenberg, B. S. (1986). Epidemiology of clinically diagnosed Alzheimer's disease. *Annals of Neurology, 19,* 415–424.

Rosen, W. G., Mohs, R.C., & Davis, K. L. (1984). A new rating scale for Alzheimer's disease. *American Journal of Psychiatry, 141,* 1356–1364.

Russell, E. W. (1979). Three patterns of brain damage on the WAIS. *Journal of Clinical Psychology, 35,* 611–620.

Scheibel, A. B., & Wechsler, A. F. (1986). *The biological substrates of Alzheimer's disease.* Orlando: Academic Press.

Schoenberg, B. S., Anderson, D. W., & Haerer, A. F. (1985). Severe dementia: Prevalence and clinical features in a biracial US population. *Archives of Neurology, 42,* 740–743.

Schwartz, M., Crascy, H., Grady, C. L., DeLeo, J. M., Frederickson, H. A., Cutler, N. R., & Rapoport, S. I. (1985). Computed tomographic analysis of brain morphometrics in 30 healthy men, aged 21 to 81 years. *Annals of Neurology, 17,* 146–157.

Seguerra, J. M. (1984). Alzheimer's disease: A single entity? *Archives of Neurology, 41,* 362.

Seltzer, B., & Sherwin, I. (1983). A comparison of clinical features in early and late onset primary degenerative dementia. *Archives of Neurology, 40,* 143–146.

Semple, A. A., Smith, C. M., & Swash, M. (1982). The Alzheimer disease syndrome. In Corkin (Ed.), *Alzheimer's disease: A report of progress* (pp. 93–107). New York: Raven Press.

Spath, H. (1980). *Cluster analysis algorithms.* Chichester: Ellis Harwood.

Springer, S. P., & Deutsch, G. (1981). *Left brain, right brain.* San Francisco: W. H. Freeman.

Storandt, M., Botwinick, J., Danziger, W. L., Berg, L., & Hughes, C. P. (1984). Psychometric differentiation of mild senile dementia of the Alzheimer's type. *Archives of Neurology, 41,* 497–499.

Strub, R. L., & Black, F. W. (1981). *Organic brain syndromes: An introduction to neurobehavioral disorders.* Philadelphia: F. A. Davis.

Terry, R. D. (1978). Aging, senile dementia and Alzheimer's disease. In R. Katzman, R. D. Terry, & K. L. Bick (Eds.), *Alzheimer's disease: Senile dementia and related disorders* (pp. 11–14). New York: Raven Press.

Terry, R. D., & Katzman, R. (1983). Senile dementia of the Alzheimer's type. *Annals of Neurology, 14,* 497–506.

Turkheimer, E., Yeo, R., & Bigler, E. D. (1983). Digital planimetry in APLSF. *Behavioral Research Methods and Instrumentation, 15,* 471–473.

Turkheimer, E., Cullum, C. M., Hubler, D. W., Paver, S. W., Yeo, R. A., & Bigler, E. D. (1984). Quantifying cortical atrophy. *Journal of Neurology, Neurosurgery and Psychiatry, 47,* 1314–1318.

Wechsler, A. F. (1977). Presenile dementia presenting as aphasia. *Journal of Neurology, Neurosurgery and Psychiatry, 40,* 303–305.

Wechsler, A. F., Verity, M. A., Rosenschein, S., Fried, I., & Scheibel, A. B. (1982). Pick's Disease: A clinical, computed tomographic and histologic study with Golgi impregnation observations. *Archives of Neurology, 39,* 287–290.

Wechsler, D. (1945). A standardized memory scale for clinical use. *Journal of Psychology, 19,* 87–93.

Wechsler, D. (1955). *Wechsler Adult Intelligence Scale manual.* New York: Psychological Corporation.

Wechsler, D. (1958). *The Measurement and Appraisal of Adult Intelligence* (4th ed.). Baltimore: Williams & Wilkins.

Wells, C. E. (1984). Diagnosis of dementia: A reassessment. *Psychosomatics, 25,* 183–190.

Wisniewski, K. E., Wisniewski, H. M., & Wen, G. Y. (1985). Occurrence of neuropathological changes and dementia of Alzheimer's disease in Down's syndrome. *Annals of Neurology, 17,* 278–282.

Wolpert, S. N. (1977). The ventricular size on computed tomography. *Journal of Computer Assisted Tomography, 2,* 22–226.

Yeo, R. A., Turkheimer, E., & Bigler, E. D. (1984). The influence of sex and age on unilateral cerebral lesion sequelae. *International Journal of Neuroscience, 24,* 299–301.

8

Neuropsychological and Neuroanatomic Aspects of Complex Motor Control

K. Y. HAALAND and RONALD A. YEO

INTRODUCTION

The different roles of the right and left hemispheres have been of great interest to investigators examining the organization and control of voluntary movements. Although the role of the left hemisphere traditionally has been emphasized (Liepmann, 1913), more recent studies provide some evidence for special right hemisphere competencies (Watson, Fleet, Gonzalez-Rothi, and Heilman, 1986). Just as the appreciation of the right hemisphere's independent role in language (Perecman, 1983) has improved our understanding of language processing and its neural correlates, differentiation of the individual and coordinated roles of the two hemispheres in the control of voluntary movement should improve our understanding of movement control. In addition, very little is known about how different parts of each hemisphere control movement. Early theories placed equal emphasis on the roles of frontal and parietal areas (Liepmann, 1913; Geschwind, 1965), but more recent data have suggested that the parietal lobe is particularly important (Heilman & Gonzalez-Rothi, 1985; DeRenzi, Faglioni, Lodesoni, & Vecchi, 1983), at least with regard to limb praxis and hand posturing.

In most of the neurological literature the emphasis has been on how a particular hemisphere controls contralateral movement. When complex motor skills are examined, however, the cognitive aspects of the task, such as programming, decision making, and retrieval, become relatively more important. Such cognitive skills may be controlled by a single hemisphere, in contrast to each hemisphere's contralateral control of more simple movements. For complex tasks the approach has been to examine the limb ipsilateral to lesion. The logic underlying this approach is that if one hemisphere is more important for the control of the cognitive aspects of the motor task, then deficits should be seen in the ipsilateral as well as the contralateral limb. The essential advantage of examining the ipsilateral limb is that the effects of primary motor deficits are minimized. This review focuses on the behavioral characteristics of movement deficits in the limb ipsilateral to right or left hemisphere damage. Although limb apraxia is discussed as it relates to

K. Y. HAALAND ● Psychology Service, Veterans Administration Medical Center, and University of New Mexico, Albuquerque, New Mexico 87108. RONALD A. YEO ● Department of Psychology, University of New Mexico, Albuquerque, New Mexico 87131.

other motor deficits, a recent paper (Faglioni & Basso, 1985) does an excellent and thorough job of reviewing the intricacies of the limb apraxia data from both case studies and group studies.

Limb apraxia is a deficit that has been observed in the ipsilateral arm more frequently after left than right hemisphere damage (Liepmann, 1913; Geschwind, 1965, 1975; Heilman & Gonzalez-Rothi, 1985; DeRenzi, Motti, & Nichelli, 1980). Liepmann (1913) concluded from this finding that the left hemisphere is more important for the control of skilled motor movement; the limits of skilled movement, however, were never well defined.

Investigators basically have used two approaches toward specifying the differential roles of the left and right hemispheres in motor control. First, movement deficits in the limb ipsilateral to lesion have been studied in patients with left versus right hemisphere damage (DeRenzi et al., 1980; Haaland & Flaherty, 1984; Jason, 1985, 1986; Kimura, 1977, 1982; Kimura & Archibald, 1974; Kolb & Milner, 1981; Wyke, 1966, 1967, 1968, 1971a,b). Second, left-hemisphere-lesioned patients with and without limb apraxia have been compared on a variety of different movements (Heilman, 1975; Haaland, Porch, & Delaney, 1980; Haaland, 1984) to assess more directly the relationship of apraxia to other motor skills. Recently, the role of intrahemispheric lesion location has been studied in limb apraxia (Kertesz & Ferro, 1984; Basso, Luzzatti, & Spinnler, 1980; Haaland, Yeo, & Koditawakku, 1986), experimental hand postures (Basso, Faglioni, & Luzzatti, 1985; DeRenzi et al., 1980, 1983; Kolb & Milner, 1981; Jason, 1985, 1986), and other motor skills (Haaland, Harrington, & Yeo, 1987; Jeannerod, 1986).

LIMB APRAXIA: BEHAVIORAL CHARACTERIZATION AND LESION LOCATION

Although limb apraxia appears to be more common after left hemisphere damage, much of the research leading to this conclusion has been in the form of clinical case studies with poor methodological control. For example, control groups are sometimes not used. Because the arm ipsilateral to the lesion is usually tested (because of contralateral hemiplegia), the left-hemisphere patient uses the left, nonpreferred hand, and the right-hemisphere patient uses the right, preferred hand. If control groups only use their right hand (or if no control group is used), it is possible that what is interpreted as a hemisphere-of-lesion effect is actually a hand effect. There is no question that right hand performance surpasses left hand performance on most motor tasks in normals (Todor and Smiley, 1985), so there is a greater chance for the group using their left, nonpreferred hand (left-hemisphere-damaged group) to look as if they are more impaired. Although some investigators find that right and left hand performance is similar in control subjects when using an overall limb apraxia score (DeRenzi et al., 1980), others (Haaland & Flaherty, 1984) have found hand differences when the types of errors made by the right or left hand of normals are examined. These differences were critical in demonstrating that the left- and right-hemisphere groups made different types of errors.

Other factors that vary considerably across studies are the types of movements examined, the definition of errors, the reliability of scoring, and the criteria for designating a patient apraxic. Limb apraxia traditionally has been assessed to verbal command or, if the patient has difficulty, to imitation. A variety of gestures are usually examined. Some have suggested that single-hand postures and sequences of hand postures are directly analogous to limb praxis testing on the basis of similar incidence of deficits in large groups of patients (DeRenzi et al., 1980), but others have argued that limb apraxia test procedures are relatively less sensitive to brain dysfunction (Kolb & Milner, 1981). The majority of investigators have continued to use the more traditional approach, and this review generally deals separately with traditional limb apraxia data and experimental hand-posturing tasks.

The traditional apraxia examination typically uses nonrepresentational, intransitive (e.g., wave goodbye), and transitive (e.g., brush teeth) movements to command and imitation. Movements with an object present are also assessed. It has not been clear, however, whether the greater incidence of limb apraxia after left hemisphere damage results from poorer performance on all three movement types and also whether the types of errors made by patients with left or right hemisphere damage differ. When this issue was examined, it was found that the left-hemisphere group performed more poorly than their control group on all three ipsilateral movements (nonrepresentative, intransitive, and transitive), whereas the right-hemisphere group performed more poorly than their control group only on the nonrepresentative and intransitive movements (Haaland & Flaherty, 1984). These results imply that limb apraxia is more common after left hemisphere damage, but the effect is largely related to the left hemisphere's special role in controlling transitive movements. It is not yet clear what characteristics of the transitive movements require left hemisphere control. However, these results suggest the importance of using transitive movements in limb apraxia assessment, as further investigation may help elucidate hemispheric differences. Although direct comparisons with hand posture sequencing were not done in this study, left/right hemisphere differences were not seen for single hand/arm movements, in disagreement with some (DeRenzi *et al.*, 1980; Kimura, 1982) but not other (Kimura & Archibald, 1974; Jason, 1983a,b) researchers.

This study also found differences in the type of body-part-as-object errors made by left-versus right hemisphere-damaged patients, suggesting that the mechanisms producing their deficits may be quite different, analogous to interhemispheric differences in lesion sequelae in dyscalculia (Levin, 1979). Contrary to expectation, right hemisphere damage did not product spatial errors, such as poor orientation of the hand or limb. Rather, the right-hemisphere group made more body-part-as object errors in which the size of the object the patient was pretending to use was not taken into account (e.g., the fist holding the pretended toothbrush moved from one side of the face to the other rather than staying on one side as if the object were actually present). These errors occur more commonly in normal 12-year-olds (Kaplan, 1968). In contrast, the left-hemisphere group made more traditional body-part-as-object errors (e.g., touching the teeth with an extended finger indicating a toothbrush), which are more commonly made by 4-year-olds (Kaplan, 1968). Other laboratories (DeRenzi, Motti, & Nichelli, 1980; Lehmberhl, Poech, & Willmes, 1983; Heilman & Gonzalez-Rothi, 1985; Rapcsak, Gonzalez-Rothi, & Heilman, 1988) are also attempting to define more accurately the types of errors that characterize limb apraxia and to control for hand preference effects and reliability issues. Recently, Poizner has developed computer and video techniques to measure the arm's position and velocity over time as patients are performing items from a limb apraxia assessment. These techniques should be very helpful in better characterizing the deficits of different apraxic patients and left- versus right-hemisphere-damaged patients (Poizner, Heilman, Gonzalez-Rothi, Verfaellie, & Mack, 1987). One very intriguing issue is whether right-hemisphere patients who are apraxic (as many as 20%) and left-hemisphere patients are failing apraxia tasks for different reasons. This information could facilitate understanding of how the brain controls movements in general.

Another approach toward specifying the neural and cognitive processes in skilled movement is to examine the relationship between limb apraxia and other motor skills. Unfortunately, none of these studies have examined a wide variety of motor skills to determine if impairment of certain movements but not others was associated with limb apraxia. Also, lesion size and intrahemispheric location have been largely ignored as determinants of performance. Heilman (1975) compared the ipsilateral finger-tapping skills of left-hemisphere-damaged patients with and without limb apraxia. He found that the apraxic patients demonstrated impaired finger tapping. When we examined left-hemisphere-damaged patients with and without limb apraxia (Haaland, Porch, & Delaney, 1980), we found very subtle differences in the apraxic patients' abilities to inhibit ipsilateral responses but no differences in their ipsilateral finger-tapping speed. Some investigators (Heilman, Schwartz, & Geschwind, 1975) have suggested that patients with limb apraxia did not retain skills learned on a pursuit rotor task as effectively as did nonpraxics. This is a fascinating issue relevant to Liepmann's idea that the left hemisphere is important for learning new motor skills, but additional work is necessary to replicate these findings and determine which aspects of motor learning are affected (Annett, 1985). This issue is quite important from a rehabilitation standpoint. If apraxic patients do have difficulty learning new motor skills, their ability to compensate for their hemiplegia with the ipsilateral arm and to use gestural compensation for aphasia would be significantly reduced (Helm-Estabrooks, Fitzpatrick, & Barresi, 1982).

Despite some progress in specification of the construct of apraxia, understanding complex motor control also requires integration of behavioral data with more precise anatomic data. Besides assessing the differential roles of the right and left hemispheres, the roles of different areas within each hemisphere and various subcortical areas must be investigated to understand how the brain controls gestures.

Liepmann (1913) differentiated several types of limb apraxia based on his theory of left hemisphere organization: ideational, ideomotor, and limb kinetic. There has been considerable controversy about the definition and existence of these subtypes and whether they are actually associated with different lesions within the left hemisphere. The differentiation of these subtypes is not discussed here, but reviews are available elsewhere (Heilman, 1979; Heilman & Gonzalez-Rothi, 1985). Based largely on a group study in which he observed limb apraxia in 20 of 41 right hemiplegics and 0 of 42 left hemiplegics, Liepmann believed that the left hemisphere controls complex motor skills. He emphasized the role of left frontal areas and their connections to the posterior left hemisphere. In his view, the left parietal cortex did not play a particularly important role except that lesions to parietal cortex frequently damaged superficial fibers carrying information to the frontal lobes from occipital, parietal, and temporal lobes.

Liepmann also identified and provided the first neuropathological confirmation of callosal apraxia (Liepmann & Maas, 1907). In this syndrome, the mid-corpus-callosum lesion is purported to disconnect the left hemisphere, which Liepmann hypothesized controls complex movements of both arms, and the right hemisphere, which controls the left limb. This disconnection results in limb apraxia of the left limb even to imitation. Thus, the left arm deficit cannot be explained by left hemisphere damage leading to comprehension deficits. However, Liepmann's patient, Ochs, had a coexistent lesion of the left basis pontis, which complicated interpretation of the case. Liepmann believed that this lesion accounted for his right hemiplegia but not his left-sided apraxia.

Geschwind (1965) also believed that callosal apraxia was related to the left hemisphere's special role in movement control rather than its role in language comprehension, which would mean that patients with callosal apraxia should demonstrate limb apraxia to imitation as well as to command. Interestingly, his famous patient (Geschwind & Kaplan, 1962) with callosal disconnection showed left-sided limb apraxia to command but not to imitation. Other cases of callosal apraxia have been published (see Faglioni & Basso, 1985, for review), but there is still considerable controversy regarding callosal apraxia, as the commissurotomy literature has not shown any consistent evidence of left-sided limb apraxia except to verbal command, in which case the effect can be explained by a disconnection between the left hemisphere's language-processing facilities and the right hemisphere's motor control of the left hand (Zaidel & Sperry, 1977; Volpe, Sidtis, Holtzmann, Wilson, & Gazzaniga, 1982). These results do not support a special role for the left hemisphere in praxic functions, and various explanations (e.g., extracallosal lesions, the older age of vascular and tumor cases, or developmental neural abnormalities in the commissurotomy patients) have been suggested as reasons why callosal patients less frequently show left arm apraxia.

Ideomotor apraxia has been the best studied of Liepmann's apraxia subtypes. Liepmann (1913) and later Geschwind (1965) attributed this imitation deficit to disconnection among posterior areas that process the visual cue or the auditory command, the premotor area, which programs the movement, and the motor area, which controls movement execution. Therefore, lesions to the left premotor area should result in greater or at least similar incidence of limb apraxia. This does not appear to be the case, however, based on limb apraxia case studies (Faglioni & Basso, 1985) and experimental hand posture studies (Kolb & Milner, 1981; DeRenzi et al., 1983). Heilman

(1979) has suggested an alternative theory in which the parietal cortex, especially the inferior parietal cortex, plays a critical role as the repository of visuokinesthetic engrams that must be stored or retrieved to perform gestures accurately. Some studies (Heilman, Rothi, & Valenstein, 1982) support this role, but all of these experiments have provided descriptive rather than quantitative specification of lesion location. In fact, in one study (Heilman et al., 1982), lesion location (anterior versus posterior) was inferred from language data although three of the four fluent aphasics had lesions including the left parietal lobe from CT scan. These results were later replicated using a task that was less dependent on auditory comprehension in order to assess more directly the apraxic patient's knowledge of the meaning of gestures (Gonzalez-Rothi, Heilman, & Watson, 1985).

The widespread availability of computerized axial tomography (CT scans) has allowed the possibility of precise delineation of lesion size and locus, allowing a more differentiated analysis of the role of various left hemisphere structures and concurrently allowing neuroanatomic investigations to go beyond individual case studies. Several important studies have attempted to examine the relationship between CT scan parameters and apraxia. Basso et al. (1980) obtained CT scans and data on ipsilateral imitation of hand postures for 123 patients with hemorrhagic or ischemic lesions. The CT parameters were quantified as follows. Based on the film image, the extent of the lesion was drawn on a lateral view of the left hemisphere. A 60-column grid was superimposed over this lateral view, each column being oriented perpendicular to the anterior–posterior axis of the brain. Each lesion was thus mapped by two numbers: one indicating its most anterior extent and one its most posterior. These measures served to describe the locus of the lesion along a single anterior–posterior axis. The relationship between lesion locus and apraxia was examined by constructing histograms, with lesion site being represented on the abscissa (anterior lesions to the left, posterior to the right) and the proportion of patients with a lesion in that area being represented on the ordinate. Comparison of the histograms drawn for subjects with and without apraxia could thus potentially reveal which lesion sites were characteristically associated with apraxia. Essentially, there was no difference in the cortical regions lesioned in the apraxic versus nonapraxic groups.

Subcortical lesion extent (defined as 2 cm or more below the cortex) was examined in this study in a similar fashion. The authors stated that nonapraxic subjects tended to show more small, deep lesions than did apraxic subjects. It is not clear, however, whether the absence of apraxia with such lesions reflects their small size, their subcortical locus, or both. Finally, the relationship between lesion volume and apraxia was examined. Lesion volume was determined by calculating the number of pixels within the perimeter of a clinically delineated lesion. Although no correlation of lesion size and apraxia was reported, the scatterplot of these two variables suggested little if any relationship.

The Basso et al. study thus provides little evidence for regions within the left hemisphere especially relevant for apraxia. Although this outcome may be consistent with some sort of "mass action" principle, two issues argue against such an interpretation. First, some patients with very small lesions were found to be apraxic. Second, no relationship was observed between lesion volume and apraxia, as would be predicted by the mass action hypothesis. We are left with the conclusion that there may well be areas within the left hemisphere essential for praxis but that the CT quantification methods used were unable to discover such regions.

As this investigation was the first in the apraxia literature to deal with quantification of lesion locus, let us look closely at their methodology. First, the reliance on visual comparison of histograms, rather than methods of statistical inference, raises questions of internal validity. Second, as Kertesz and Ferro (1984) point out, superimposing a large number of lesions from patients with neurobehavioral problems essentially outlines the distribution of the middle cerebral artery. The contributions of specific portions of the middle cerebral artery territory to praxis is thus obscured. Finally, it should be recognized that describing lesion locus in terms of a single anterior–posterior dimension means collapsing across the superior–inferior dimension of the brain.

That is, this lesion locus quantification method implicitly assumes that lesions that are similarly placed on the anterior–posterior axis produce similar effects regardless of their superior–inferior position. For example, supplementary motor and anterior temporal regions, which are approximately in the same position along the anterior–posterior axis, would be quantitatively described as having the same locus. As regional differences in motor control may be obscured by such a quantification strategy, it is important to supplement such an analysis with other approaches to quantification of lesion locus.

A more recent investigation by Basso and colleagues (Basso *et al.*, 1985) adopted a different approach to the investigation of left hemisphere sites crucial for apraxia. First, they divided their sample of 152 stroke patients into those with large versus small lesions. Large lesions were defined as involving more than 50% of the maximal anteroposterior distance of the brain, which essentially means the entire distribution of the middle cerebral artery. Patients with small lesions were divided into eight groups (frontal, frontocentral, parietal, temporoparietooccipital, temporal, occipital, anterior mesial, and deep/subcortical), although not all of these groups contained sufficient numbers of patients for analysis. Patients with temporal and occipital lesions were generally nonapraxic. Comparison of frontal and parietal patients revealed a significantly greater incidence of apraxia in the latter; 24% of the frontal patients were apraxic, as compared to 62% of the parietal patients. The deep/subcortical lesions did not typically produce apraxia. Of 21 patients with such lesions, only 3 were apraxic. Also, in contrast to their earlier study, an association was noted between lesion size and apraxia. Patients with large lesions were more frequently apraxic than those with small lesions.

Kertesz and Ferro (1984) also investigated the role of lesion site and size in ideomotor apraxia. Their apraxia examination differed from that of Basso *et al.* (1980), however, in that items from the apraxia battery were first administered by verbal command. Only if the subjects failed to command were the items administered to imitation. One hundred seventy-seven ischemic stroke patients were studied, and for each, CT scans were obtained. Lesion size was determined as follows. On each slice, the areas of the brain and lesion were determined by planimetry. Integrating across slices, volumetric expressions of brain and lesion size were obtained; lesion size was expressed as a proportion of lesion volume to hemisphere volume for all analyses. Overall lesion size was positively correlated with severity of apraxia ($r = 0.39$). Three groups of patients were then composed: those with "small" (ratios less than 0.10), "moderate" (ratios between 0.10 and 0.30), and "large" (ratios greater than 0.30) lesions.

To investigate the role of various regions in the left hemisphere, Kertesz and Ferro studied two subgroups, those patients with small lesions who were apraxic and those with large lesions who were not apraxic. The authors reasoned that small lesions producing apraxia would indicate critical left hemisphere sites. Nine patients were identified as having small lesions and apraxia, and the most commonly affected site was the anterior half of periventricular white matter. Such a locus was considered consistent with the Liepmann (1913) and Geschwind (1965) formulations. This lesion could cause disconnection of the frontal and parietal lobes, preventing verbal or visual gesture cues from reaching the motor regions of the frontal cortex.

There are, however, significant difficulties with the authors' interpretations of their data. Sixty patients (Table 2 of Kertesz & Ferro, 1984, p. 925) had "small" lesions. To conclude that the anterior periventricular region is of central importance, we need to know the anatomic distribution of the 51 patients with small lesions who were not apraxic. If some of these patients had similarly located lesions, as seems possible, concluding that lesions of the anterior periventricular region are especially important for apraxia seems premature. With respect to their subjects with large lesions and no apraxia, these subjects were noted to have atypical anatomic asymmetries, which the authors speculated might be related to bilateral representation of motor control.

We (Haaland *et al.*, 1986) recently attempted to examine the relationship between left hemisphere lesion locus and apraxia in 28 ischemic stroke patients. All apraxic tasks were performed by the left (ipsilateral) hand. Three different types of gesture imitation tasks were administered: transitive, intransitive, and nonrepresentative. Lesion parameters were quantified from CT scans

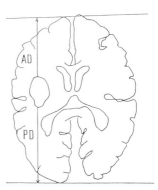

FIGURE 1. Determination of linear expressions of lesion locus: distance from the anterior (AD) and posterior (PD) poles (see text for explanation).

as follows. The CT scans were first traced, carefully outlining the brain, ventricles, and lesion. The tracing was then digitized on a digitizing tablet, which writes X and Y coordinates at the rate of 200 per second. Using these data, a computer program determined the areas of various structures on each slice and the volumes of various structures across slices. Lesion size was expressed as a ratio of lesion volume to brain volume.

Two different classes of procedures were used to describe the anatomic locus of the lesion. First, lesion locus was described in terms of distance from the frontal pole, as described by Robinson, Kubos, Starr, Rao, and Price (1984) (see Figure 1). The distance of the anterior border of the lesion from the frontal pole was measured on every slice on which a lesion appeared; distances were averaged across slices to provide an overall index of lesion location. The smaller this value, the more anterior was the lesion. An analogous procedure was used to measure the distance of the most posterior extent of the lesion from the occipital pole. The smaller this distance, the more posterior was the lesion.

The second approach to lesion locus quantification involved calculation of lesion size in various brain regions. Essentially, the left hemisphere was divided into halves, thirds, and quarters, resulting in, respectively, two-, three-, and four-variable descriptions of lesion locus. The simplest method, dividing the left hemisphere into halves, involved separate determination of the proportion of the anterior and posterior halves that were lesioned. To do this, a line was derived bisecting the interhemispheric fissure. The bisector was determined as follows (see Figure 2): One line segment was drawn connecting the two frontal poles; another was drawn connecting the two occipital poles. The interhemispheric fissure was extended so that it intersected anteriorly and

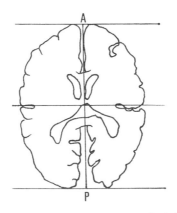

FIGURE 2. Determination of anterior and posterior halves of the left hemisphere.

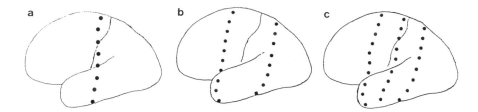

FIGURE 3. Three approaches to dividing the left hemisphere into separate anatomic regions (see text for explanation).

posteriorly with these line segments, at points A and P as shown in Figure 2. The perpendicular bisector of segment AP divided the left hemisphere into halves for every CT slice. Lesion locus could thus be described by two variables: the proportion of the anterior half of the hemisphere damaged and the proportion of the posterior half damaged (see Figure 3a).

Because one of the goals of this project was to investigate the utility of different ways to quantify lesion locus, we also divided each hemisphere into thirds. To accomplish this, segment AP (Figure 2) was divided at two points, one-third and two-thirds of the total AP distance. Lesion size could thus be expressed in terms of three variables: the proportion of each third of the left hemisphere lesioned. Figure 3b shows that this quantification system approximately divides the left hemisphere into frontal, sensory–motor, and parietooccipital regions.

The data from the two volumetric approaches to lesion locus described above were used to derive a four-variable system to express locus. Essentially, the middle third in the three-variable approach was divided into two segments, one approximately the motor strip and one approximately the sensory strip and anterior parietal area (see Figure 3c). This was accomplished by subtracting from the amount of lesion in the anterior half of the brain the amount in the anterior third; the remainder reflects lesion volume in the anterior half of the middle third. Similarly, subtracting the lesion volume in the posterior third of the brain from the lesion volume in the posterior half provides a measure of the amount of lesion in the posterior portion of the middle third of the left hemisphere.

The subcortical extent of each lesion was also determined. In each slice the area of lesion extending into subcortical areas was quantified. For the present study, the extreme capsule marked the lateral extent of subcortical tissue, and white matter just anterior to the most anterior extension of the lateral ventricles marked the anterior extent of cortical tissue.

Before reviewing the results of this investigation, let us comment on the various methods of quantifying lesion locus. Determining the appropriate way to quantify locus is an interesting methodological issue in attempts to elucidate brain–behavior relationships. As the number of separate regions quantified increases, both advantages and disadvantages can be noted. A definite advantage is increasing anatomic specificity; smaller brain regions more closely approximate functional anatomic units. As the number of regions quantified increases, however, the psychometric properties of the variables are adversely affected. Specifically, they tend towards dichotomous (i.e., lesioned or not lesioned) rather than continuously distributed variables. Further, the problem of collinearity arises. Two small, adjacent structures will tend either to be both lesioned or both intact. Multivariate procedures have difficulty in determining the independent effects of highly correlated variables. Also, as the number of predictor variables increases, statistical power decreases, so ideally, sample size should increase as the number of predictor variables increases. All researchers in the field know the difficulty of obtaining large numbers of "clean" unilateral stroke patients with small focal lesions.

Lesion locus quantification methods can be placed on a continuum in terms of the number of separate sites examined. At one extreme of this continuum is the structure checklist approach in which the status of a large number of regions is specified. At the other end of this continuum

TABLE 1. Intercorrelations among Apraxia Scores[a]

	Total	Nonrepresentative	Transitive
Nonrepresentative	0.75[+++]	—	—
Transitive	0.81[+++]	0.50[++]	—
Intransitive	0.83[+++]	0.41[+]	0.48[++]

[a]Significance: [+]$P<0.05$; [++]$P<0.01$; [+++]$P<0.001$.

is a single-variable description of locus, such as the technique of (Robinson et al., 1984) describing locus in terms of distance from the frontal pole. The approaches we have emphasized fall toward the middle of this continuum, attempting to balance anatomic specificity with statistical rigor within the constraints of the typical sample size in human brain lesion studies. In this review we present results from the "simpler" quantification approaches, those describing locus in terms of distance of the lesion from the poles, and a three-variable volume system: extent of frontal half involvement, extent of posterior half involvement, and extent of subcortical involvement. The quantification approaches using additional volumetric expressions of lesion size (i.e., dividing the brain into thirds or quarters) revealed the same pattern of results, though statistical significance was less frequently achieved, probably because of the greater adjustment in the total explained variance with a greater number of predictors as well as the more dichotomous nature of these lesion variables (i.e., more frequent occurrence of "zero" lesion values).

Correlations among the apraxia scores are shown in Table 1. The generally low intercorrelations among nonrepresentative, transitive, and intransitive imitative movements along with high interrater reliabilities of the three variables (Haaland & Flaherty, 1984) suggests separate analyses of the different types of movement in order to evaluate possible differences in anatomic sites crucial for these functions. Table 2 shows the pattern of correlations between apraxia variables and volumetric expressions of lesion size in the anterior and posterior halves of the brain and in subcortical regions. Across types of apraxia, posterior involvement appears to produce greater impairment. For transitive movements, the correlation with posterior involvement did not reach significance, although extent of subcortical involvement was significantly related to severity of deficits. This general pattern of results is also reflected in the correlations between apraxia error scores and the anterior–posterior distance expressions of lesion locus (see Table 3). With the exception of transitive movements, the more posterior the lesion the greater was the deficit.

Multiple regression analyses were performed to reveal the amount of variance accounted for by the three volumetric expressions of lesion locus; these are shown in Table 4. The magnitude of the R^2 values shown in this table clearly suggest that other determinants of apraxia than those we investigated must also be important. Overall lesion size (a ratio of lesion volume to brain volume) was not significantly related to any apraxia score.

The general thrust of these data confirm the conclusion of Basso et al. (1985) regarding the importance of posterior regions for the imitation of limb movements. Of note, however, is the

TABLE 2. Intercorrelations of Apraxia Scores and Volumetric Expressions of Lesion Locus[a]

	Total	Nonrepresentative	Intransitive	Transitive
Anterior	−0.07	−0.22	0.04	0.27
Posterior	0.47[++]	0.38[+]	0.43[+]	0.35
Subcortical	0.42[+]	0.23	0.36[+]	0.38[+]

[a]Significance: [+]$P<0.05$; [++]$P<0.01$.

TABLE 3. Intercorrelations of Apraxia Scores with Linear Expressions of
Lesion Locus[a]

	Total	Nonrepresentative	Intransitive	Transitive
Distance from frontal pole	0.36	0.62[++]	0.31	0.04
Distance from occipital pole	−0.47[++]	−0.62[++]	−0.41[+]	−0.18

[a]Significance: [+]$P<0.05$; [++]$P<0.01$.

fact that this conclusion does not appear to hold for transitive movements. Given our earlier observation (Haaland & Flaherty, 1984) of transitive movements being the only type on which left-lesioned patients perform worse, this suggests the possibility that the neural control of transitive movements is relatively more diffusely represented in the left hemisphere than are other types of movement control.

With the exception of one patient with an anterior cerebral artery lesion, all of our patients suffered occlusions of part or all of the middle cerebral artery. Thus, our sample generally did not have lesions very anterior or very posterior. Neither did they have lesions in the supplementary motor area, an important motor control center. These issues need to be kept in mind in interpreting our results regarding the importance of "posterior" regions. "Posterior," in this sample, essentially means parietal, as only a few patients had involvement of the more anterior occipital regions and none had lesions that encroached on the occipital pole.

The mechanisms Geschwind and Heilman propose as the anatomic basis of imitation limb apraxia of the ipsilateral arm are in agreement as to the role of anterior brain structures. Motor association area lesions, motor strip lesions, and subcortical lesions each lead to apraxia via disconnection of the right hemisphere (which controls the ipsilateral arm) from left hemisphere control systems. In this regard, it is somewhat surprising that our investigation as well as others (Basso et al., 1985; Kolb & Milner, 1981; DeRenzi et al., 1983; Kimura, 1982) all suggest that frontal lesions are less often associated with apraxia. This seems to suggest that alternative pathways may be available for connecting left parietal and right hemisphere motor regions. Perhaps connections existing between the supplementary motor area and posterior neocortical regions (Haaxma & Kuypers, 1975) are only infrequently destroyed by frontal lesions. The supplementary motor area has been associated with movement sequencing (Roland, Larsen, Lassen, & Skinhoj, 1980) and imitative transitive limb movements (Watson et al., 1986) by some but not others (Damasio & Van Hoesen, 1980). Afferents of the supplementary motor area include regions of the parietal lobe, thalamus, and basal ganglia; the heaviest projection of efferents is to the putamen (Jurgens, 1984). Because the supplementary motor area is rarely damaged in large series of stroke patients, being in the distribution of the anterior cerebral artery, its role in the guidance of movement would tend to be compromised largely by disconnection, i.e., deep parietal or basal ganglia lesions. Our data suggests that each of these types of lesions was associated with apraxia (see Table 2). Thus, frontal lesions that do not encroach on the supplementary motor area or basal

TABLE 4. Multiple Regression Analyses of Apraxia Scores Predicted by
Volumetric Expressions of Lesion Locus

	Total	Nonrepresentative	Intransitive	Transitive
R^2	0.24	0.26	0.24	0.28
P	0.02	0.02	0.08	0.04

ganglia may not produce striking apraxia, as the connected and intact supplementary motor area may be able to control visually guided movements of the ipsilateral hand via intact homologous callosal projections.

The comparability of these quantitative approaches to lesion location with the more typical descriptive approach might be enhanced by dividing the hemisphere by appropriate anatomic markings (e.g., central sulcus) and applying the quantitative approach within those boundaries. Another approach that would be helpful in characterizing limb apraxia and its relationship to the brain would be careful case studies that document behavioral and lesion locus data (Marshall & Newcomb, 1984).

EXPERIMENTAL HAND POSTURE TASKS: BEHAVIORAL CHARACTERIZATION AND LESION LOCATION

Some of the most extensive and careful studies of movement control have examined specific limb postures (See Table 5 for summary). Such movements are similar to those examined in limb apraxia, and deficits in the two types of movements are highly correlated (DeRenzi et al., 1980) to the point that these investigators discuss the movement deficits as limb apraxia. These movements have several advantages over the typical apraxia assessment, including the fact that they are meaningless, the body part used (hand, arm, or hand and arm) can be varied, and they can be combined to produce sequences of differing lengths and complexities.

The first study of this type was conducted by Kimura and Archibald (1974), who showed that single arm/hand postures of the ipsilateral limb were performed similarly by patients with right or left hemisphere damage, whereas sequences of arm/hand postures and gestures used in traditional limb apraxia testing were performed more poorly after left hemisphere damage. No control group was used in this study, so it was not clear whether the single movements were actually impaired in both patient groups. The authors interpreted these results as supporting the role of the left hemisphere in motor sequencing.

Unfortunately, this study had several problems related to comparability of the two brain-damaged groups, problems that could reflect different lesion locations in the two groups. For example, the right-hemisphere group was selected on the basis of hemiplegia (right-hemisphere group, 11 of 14 hemiplegic; left-hemisphere group, 9 of 16 hemiplegic), whereas the left-hemisphere group was selected if they had aphasia (14 of 16 were aphasic). This difference could obviously result in a greater incidence of anterior cortical or capsular lesions in the right-hemisphere group and a greater proportion of cortical lesions in the left-hemisphere group. These differences in themselves could result in poorer performance of the left-hemisphere group, independent of the hemisphere of lesion. Anatomic data were not presented to evaluate this concern. In addition, if patients used verbal mediation strategies to remember the items on the sequencing task, it would be expected that the left-hemisphere patients, who are more often aphasic, could show difficulties strictly on this basis. Correlations between language and sequencing performance, however, were relatively low (r ranged from 0.3 to 0.4), which suggested that the motor deficits cannot be explained entirely by the language deficits.

This same conclusion was reached by Goodglass and Kaplan (1963) in their classic study of limb apraxia. In another study (DeRenzi et al., 1980), left hemisphere patients were divided into aphasic and nonaphasic groups. Two of 40 nonaphasic patients and 48 of 60 aphasic patients showed hand-posture deficits. However, this comparison is not entirely appropriate because the aphasic patient may be the most severely impaired on motor as well as all other tasks; nonmotor measures were not presented to demonstrate the specificity of this deficit. In addition, the correlation between the movement-copying score and auditory comprehension (Token Test) in the aphasic patients was 0.56 and accounted for only 31% of the variance. These results suggest that language deficits cannot entirely account for movement-sequencing deficits, and most investiga-

TABLE 5. Ipsilateral Sequencing after Unilateral Hemisphere Damage

Reference	Patients[a]	Anatomic verification	Interhemisphere control			Intrahemispheric control
			Left hemisphere dominance	Right hemisphere dominance	Bilateral control	
Kimura and Archibald (1974)	$N=30$; LHL $=16$ RHL $=14$ Vascular primarily	None reported	Sequential arm/hand postures Limb apraxia	—	Single hand postures from memory	None reported
Kimura (1977)	$N=45$; LHL $=29$ RHL $=16$ Vascular	None reported	Sequential hand postures on box	—	—	None reported
Kimura (1982)	$N=118$; LHL $=72$ RHL $=46$ Vascular and tumor	Method not specified; anterior or posterior; damaged lobe	Single and sequential arm/hand postures	—	—	Single: left parietal Sequential: left anterior and left parietal
DeRenzi et al. (1980)	$N=280$; LHL $=100$ RHL $=80$ C $=100$ Vascular, tumor, trauma	None reported	Single and sequential finger/hand postures	—	—	None reported
DeRenzi et al. (1983)	$N=120$; LHL $=60$ C $=60$ Frontal $=13$ Parietal $=20$ Vascular, tumor, trauma	CT scan	—	—	—	Single: left parietal Sequential: left parietal, not left frontal
Kolb and Milner (1981)	$N=123$; LHL $=58$ RHL $=47$ C $=18$ Lobectomies	Surgical data	Sequences of arm/hand postures (left parietal most impaired)	—	Sequences of arm/hand postures (right and left frontal)	Sequential: left and right frontal and left parietal

Jason (1983a)	Exp. I $N=32$ LHL = 17 RHL = 15 Vascular Exp. II $N=38$ LHL = 21 RHL = 17 Vascular	None reported	Memory for sequences of hand postures	—	Sequences of hand postures without memory component	None reported
Jason (1983b)	Same group Exp. II (1983a)	None reported	Memory for sequences of hand postures even if order not critical	—	Sequences of hand postures without memory component	None reported
Jason (1985)	$N=90$; LHL = 35 RHL = 40 C = 15 Lobectomies	Surgical data	Acquisition of hand posture sequences	—	—	Sequential hand postures: left frontal and temporal; no parietal group
Jason (1986)	$N=90$; Left = 35 Right = 40 Controls = 15 Lobectomies	Surgical data	—	—	Single arm, hand, or arm/hand postures without memory component Sequences of arm, hand, or arm/hand postures without memory	Single postures: left and right frontal and temporal lobes Sequential postures: left and right frontal and temporal but especially frontal

[a]LHL, left hemisphere lesion; RHL, right hemisphere lesion; C, control.

tors feel that overlap of anatomic control for language and praxis or the possible similar cognitive requirements for language and praxis are the reasons these two skills are related (Kertesz, 1985). It should be noted that in one investigation (Haaland, 1984) correlations between limb apraxia and various motor tasks actually were somewhat higher than correlations between limb apraxia and language tasks, but not significantly so.

The symbolic nature of the movement has been shown not to be important in traditional limb apraxia (Dee, Benton, & VanAllen, 1970; Haaland & Flaherty, 1984). Jason (1986) and DeRenzi et al. (1980) have also investigated the effect of varying the specific body part used on sequencing tests and found that the pattern of results is similar whether arm, hand, or arm and hand positions are used. These data make it less likely that discrepancies across studies can be associated with differences in the part of the upper limb being used, in contrast to suggestions that axial movements are preserved in apraxia while distal movements are not (Geschwind, 1965).

The issue of whether single hand postures as well as sequences of hand postures are primarily controlled by the left hemisphere has still not been resolved. Although some have linked the left hemisphere strictly to the sequential task (Kimura & Archibald, 1974), others (DeRenzi et al., 1980, 1983; Kimura, 1982) have found deficits in the imitation of single and sequential hand postures after left hemisphere damage. These studies suggest that the sequential requirements per se were not crucial for the left hemisphere to exert predominant control (see also Roy, 1981).

The fact that neither Kimura (1977), using a sequencing box, or Jason (1983b), using sequential hand postures, showed any significant order errors in left-hemisphere-damaged patients further confirmed this notion. Jason (1983b) also showed that the left-hemisphere group performed more poorly than the right-hemisphere group, regardless of whether sequential order was considered crucial. Although sequential requirements per se may not be critical in inducing left hemisphere control, the additional difficulty of the task may influence the asymmetry observed between the hemispheres. In addition, Jason has manipulated the memory component of these tasks, essentially by presenting single hand postures for imitation under time pressure for performance. When this is done, both right- and left-hemisphere-damaged groups demonstrate similar deficits. These results support the idea that the performance of single postures can be controlled by the left or right hemisphere, whereas the performance of sequential hand postures with a memory component is differentially controlled by the left hemisphere. In the context of studying the relative importance of the left frontal and parietal areas, it has been shown that the performance of single and sequential postures are highly correlated and are either both impaired, as in left-parietal patients, or both unimpaired, as in left-frontal patients (DeRenzi et al., 1983). Thus, this study supports the idea that single hand postures and sequences are similarly controlled within the left hemisphere. This issue is still not resolved and is actually quite important because the factors that influence sequential hand postures are much more complex than those for single postures. If the left hemisphere controls single postures as well as sequences of postures, there are significant implications for the nature of the left hemisphere's special role in movement.

Jason (1983a,b, 1985, 1986) has provided the most careful look at the possible reasons that the left hemisphere would be specialized for controlling limb postures, whether sequences or single movements. He suggested that the difficulty of the task affects its sensitivity to brain dysfunction; one possible reason for the discrepancy in the literature regarding single movements is that simple movements are easier, especially in that there is no significant memory component, and subtle variations in administration may influence whether the left-hemisphere group shows greater impairment. He examined the influence of memory on the sequencing task by presenting it in two ways (Jason, 1983a). First, he demonstrated the sequence, and when finished, the patient performed the task. This approach introduces a significant memory component. Second, he presented the sequences with only a minimal memory component by presenting each posture in the sequence at a particular rate and requiring the patient to duplicate each posture as it was demonstrated. When memory was a significant factor, the left-hemisphere group performed more poorly; in the second case, when memory was reduced and efficient processing was rendered more important, there were no group differences. No control group was used, so it is not known whether the right- and left-hemisphere groups were normal or impaired when memory was not a factor. When recall was examined using a Brown–Peterson paradigm in which the posture had to be performed after delays varying from 0 to 20 s, he found the left-hemisphere-damaged patients to be more impaired than the right-hemisphere patients. These results also emphasize the importance of memory factors. The specific aspects of memory involved (e.g., encoding, consolidation, storage or retrieval, including response selection) are not clear.

Several of these studies have addressed the issue of which regions of the left hemisphere are critical for generating these movements. Even though early theories of limb apraxia emphasized the equal roles of the frontal and parietal lobes of the left hemisphere, recent data suggests that the left parietal lobe plays a more central role in controlling praxis (Heilman & Gonzalez-Rothi, 1985; Gonzalez-Rothi & Heilman, 1985; DeRenzi et al., 1983). Therefore, one might expect the parietal lobe to be particularly important for controlling single hand postures and sequences, since they seem to be quite similar to apraxia tests (DeRenzi et al., 1980, 1983).

As can be seen in Table 6, across studies the performance of patients with left parietal lobe damage is consistently impaired for sequences of hand postures or single hand postures. In contrast, the left frontal lobe and left temporal lobe are inconsistently involved for sequences and single hand postures. There is no single explanation for this discrepancy across studies, as they differ in many respects, including the patient population used, whether a control group was used, and the exact task requirements. For example, some studies (Kolb & Milner, 1981; Jason, 1985, 1986) used postsurgical epileptic patients (some with childhood onset of seizures), whereas others (Kimura, 1982; DeRenzi et al., 1983) used patients with focal vascular or neoplastic etiologies. These groups very likely vary in lesion size, age of onset of abnormalities, and age when tested, with the epileptic mean ages in their 20s and the focal lesion groups' mean ages in their 50s. In addition, some investigators repeated the sequence once (Kolb & Milner, 1981), and others repeated it three times (Kimura, 1982); DeRenzi et al. (1983), who found no deficits in their frontal group, repeated the sequence five times. It is possible the deficit is subtle enough in the left frontal patient that five stimulus repetitions eliminates the performance deficit. What is not apparent from Table 6 is that although Jason (1986) finds deficits in sequencing and single hand postures for the frontal and temporal lobe groups, it is only on the sequencing task that the frontal group performs more poorly than the temporal group. This suggests that the frontal lobes may play a role in the rapid execution of multiple hand positions even when a memory component is not present.

One of Jason's (1985) sequencing tasks uses four sets of five hand positions, whereas others (Kolb & Milner, 1981; DeRenzi et al., 1983; Kimura, 1982) use sequences with about three movements. This difference may explain the incidence of deficits in the left-temporal group when compared to the Kolb & Milner (1981) study, which used a similar population. Jason's later study (1986) differed from the others by its use of metronome-paced presentation of each posture. This time pressure, the increased number of hand postures in each sequence, and the focus on performance without a memory component may explain the incidence of deficits in all groups tested, including the right-frontal and temporal groups. One might expect, similar to other complex motor tasks (Haaland & Delaney, 1981), that these time pressures could increase the requirement of intra- and interhemispheric collaboration so that ipsilateral deficits are seen after unilateral right- or left-hemisphere damage regardless of intrahemispheric location. Jason (1986), however, also found a similar pattern of deficits for imitation of single postures that cannot be explained in this fashion and is contradictory to the Kimura (1982) and DeRenzi et al. (1983) studies. Whereas these authors suggest that the left parietal area is critical for controlling the generation of single postures, Jason finds deficits after frontal or temporal excisions of either hemisphere. This discrepancy could be associated with Jason's more difficult task and/or lesion size, which is likely to be larger in the epileptic groups used by Jason (1986). At this point it is impossible to definitely explain these discrepancies.

TABLE 6. Ipsilateral Motor Deficits after Unilateral Hemisphere Damage

Reference	Patient[a]	Anatomic verification	Interhemisphere control			Intrahemispheric control
			Left hemisphere dominance	Right hemisphere dominance	Bilateral control	
Wyke (1966)	N = 50; LHL = 15, RHL = 15, C = 20; Tumor and temporal lobectomies	Surgical data	Maintenance of static arm posture without vision	—	—	Left parietal: 100% impaired; Left frontal: 67% impaired; Left temporal: 50% impaired
Wyke (1967)	N = 68; LHL = 18, RHL = 20, C = 30; Tumors and temporal lobectomies	Surgical data	Single arm RT and MT; repetitive arm movements to two targets	—	—	Left parietal: 75% impaired; Left frontal: 50% impaired; Left temporal: 38% impaired
Wyke (1968)	N = 67; LHL = 17, RHL = 20, C = 30; Tumors and temporal lobectomies	Surgical data	Target aiming if patients with VFD excluded	—	Target aiming if patients with VFD included	Frontal: 76% impaired; Parietal: 67% impaired; Temporal: 68% impaired
Wyke (1971a)	N = 60; LHL = 20, RHL = 20, C = 20; Tumor and temporal lobectomies	Surgical data	Repetitive arm movements to two targets (10-cm square)	—	Purdue pegboard	
Wyke (1971b)	N = 60; LHL = 20, RHL = 20, C = 20	Surgical data	Design-tracing task (bimanual)	—	—	Frontal: 56% impaired; Parietal: 60% impaired; Temporal: 41% impaired

Study	Subjects[a]	Lesion documentation				Results
Haaland et al. (1987)	Tumor and temporal lobectomies N=40; LHL=10 RHL=10 C=20 Vascular	CT scan; lesion size controlled across groups	Repetitive arm movements, especially to wide targets	—	—	Left-hemisphere lesions more anterior
Carmon (1971)	Exp. 1: N=78 LHL=19 RHL=19 C=40 Vascular and tumor	None reported	Rapid tapping	Paced tapping	—	—
Haaland and Delaney (1981)	N=112; LHL=40 RHL=32 C=40 Vascular and tumor	None	—	—	Grooved pegboard Maze coordination Vertical groove steadiness Static steadiness	—
Heap and Wyke (1972)	N=60; LHL=20 RHL=20 C=20 Tumor and temporal lobectomies	Surgical data	—	—	Pursuit rotor	Patients with impaired performance Frontal: 89% impaired Parietal: 80% impaired Temporal: 71% impaired
Vaughan and Costa (1962)	N=53; LHL=17 RHL=18 C=18 Vascular, tumor, trauma, focal atrophy	EEG data	Combined score from pressure, two-point, finger tap & Purdue dexterity test	—	—	—

[a]LHL, left hemisphere lesion; RHL, right hemisphere lesion; C, control.

These results are interesting, especially if one accepts the possibility that the hand posture tasks are comparable to limb apraxia testing. Both Liepmann's (1913) and Geschwind's (1965) disconnection hypotheses suggest that the left frontal area is more important than the left parietal area in controlling gestures. These results (summarized in Table 5 and 6) indicate, contrary to the disconnection hypothesis, that the left parietal area is more important than the left frontal area in controlling hand postures, a conclusion suggested earlier in our discussion of lesion locus and apraxia. If the left parietal region plays an important role in control of the left limb, connections must be made with right frontal motor systems. As discussed earlier, it is possible that connections are made via the left supplementary motor area. Alternatively, one can hypothesize direct connections of the left parietal lobe to the right premotor area via the right parietal lobe. This hypothesis seems unlikely because it would predict that the incidence of limb apraxia or sequencing deficits after right parietal lesion should be similar to that following left parietal lesion, a prediction contrary to the data (Kolb & Milner, 1981).

Although the incidence of limb apraxia and sequencing deficits is clearly less in patients with right- than with left-hemisphere damage, thorough studies examining left- and right-hemisphere patients, with comparison of parietal as well as frontal damage, have not been conducted. One can also hypothesize a direct connection between left parietal and right premotor areas, a hypothesis that requires the anatomic demonstration of diagonally crossing fibers in the corpus callosum. In any case, the effect of anterior right-hemisphere lesions on left-sided sequencing or limb apraxia cannot be tested because of the high incidence of left hemiplegia in this group. The central importance of the left parietal region seems clear, consistent with the idea that the left parietal lobe plays a central role in programming limb movements, possibly as a location of visuokinesthetic engrams for those movements (Heilman & Gonzalez-Rothi, 1985). Indeed, these same investigators have suggested that the roles of the left parietal and frontal areas in movement control may be quite different. Fluent aphasics (with presumed posterior damage) were more impaired than nonfluent aphasics (with presumed frontal damage) on a gesture recognition task. Even though these findings are intriguing, substantive conclusions await replication with better anatomic verification of lesion location and size.

MOTOR SKILLS: BEHAVIORAL CHARACTERIZATION AND LESION LOCATION

Numerous investigators have examined ipsilateral motor performance in left- and right-hemisphere-damaged patients to determine the nature of left hemisphere motor control. Lesions of the left, but not the right, hemisphere have been associated with ipsilateral deficits in maintaining static arm position (Wyke, 1966), single and repetitive arm movements (Wyke, 1967, 1971a; Haaland et al., 1987), target aiming on a pursuit rotor device (Wyke, 1968), limb sequencing (Jason, 1983a,b, 1985; Kimura & Archibald, 1974; Kimura, 1982; Kolb & Milner, 1981; De-Renzi et al., 1980) if a memory component is present (Jason, 1983a,b, 1985, 1986), generating single hand postures (DeRenzi et al., 1980; Kimura, 1982) and gesturing (Geschwind, 1965; Haaland & Flaherty, 1984; Heilman & Gonzalez-Rothi, 1985).

In contrast, other motor tasks that emphasize spatial localization (Nachson & Carmon, 1975), rapid movements in the left hemispace (Guiard, Diaz, & Beaubaton, 1983) and position localization (Roy & MacKenzie, 1978), have been linked to the right hemisphere and its spatial processing capabilities. It should be noted, however, that the three studies cited above have inferred right hemisphere control advantages from normal studies in which the left hand was observed to perform better than the right. One experiment (Carmon, 1971) has demonstrated differential hemispheric control as a function of finger-tapping speed. Rapid tapping was impaired with left hemisphere damage (movement times averaging 149 ms), whereas slower tapping (movement times averaging 250 and 417 ms) was impaired after right hemisphere damage.

Hypokinesia, assessed clinically by reaction time (Heilman, Bowers, Coslett, Whelan, & Watson, 1985; DeRenzi & Faglioni, 1965; Howes & Boller, 1975), has been associated with right hemisphere damage. Some have associated these deficits with the right hemisphere's role in attention and intention. The right supplementary motor area, in particular, has been associated with initiation and amplitude control on the basis of a case study (Watson et al., 1986), but this hypothesis obviously requires further verification.

The reaction time (RT) data (DeRenzi & Faglioni, 1965; Howes & Boller, 1975) are confusing because of various methodological difficulties. The major problem has been poor control of hand use, such that both left- and right-hemisphere-damaged patients used the hand ipsilateral to the lesion, but the control group used only their right hand. Since simple RTs of the left hand can be faster than those of the right hand in normals (Klapp, Greim, Mendicino, & Koenig, 1979), using only right-hand controls automatically gives the left-hemisphere group an advantage and the right-hemisphere group a disadvantage, dependent solely on the performing hand. To our knowledge only two studies have been published with brain-damaged patients in which the hand factor was properly controlled. In the first, both right- and left-hemisphere patients showed slower RTs relative to controls (Benton & Joynt, 1958). In the second study the right-hemisphere group had longer RTs than the left hemisphere group, but only right-hemisphere patients with neglect were studied, and these patients are obviously not representative of all right-hemisphere patients (Heilman, Bowers, et al., 1985). This could profitably be extended to include data on lesion size and location as well as hemispace to determine if these findings are attributable strictly to the right parietal lobe. No one has yet examined changes in reaction time of brain-damaged patients as a function of the characteristics of the movement, a study that would be quite useful since RT has been associated with movement-programming speed.

Still other studies (See Tables 5 and 6) have shown that both hemispheres are equally important in controlling ipsilateral movements. Tracking on pursuit rotor (Heap & Wyke, 1972), maze coordination (Haaland & Delaney, 1981), peg insertion (Haaland & Delaney, 1981; Wyke, 1971a), screw rotation (Kimura, 1979), sequencing without memory demands (Jason, 1983a,b, 1986), and control of single hand postures (Kimura & Archibald, 1982; DeRenzi et al., 1981; Jason, 1986) have been shown to be impaired ipsilaterally after damage to the left or right hemisphere. These results suggest that both hemispheres are important in controlling these tasks.

Therefore, Liepmann's original contention that the left hemisphere is important for the control of skilled movement cannot be entirely true. Some tasks are primarily controlled by the left hemisphere, but others are controlled by the right hemisphere or equally by both hemispheres. Despite the large number of studies done in this area, there is still no single theory of hemispheric control of movement. There are two important reasons for this. Very few studies have examined the performance of patient groups on a large number of motor tasks that methodically vary certain factors, such as type of movement required or task complexity. In addition, few of the motor studies have included adequate anatomic data regarding lesion location. When lesion location has been examined, patients with overlapping lesions (i.e., frontal, frontoparietal, and frontotemporal are included in the frontal group, and parietal, parietofrontal, and parietotemporal locations are included in the parietal group) have been used (Wyke, 1966, 1967, 1968, 1971b), an approach rendering localization quite difficult. Others have approached lesion localization in various ways. Some (Kolb & Milner, 1981; Jason, 1985, 1986) have used epileptic patients who have had focal cortical excisions, but generalizations to the normal brain may not be accurate. Others (DeRenzi et al., 1980, 1983; Basso et al., 1985) have examined larger series of patients to identify a significant number with lesions restricted to a particular lobe. Case studies have, of course, been reported (see Faglioni & Basso, 1985, for review), but not with detailed analysis as suggested by Marshall and Newcomb (1984). Quantitative analysis of CT scan parameters, as discussed earlier in our discussion of apraxia, may facilitate derivation of neuroanatomic correlates.

The work in our laboratories has been designed to characterize motor deficits systematically in unilaterally damaged patients, to correlate lesion location within each hemisphere with the pattern of motor deficits, and to examine directly the interrelationship among motor deficits, limb apraxia, and lesion location. Unfortunately, our sample is too small to allow sole focus on patients with lesions restricted to a particular lobe, but quantification of the CT scan has provided some information regarding the relationship between lesion size and location and motor deficits.

With regard to the nature of specific task factors, we examined stroke and tumor patients on five different motor tasks and found that the tasks that appeared to require greater sensory–motor integration were impaired ipsilaterally after right or left hemisphere damage (Haaland & Delaney, 1981), in agreement with others (Heap and Wyke, 1972; Wyke, 1971a). Several explanations were considered. One possibility was that the reasons for impairment were different for the right- and left-hemisphere groups. For example, the left-hemisphere group may have difficulty with the sequential characteristics of the task, and the right-hemisphere group may have difficulty with the visuospatial requirements. If sequencing was particularly difficult for the left-hemisphere group, then the performance on the grooved pegboard task, which has the greatest sequential requirements, should be most impaired for the left-hemisphere group. This was not observed. Visuospatial factors were examined by determining if correlations between a spatial test (Block Design) and the four motor tasks that were impaired in the right-hemisphere group were greater than those between a spatial task and the two tasks that showed no ipsilateral impairment. This was not the case. Thus, these two different explanations for the impairment of the left- and right-hemisphere groups are not supported. Our initial explanation of these data was the same for each lesion group and centered on the extent of sensory–motor integration demands. Movements that require greater sensory–motor integration require both hemispheres, and, therefore, unilateral damage to either hemisphere should produce ipsilateral deficits.

However, this explanation has never been examined directly by designing tasks that have similar requirements but vary in their dependence on sensory feedback. This was done recently in the context of a modified aiming task (Haaland & Harrington, in press) in which the patient was asked to move a stylus between two parallel lines that were separated by 5 or 30 mm. The narrow task was shown to be more dependent on visual feedback in normals, but performance on both tasks deteriorated when visual feedback was not available, implying both were at least somewhat dependent on visual feedback (closed loop). Nonetheless, it was predicted that if both hemispheres were required for tasks that required greater sensory–motor integration, then greater ipsilateral deficits should be seen on the narrow task than the wide in both right- and left-hemisphere-damaged patients. This was not the case; the differences between the control groups and the two brain-damaged groups were similar for both tasks, regardless of the degree of visual–motor integration requirements. Only the right-hemisphere group demonstrated an ipsilateral deficit in movement accuracy, although they demonstrated no deficit in reaction time. The left-hemisphere group showed no deficits. This suggests that sensory–motor requirements are not the only factors to explain our previous results. Sensory–motor requirements may be particularly important in explaining the deficits of the right-hemisphere group, and other (as yet unknown) factors may be critical in producing the deficits of the left-hemisphere group. Examining the quality of performance (in this case, the trajectories of the movement) may help determine if different factors affect the motor performance of right- and left-hemisphere patients. When the movement trajectories to the narrow and wide targets were examined, they appeared to be quite similar in speed and accuracy in individual patients. This observation suggested that similar strategies were used for both tasks, but there was no direct quantitative way to assess these strategies within this experiment.

Simple aiming tasks may offer a way to assess directly strategies of movement. Many investigators (Woodworth, 1899; Keele, 1981; Flowers, 1975, 1976) have differentiated two components in the aiming movement, and these two components relate to the sensory–motor integration requirements of the task as well as the motor programming requirements. Their data indicate that a rapid, preprogrammed component that is largely independent of sensory feedback (open loop) is usually followed by a slower, more sensory-dependent component (closed loop) to hit the target

(Keele, 1981). If the target is large enough or the amplitude of the movement is short enough, the closed-loop component may not be necessary; in this case the open-loop component, which gets near the vicinity of the target, can be accurate enough to hit the target.

Thus, the importance of each of these components to a particular movement can be manipulated by varying the target size and/or the movement amplitude. Although proprioceptive feedback has not been assessed, Wallace and Newell (1983) have shown that if the index of difficulty ($ID = \log_2 2A/W$, where A is the movement amplitude and W is the target width) is 3.58 or below, the aiming movement is independent of visual feedback.

In addition, the trajectory of aiming movements can be directly measured (Flowers, 1975, 1976). An initial rapid open-loop component can be separated from the subsequent closed-loop component based on changes in the velocity of the movement. In the context of the previous experiment (Haaland & Harrington, 1988), both tasks largely required closed-loop movements even though they varied somewhat in their dependence on visual feedback. The next series of experiments were designed to assess the open-and closed-loop components of movement within the same task. This was done in two ways.

In the first experiment (Haaland et al., 1987), the Fitts Tapping Task (Fitts, 1954) was used. The subject was asked to move a stylus as rapidly and accurately as possible between two targets of varying size. The open- and closed-loop components were inferred from the index of difficulty (Wallace & Newell, 1983) because when the target width increases and the movement amplitude decreases, resulting in an index of difficulty less than 3.58, the movement is largely independent of visual feedback and thus primarily open loop. When patients with left- or right-hemisphere damage were examined on this task, the left-hemisphere patients had difficulty with both tasks, but their performance on the open-loop task was particularly impaired relative to control subjects. In contrast, even though total lesion size was comparable across both groups, the right-hemisphere group showed no such ipsilateral deficits. Without CT scan data, these results suggested that the left hemisphere plays a stronger role in repetitive arm movements, especially when open-loop control is emphasized. Wyke (1967, 1971a) has also found greater deficits after left-hemisphere damage for single arm movements and repetitive arm movements. In both of her studies the index of difficulty was less than 2, certainly within the range of open-loop movements. This finding has implications for limb apraxia, which is more common after left-hemisphere damage and has been shown to be independent of sensory feedback and primarily open loop (Heilman, Mack, Gonzalez-Rothi, & Watson, 1988). These results must be extended over a wider range of ID levels to assess more directly the open/closed-loop hypothesis and to determine if the right-hemisphere patients show any closed-loop deficits.

However, in the Haaland et al. (1987) study, CT data showed that although total lesion size was similar between the left- and right-hemisphere groups, lesion location within each hemisphere was different (see Figure 4). The lesions of the left-hemisphere group were more anterior, so it is possible that these results can be attributed to intra- rather than interhemispheric lesion location. It may be frontal areas rather than the left hemisphere per se that are most important for open-loop control. These data emphasize the need to assess intrahemispheric lesion location before concluding that the left and right hemispheres play different roles. They also emphasize the potential utility of quantitative analysis because when CT scans were visually inspected, the two groups looked quite comparable. The group differences were apparent only when the CT scans were quantitatively examined.

The direct measurement approach (Flowers, 1975) is also being used in our laboratory. These results have shown that the left hemisphere group's ipsilateral open- but not its closed-loop performance was impaired, while the right hemisphere group demonstrated no deficits relative to a control group (Haaland & Harrington, 1988). In addition, if the programming necessary to perform gestures is at all comparable to the programming of simple aiming, one might expect apraxic patients to have particular difficulty with the open-loop component, even of simple aiming.

The research reviewed in this chapter clearly supports the notion of hemispheric differences

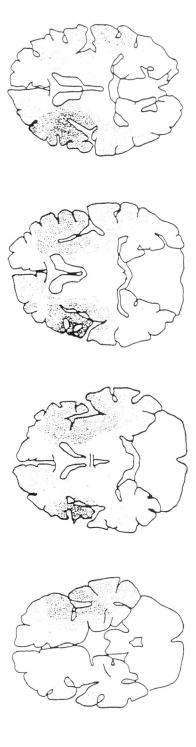
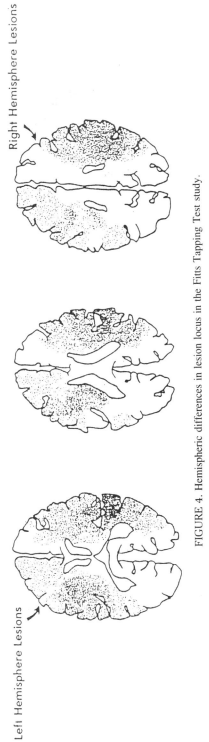

FIGURE 4. Hemispheric differences in lesion locus in the Fitts Tapping Test study.

in motor control, though the precise nature of these differences is a matter of much discussion. We (Yeo, Melka, & Haaland, 1988) recently explored possible anatomic differences between the hemispheres that may underlie behavioral differences. In 1967, Josephine Semmes (Semmes, 1967) presented an intriguing paper on this issue. Based on analysis of sensory and motor data in patients with missile wounds, Semmes hypothesized that the hemispheres differed in the degree of "focal organization." That is, a given motor control system, for example, might be more diffusely represented in the right hemisphere than the left. In her study, lesions outside the classical contralateral tactile and motor areas were more apt to be associated with deficits in patients with right hemisphere lesions than those with left lesions. Despite the potential utility of Semmes's hypothesis in accounting for hemispheric differences, there have been no direct tests of the hypothesis since her original work.

We attempted to test this hypothesis in a sample of unilateral stroke patients (34 left, 22 right) for whom quantitative expressions of lesion locus and size were obtained (Yeo, Melka, & Haaland, 1988). In this investigation lesion locus was again described in terms of distance of the lesion from frontal and occipital poles and also in terms of proportions of three large cortical regions that were damaged (see Figure 3b). Motor performance of the contralateral hand was examined by finger-tapping speed and strength of grip, two tasks that are impaired contralateral but not ipsilateral to the lesion (Haaland & Delaney, 1981). Consistent with the Semmes hypothesis, the amount of lesion outside the sensorimotor areas of the right hemisphere was associated with motor deficits ($r = 0.54$, $P < 0.001$), but lesions outside the sensorimotor area of the left hemisphere were not associated with motor deficits ($r = -0.16$, NS). These results were complicated, however, by the presence of somewhat larger lesions in the right-hemisphere group. Partial correlational analyses revealed that lesion-size differences could not account for the different pattern of results in the two groups. The broad thrust of this study supported the Semmes hypothesis, though of course limited to simple movements of the contralateral arm. Future investigations will necessarily focus on a broader array of ipsilateral and contralateral motor skills, but our recent data and those of Semmes (1967) offer one perspective on the anatomic basis for hemispheric differences in motor control.

SUMMARY AND CONCLUSIONS

This chapter has attempted to summarize selectively the behavioral and anatomic data related to developing a better understanding of the cognitive aspects of motor deficits after unilateral brain damage. For this purpose, movements ipsilateral rather than contralateral to the lesion have traditionally been examined to minimize the impact of direct damage to the primary motor system. Although some motor skills are impaired after left-hemisphere damage, others are impaired after right-hemisphere damage or damage to either hemisphere. There is, however, no comprehensive theory to conceptualize the hemispheric control of complex movement. In addition, from a methodological standpoint, experiments must begin to examine either groups of patients or individual cases with a wide variety of motor tasks in order to specify differential impairment related to lesion locations within and across the two hemispheres. With the exception of hand-posturing tasks, the selection of motor tests done with brain-damaged patients has not been derived from theory. Adoption of a conceptual framework may be advantageous, and this chapter suggests that one approach that may be helpful as a first step is the distinction between open- and closed-loop movements (Keele, 1981).

ACKNOWLEDGMENTS. The authors wish to thank Shirley Sparks for preparation of the manuscript. This work was partially supported by the Research Service, Veterans Administration Medical Center and completed while the first author was a visiting scientist at the University of Washington.

REFERENCES

Annett, J . (1985). Motor learning: A review. In H. Heuer, U. Kleinbeck, K. H. Schmidt (Eds.), *Motor behavior: Programming, control and acquisition* (pp. 189–212). Berlin: Springer Verlag.

Basso, A., Luzzatti, C., & Spinnler, H. (1980). Is ideomotor apraxia the outcome of damage to well defined regions of the left hemisphere? *Journal of Neurology, Neurosurgery and Psychiatry, 43*, 118–126.

Basso, A., Faglioni, P., & Luzzatti, C. (1985). Methods in neuroanatomical research and an experimental study of limb apraxia. In E. A. Roy (Ed), *Neuropsychological studies of apraxia and related disorders* (pp. 179–202.) Amsterdam: North Holland.

Benton, A. L., & Joynt, R. J. (1958). RT in unilateral cerebral disease. *Confina Neurologica, 19*, 247–256.

Carmon, A. (1971). Sequenced motor performance in patients with unilateral cerebral lesions. *Neuropsychologia, 9*, 445–449.

Damasio, A. R. & Van Hoesen, G. W. (1980). Structure and function of the supplementary motor area. *Neurology, 30*, 359.

Dee, H. L., Benton, A. L., & VanAllen, M. W. (1970). Apraxia in relation to hemispheric locus of lesion and aphasia. *Transactions of the American Academy of Neurology, 95*, 147–150.

DeRenzi, E., & Faglioni, P. (1965). The comparative efficiency of intelligence and vigilance tests in detecting hemispheric cerebral damage *Cortex, 1*, 410–433.

DeRenzi, E., Motti, F., & Nichelli, P. (1980). Imitating gestures: A quantitative approach to ideomotor apraxia. *Archives of Neurology, 37*, 6–10.

DeRenzi, E., Faglioni, P., Lodesoni, M., & Vecchi, A. (1983). Performance of left brain-damaged patients on imitation of single movements and motor sequences: Frontal and parietal-injured patients compared. *Cortex, 19*, 333–343.

Faglioni, P., and Basso, A. (1985). Historical perspectives on neuroanatomical correlates of limb apraxia. In E. A. Roy (Ed.), *Neuropsychological studies of apraxia and related disorders* (pp. 3–44). Amsterdam: North Holland.

Fitts, P. M. (1954). The information capacity of the human motor system controlling the amplitude of movements. *Journal of Experimental Psychology, 47*, 381–391.

Flowers, K. A. (1975). Ballistic and corrective movements on an aiming task: Intention tremor and parkinsonian movement disorders compared. *Neurology, 25*, 413–421.

Flowers, K. A. (1976). Visual "closed loop" and "open loop" characteristics of voluntary movement in patients with parkinsonism and intention tremor. *Brain, 99*, 269–310.

Geschwind, N. (1965). Disconnexion syndromes in animals and man. *Brain, 88*, 237–294, 585–644.

Geschwind, N. (1975). The apraxias: Neural mechanisms of disorders of learned movement. *American Scientist, 63*, 188–195.

Geschwind, N., & Kaplan, E. F. (1962). A human cerebral disconnection syndrome. *Neurology, 12*, 675–685.

Gonzalez-Rothi, L. J., & Heilman, K. M. (1985). Ideomotor apraxia: Gestural discrimination, comprehension and memory. In E. A. Roy (Ed.), *Neuropsychological studies of limb apraxia and related disorders* (pp. 65–74). Amsterdam: Elsevier.

Gonzalez-Rothi, L. J., Heilman, K. M., & Watson, R. T. (1985). Pantomine comprehension and ideomotor apraxia. *Journal of Neurology Neurosurgery and Psychiatry, 48*, 207–210.

Goodglass, H. J., & Kaplan, E. (1963). Disturbance of gesture and pantomine in aphasia. *Brain, 86*, 703–720.

Guiard, Y., Diaz, G., & Beaubaton, D. (1983). Left-hand advantage in right-handers for spatial constant error: Preliminary evidence in a unimanual ballistic aimed movement. *Neuropsychologia, 21 (1)*, 111–115.

Haaland, K. Y. (1984). The relationship of limb apraxia severity to motor and language deficits. *Brain and Cognition, 3*, 307–316.

Haaland, K. Y., & Delaney, H. D. (1981). Motor deficits after left or right hemisphere damage due to stroke or tumor. *Neuropsychologia, 19*, 17–27.

Haaland, K. Y., & Flaherty, D. (1984). The different types of limb apraxia errors made by patients with left vs. right hemisphere damage. *Brain and Cognition, 3*, 370–384.

Haaland, K. Y., & Harrington, D. L. (1988). Hemispheric asymmetry of open- and closed-loop movement. *Journal of Clinical and Experimental Neuropsychology, 10*, 23.

Haaland, K. Y., & Harrington, D. L. (1988). The role of the hemispheres in closed loop movements. *Brain and Cognition*, (in press).

Haaland, K. Y., Porch, B. E., & Delaney, H. D. (1980). Limb apraxia and motor performance. *Brain and Language, 9*, 315–323.

Haaland, K. Y., Yeo, R. A., & Koditawakku, P. W. (1986). Limb apraxia and intrahemispheric lesion locus. *Journal of Clinical and Experimental Neuropsychology, 9*, 46.

Haaland, K. Y., Harrington, D. L., & Yeo, R. (1987). The effects of task complexity on motor performance in left and right CVA patients. *Neuropsychologia, 25*, 783–794.

Heap, M., & Wyke, M. (1972). Learning of a unimanual skill by patients with brain lesions: An experimental study. *Cortex, 8*, 1–18.

Heilman, K. M. (1975). A tapping test in apraxia. *Cortex, 11*, 259–263.

Heilman, K. M. (1979). Apraxia. In K. M. Heilman & E. Valenstein (Eds.), *Clinical neuropsychology* (pp. 159–185). New York: Oxford University Press.

Heilman, K. M., & Gonzalez-Rothi, L. J. (1985). Apraxia. In K. M. Heilman & E. Valenstein (Eds.), *Clinical neuropsychology* (pp. 131–150). New York: Oxford University Press.

Heilman, K. M., Schwartz, H. D., and Geschwind, N. (1975). Defective motor learning in ideomotor apraxia. *Neurology, 25*, 1018–1020.

Heilman, K. M., Rothi, L. J., & Valenstein, E. (1982). Two forms of ideomotor apraxia. *Neurology, 32*, 342–346.

Heilman, K. M., Bowers, D., Coslett, H. D., Whelan, H., & Watson, R. T. (1985). Directional hypokinesia: Prolonged reaction time for leftward movements in patients with right hemisphere lesions and neglect. *Neurology, 35*, 855–859.

Heilman, K M., Watson, R. T., & Valenstein, E. (1985). Neglect and related disorders. In K. M. Heilman & E. Valenstein (Eds.), *Clinical neuropsychology* (pp. 243–293). New York: Oxford University Press.

Heilman, K. M., Mack, L., Gonzalez-Rothi, L. J., & Watson, R. T. (1987). Transitive movements in a deafferented man. *Cortex, 23*, 525 530

Helm-Estabrooks, N., Fitzpatrick, P. M., & Barresi, B. (1982). Visual action therapy for global aphasia. *Journal of Speech and Hearing Disorders, 47*, 385–389.

Howes, D., & Boller, F. (1975). Simple reaction time: Evidence for focal impairment from lesions of the right hemisphere. *Brain, 98*, 317–332.

Jason, G. W. (1983a). Hemispheric asymmetries in motor function. I. Left hemisphere specialization for memory but not performance. *Neuropsychologia, 21*, 35–45.

Jason, G. W. (1983b). Hemispheric asymmetries in motor function. II. Ordering does not contribute to left hemisphere specialization. *Neuropsychologia, 21*, 47–58.

Jason, G. W. (1985). Manual sequence learning after focal cortical lesions. *Neuropsychologia, 23*, 483–496.

Jason, G. W. (1986). Performance of manual copying tasks after focal cortical lesions. *Neuropsychologia, 23*, 483–496.

Jeannerod, M. (1986). Mechanisms of visuomotor coordination: A study in normal and brain-damaged subjects. *Neuropsychologia, 24*, 41–78.

Jurgens, U. (1984). The efferent and afferent connections of the supplementary motor area. *Brain Research, 300*, 63–81.

Kaplan, E. F. (1968).*Gestural representation of implement usage: An organismic development study*. Doctoral dissertation. Ann Arbor: University Microfilms International.

Keele, S. W. (1981). Behavioral analysis of movement. In V. Brooks (Ed.), *Handbook of physiology: Motor control* (pp. 1391–1414). Bethesda: American Physiological Society.

Kertesz, A. (1985). Apraxia and aphasia: Anatomical and clinical relationship. In E. A. Roy (Ed.), *Neuropsychological studies of apraxia and related disorders* (pp. 163–178). Amsterdam: Elsevier.

Kertesz, A., & Ferro, J. M. (1984). Lesion size and location in ideomotor apraxia. *Brain, 107*, 921–933.

Kimura, D. (1977). Acquisition of a motor skill after left-hemisphere damage. *Brain, 100*, 527–542.

Kimura, D. (1979). Neuromotor mechanisms in the evolution of human communication. In H. D. Stehlis & M. J. Raleigh (Eds.), *Neurobiology of social communication in primates* (pp. 197–219). New York: Academic Press.

Kimura, D. (1982). Left-hemisphere control of oral and brachial movements and their relation to communication. *Philosophical Transactions of the Royal Society of London, B298*, 135–149.

Kimura, D., & Archibald, Y. (1974). Motor functions of the left hemisphere. *Brain, 97*, 337–350.

Klapp, S. T., Greim, D. M., Mendicino, C. M., & Koenig, R. S. (1979). Anatomic and environmental dimensions of stimulus-response compatibility: Implication for theories of memory coding. *Acta Psychologia, 43*, 367–379.

Kolb, B., & Milner, B. (1981). Performance of complex arm and facial movements after focal brain lesions. *Neuropsychologia, 19*, 491–503.

Lehmberhl, G., Poech, K., & Willmes, K. (1983). Ideomotor apraxia and aphasia: An examination of types and manifestations of apraxic symptoms. *Neuropsychologia, 21*, 199–212.

Levin, H. (1979). The acalculias. In K. M. Heilman & E. Valenstein (Eds.), *Clinical neuropsychology*. New York: Oxford University Press.

Liepmann, H. (1913). Motor aphasia, anarthria and apraxia. *Transactions of the 17th International Congress of Medicine (London), XI (2)*, 97–106.

Liepmann, H., & Maas, O. (1907). Fall von linksseitiger agraphie und apraxie bei rechtsslitiger löhmung. Z. F. Psychologie. *Neurolo. 10*, 214–227.

Marshall, J. C., & Newcombe, F. (1984). Putative problems and pure progress in neuropsychological single-case studies. *Journal of Clinical Neuropsychology, 6*, 65–70.

Nachson, I., & Carmon, A. (1975). Hand preference in sequential and spatial discrimination tasks. *Cortex, 11*, 123–131.

Perecman, E. (1983). *Cognitive processing of the right hemisphere*. New York: Academic Press.

Poizner, H., Heilman, K. M., Gonzalez-Rothi, L., Verfaellie, M., & Mack, L. (1987). Apraxia: Three dimensional spatial trajectory analysis. *Journal of Clinical Experimental Neuropsychology, 9*, 46.

Rapcsak, S. Z., Gonzalez-Rothi, L. J., & Heilman, K. M. (1988). Apraxia in patient with atypical cerebral dominance. *Brain and Cognition, 6*, 450–463.

Robinson, R., Kubos, K., Starr, L., Rao, K., & Price, T. (1984). Mood disorders in stroke patients. *Brain, 107*, 81–93.

Roland, P. E., Larsen, B., Lassen, N. A., & Skinhoj, E. (1980). Supplementary motor area and other cortical areas in organization of voluntary movements in man. *Journal of Neurophysiology, 43*, 118–136.

Roy, E. (1981). Action sequencing and lateralized cerebral damage: Evidence for asymmetries in control. In J. Long & A. Baddely (Eds.), *Attention and performance IX*, Hillsdale, NJ: Lawrence Earlbaum.

Roy, E. A., & MacKenzie, C. L. (1978). Handedness effects on kinesthetic spatial location judgments. *Cortex, 14*, 250–258.

Semmes, J. (1967). Hemispheric specialization: A possible clue to mechanism. *Neuropsychologia, 6*, 11–26.

Todor, J. I., and Smiley, A. L. (1985). Performance differences between the hands: Implications for studying disruption of limb praxis. In E. A. Roy (Ed.), *Neuropsychological studies of apraxia and related disorders*. Amsterdam: Elsevier.

Vaughan, H. G., & Costa, L. D. (1962). Performance of patients with lateralized cerebral lesions. II. Sensory and motor tests. *J. Nervous and Mental Disease, 134*, 237–244.

Volpe, B. T., Sidtis, J. J., Holtzman, J. D., Wilson, D. H., & Gazzaniga, M. S. (1982). Cortical mechanisms involved in praxis: Observations following partial and complete section of the corpus callosum in man. *Neurology, 32*, 645–650.

Wallace, S. A., & Newell, K. M. (1983). Visual control of discrete aiming movements. *Quarterly Journal of Experimental Psychology, 35A*, 311–321.

Watson, R. T., Fleet, W. S., Gonzalez-Rothi, L., & Heilman, K. M. (1986). Apraxia and the supplementary motor area. *Archives of Neurology, 43*, 787–792.

Woodworth, R. S. (1899). The accuracy of voluntary movement. *Psychological Review Monographs* (Supplemental Issue 3).

Wyke, M. (1966). Postural arm drift associated with brain lesions in man. *Archives of Neurology, 15*, 329–334.

Wyke, M. (1967). Effects of brain lesions on the rapidity of arm movement. *Neurology, 17*, 1113–1120.

Wyke, M. (1968). The effects of brain lesions in the performance of an arm–hand precision task. *Neuropsycholgia, 6*, 125–134.

Wyke, M. (1971a). The effects of brain lesions on the performance of bilateral arm movements. Neuropsychologia, 9, 33–42.

Wyke, M. (1971b). The effects of brain lesions on the learning performance of a bimanual coordination task. *Cortex, 7*, 59–72.

Yeo, R. A., Melka, B. E., & Haaland, K. Y. (1988). Intrahemispheric differences in motor control: A reevaluation of the Semmes' hypothesis. *Journal of Clinical and Experimental Neuropsychology 10*, 23.

Zaidel, D., and Sperry, R. W. (1977). Some long term motor effects of cerebral commissurotomy in man. *Neuropsychologia, 15*, 193–204.

9

Structural Brain Abnormalities in the Major Psychoses

SARAH RAZ

INTRODUCTION

Since Kraepelin first described the symptoms associated with schizophrenia or "dementia prae-cox," numerous studies have attempted to elucidate its etiology. Explanatory efforts ranged from incorporating schizophrenia into a diversity of normal human experiences (Laing, 1960) to describing it as an organic brain disease (e.g., Crow, 1983, 1984). It is this latter approach that generated great interest in structural parameters of schizophrenic brains.

Four methods have been used to investigate morphological brain abnormalities in schizo-phrenia: postmortem investigations (PM), pneumoencephalography (PEG), computerized tomog-raphy (CT), and recently, magnetic resonance imaging (MRI). Postmortem studies have yielded inconsistent and nonspecific findings. Furthermore, the results of these investigations lacked clar-ity because it was impossible to establish postmortem whether a particular cerebral pathology had been an antecedent or a consequence of the schizophrenic illness and its treatment. Neverthe-less, Weinberger, Wagner, and Wyatt (1983), who reviewed recent PM studies, concluded that the results of these investigations were suggestive of a limbic focus of pathology unique to schizo-phrenia and absent in other psychiatric disorders.

Pneumoencephalography has been applied to the *in vivo* examination of brain morphology in schizophrenia. The majority of PEG studies reported ventricular dilation and widening of sulci and fissures in this group of patients (Haug, 1962). The interpretation of these findings has been questioned, though, because of methodological problems inherent in the technique. For example, as a result of the morbidity and mortality risks associated with this invasive procedure, most PEG investigations either lacked a control group or used inadequate controls. Another serious limitation of PEG is that the gas used as a contrast medium caused distention of the cerebral ventricles (Probst, 1973).

Both CT and MRI permit reliable noninvasive *in vivo* imaging of brain structures. These two radiological techniques are therefore more suitable than either PM or PEG to investigate morpho-logical differences between psychiatric patients and controls. The use of MRI for the study of brain pathology in schizophrenia is quite novel, but CT has been applied to research in this area for over a decade, since the pioneering investigation of Johnstone, Crow, Frith, Stevens, and Kreel (1976). It is the latter body of research that is the primary focus of this review.

SARAH RAZ • Department of Psychology, University of Texas at Austin, Austin, Texas 78712.

To date, based on analyses of CT data, five types of morphological brain abnormalities have been reported in schizophrenic patients. These abnormalities include:

1. Subcortical (central) pathology, inferred from enlargement of the cerebral ventricles.
2. Cortical pathology, the so-called "cortical atrophy," inferred from dilation of cortical fissures and sulci.
3. Cerebellar pathology, the so-called "cerebellar atrophy," inferred from enlargement of cerebrospinal fluid (CSF)-filled spaces neighboring the cerebellum.
4. Reversed cerebral asymmetry: reversal of the anatomic asymmetries usually observed in the normal human brain.
5. Changes in regional brain density, inferred from regional increase or decrease in x-ray attenuation values generated by CT.

Though presence of cerebellar pathology, reversed cerebral asymmetry, and changes in regional brain density have been documented in several CT investigations in schizophrenia (see Weinberger, Torrey, & Wyatt, 1979; Golden, Graber, et al., 1980; Lippmann et al., 1982; Luchins, Weinberger, & Wyatt, 1982; Dewan et al., 1983; Kanaba, Shima, Tsukumo, Masuda, & Asai, 1984), numerous failures at replication have also been reported (Coffman, Mefferd, Golden, Bloch, & Graber, 1981; Andreasen, Dennert, Olsen, & Damasio, 1982; Jernigan, Zatz, Moses, & Cardellino, 1982; Weinberger, DeLisi, Perman, Targum, & Wyatt, 1982; Luchins & Meltzer, 1983b; Coffman & Nasrallah, 1984; Boronow et al., 1985; Yates, Jacoby, & Andreasen, 1987). Ventriculomegaly in particular, and to a lesser extent cortical atrophy, however, are relatively established pathological phenomena in schizophrenia. It has been proposed that enlargement of the cerebral ventricles be regarded as the most replicable biological feature of the disorder (Farmer, Jackson, McGuffin, & Storey, 1987). In a quantitative review of the English language literature (Raz, Raz, & Bigler, 1988), the average effect obtained from 37 studies of ventricular size comparing schizophrenics with nonpsychotic controls was rather large: the mean difference between schizophrenics and controls approached 0.8 of the standard deviation unit. This abnormality has been found to be associated neither with sociodemographic variables such as sex (Weinberger, Torrey, Neophytides, & Wyatt, 1979; Bishop et al., 1983), age (Moriguchi, 1981; Jeste et al., 1982; Bishop et al., 1982), or race (Jernigan, Zatz, Moses, & Berger, 1982; Jeste, Kleinman, Potkin, Luchins, & Weinberger, 1982; Pearlson, Garbacz, Breakey Ahn, & DePaulo, 1984) nor with psychiatric treatments such as ECT or chemotherapy (Weinberger, Torrey, Neophytides, et al., 1979; Weinberger et al., 1982).

Three broad categories of pathological processes have been proposed to account for CT findings: (1) focal changes in either periventricular structures, possibly the limbic system (Weinberger, 1984), the diencephalon (Pandurangi et al., 1984), or the basal ganglia (Nyback, Wiesel, Berggren, & Hindmarsh, 1982), or in distal structures such as the prefrontal cortex that are connected to periventricular regions via anatomic projections (Shelton, Weinberger, Doran, & Pickar, 1985); (2) generalized degenerative process resembling Alzheimer's dementia (Tachiki, Kurtz, Kling, & Hullet, 1984); and (3) CSF flow disturbance (Reveley & Reveley, 1983). Whatever the cause, these abnormalities are not unique to schizophrenia, for they have been observed in other patient populations.

Ventriculomegaly has been documented in patients with systemic lupus erythematosus (Bilaniuk, Patel, & Zimmerman, 1977; Gonzales-Scarano et al., 1979) and in patients with closed-head injury (Levin, Meyers, Grossman, & Sarwar, 1981; Cullum & Bigler, 1986). Both cortical atrophy and ventricular enlargement have been observed in normal subjects in the seventh or eight decade (Jacobs, Kinkel, Painter, Murawski, & Heffner, 1978) and in patients with anorexia nervosa (Datlof, Coleman, Forbes, & Kreipe, 1986), Parkinson's disease (Becker, Schneider, Hacker, & Fischer, 1979), Cushing's syndrome (Heinz, Martinez, & Haenggeli, 1977), severe migraine headaches (Hungerford, du Boulay, & Zilkah, 1976), senile dementia (Fox, Topel, & Huckman, 1975), and protein malnutrition (El-Tatawy, Badrawi, & El-Bishlawy, 1983). Patients receiving exogenous steroids (e.g., ACTH) were also reported to develop ventriculomegaly and widened cortical sulci and fissures (Bentson, Reza, Winter, & Wilson, 1978, Okuno, Ito, Konishi, Yoshioka, & Nakano, 1980).

Widening of CSF spaces neighboring the cerebellum as well as enlargement of the ventricles and widening of cortical sulci have been reported in alcoholics (Carlen, Wortzman, Holgate, Wilkinson, & Rankin, 1978; Wilkinson, 1982). Reversed cerebral asymmetry has been reported in autistic (Hier, LeMay, & Rosenberger, 1978, 1979) and dyslexic children (Hier, LeMay, & Rosenberger, 1978). Finally, decreased brain density has also been observed in patients with dementia (Naeser, Gebhart, & Levine, 1980).

Most important for evaluating the significance of CT findings in schizophrenia were reports of structural brain abnormalities in other adult psychiatric disorders. Patients with unipolar depression (Scott, Golden, Ruedrich, & Bishop, 1983), bipolar depression (Pearlson & Veroff, 1981), and obsessive–compulsive disorder (Behar *et al.*, 1984) were all reported to have dilated cerebral ventricles. The latter finding has not been uniformly observed (see Insel, Donnelly, Lalakea, Alterman, & Murphy, 1983). Nonetheless, CT abnormalities in schizophrenia appear to be nonspecific phenomena that may characterize all psychotic disorders.

It has been suggested that structural brain abnormalities may have clinical or biological significance within the specific context of schizophrenia or, perhaps, within the unique context of each of the major psychoses. According to this approach (Crow, 1980), schizophrenics with structural brain abnormalities may constitute a distinct subgroup, differing from schizophrenics without these abnormalities in etiology, symptomatology, and course of illness. Schizophrenics with type I syndrome are characterized by an acute remitting illness with positive symptoms, favorable response to neuroleptics, and absence of intellectual deterioration. Schizophrenics with type II syndrome, on the other hand, are characterized by a progressive chronic course with negative symptoms (the "defect state"), intellectual deterioration, and poor response to neuroleptics. Whereas type I syndrome, according to Crow, is likely to be associated with changes in dopamine receptors, type II syndrome is probably related to the structural abnormalities observed on CT scans and presumed to reflect cell loss. There has been no similar model to direct CT studies of affective disorders, since at least initially these patients have been treated primarily as a group of psychiatric controls.

In the following sections, CT research on structural brain abnormalities in the major psychoses is reviewed. The goal is to examine the clinical and biological significance of CT findings in the psychoses.

CLINICOPATHOLOGICAL AND BIOCHEMICAL CHARACTERISTICS AND CT ABNORMALITIES IN THE PSYCHOSES

The investigation of relationships between clinicopathological or biochemical characteristics and morphological brain abnormalities in the major psychoses has been largely influenced by Crow's model, which proved to have considerable heuristic value, providing clearly stated and testable hypotheses. Another two-syndrome model has been initially proposed by Campbell, Hays, Russel, and Zacks (1979), later expanded by Reveley, Reveley, and Murray (1984) and Murray, Lewis, and Reveley (1985). These investigators suggested that CT abnormalities were physiological attributes of a schizophrenic subgroup with a nongenetic etiology. Enlarged CSF spaces are, therefore, the putative sequelae of environmental insult to the brain, which has presumably resulted in psychiatric illness. According to both of these models, CT abnormalities are valuable markers, as they allow subtyping of the "schizophrenias" into two distinct syndromes.

It should be noted that the two-syndrome concept of schizophrenia has been challenged since its inception. Mackay (1980), for instance, proposed a model of unitary schizophrenic illness in which reduced dopaminergic activity, behaviorally expressed in negative symptoms, was the primary underlying process. Episodes of dopamine overactivity reflected in periodic exacerbation of positive psychotic symptoms, were presumed to be superimposed on this primary disorder. This chapter provides a review of major findings on the clinicopathological correlates of CT abnormalities in the major psychoses in an attempt to establish whether the results provide sufficient support for a two-syndrome concept or whether a unitary illness view should be adopted.

Psychotic Symptomatology and CT Abnormalities

The relationships between symptomatology and CT abnormalities in the psychoses have been investigated from two perspectives. The first involves the study of positive and negative symptom clusters in psychotic patients whose CT scans were classified as normal or abnormal. The second involves comparing different diagnostic subtypes within each of the major psychoses on CT parameters.

Positive and Negative Symptoms in the Psychoses

The study of symptomatology and morphological abnormalities in schizophrenia has been largely inspired by Crow's two-syndrome hypothesis (Crow, 1980). Type I schizophrenia, the acute remitting syndrome, was presumed to be characterized by a preponderance of positive or excess symptoms (e.g., delusions, hallucinations, formal thought disorder). Type II syndrome, the "defect state," on the other hand, was thought to be characterized by a predominance of negative symptoms (e.g., psychomotor retardation, affective flattening, amotivation, poverty of speech).

As Table 1 shows, the hypothesized inverse relationship between presence of positive symptoms and CT abnormalities was demonstrated in only 3 of 16 studies (19%). Similarly, an association between CT abnormalities and presence of negative symptoms was documented in only 4 of 16 studies (25%), in 2 of which only nonsignificant trends were reported. In many of the studies, patients were on neuroleptics during assessment of symptoms. This could have masked expected relationships, because antipsychotic medication has been reported to reduce both positive and negative symptoms (Goldberg, 1985; Meltzer, Sommers, & Luchins, 1986). Inspection of Table 1 shows, however, that the findings of investigations in which medication-free patients were assessed did not provide more support to Crow's hypothesis than the findings of studies in which patients were evaluated while on neuroleptics.

With the exception of studies in which the Scale for Assessment of Negative Symptoms (SANS) was used, in most investigations scales not designed specifically for assessment of positive and negative symptoms were utilized, with groups of items being assigned to either class on the basis of face validity. Nevertheless, inspection of Table 1 did not reveal a relationship between the symptom scale chosen and the findings. Thus, the bulk of the evidence is not supportive of Crow's two-syndrome hypothesis, at least as regards the link between symptomatological features and presence of changes in brain morphology in schizophrenia subtypes. It should be mentioned that at present there is little agreement about the operational definition of negative and positive symptom clusters. There are difficulties in quantification of pathological traits that may vary during the course of illness, and the validity of these constructs or the instruments purporting to measure them has not been established (for a comprehensive review of conceptual and methodological difficulties in assessment of negative symptoms see Sommers, 1985). Thus, it can not be determined whether the nonsignificant results reported in most studies are attributable to inadequate measurement instruments or to an inadequate theoretical model.

Only three groups of investigators have studied the relationship between symptomatology and CT abnormalities in patients with affective disorders. Luchins, Lewine, and Meltzer (1984) and Nasrallah, McCalley-Whitters, and Pfohl (1984) were unable to demonstrate a link between ventricular size and positive symptomatology in patients with bipolar or unipolar disorder. Pearl-

TABLE 1. Characteristics and Results of Studies Investigating the Relationships between Brain Morphology and Symptomatology in Schizophrenia

Author	Year	N	CT measure[a]	Med.[b]	Instrument[c]	Symptoms Positive	Symptoms Negative
Andreasen, Olsen, et al.	1982	32	VBR	+	SANS, SADS	$P<0.02$	$P<0.15$
Bishop et al.	1983	46	VBR	+	BPRS	$P<0.05$	n. s
Boronow et al.	1985	30	VBR, CA	−	BPRS	n. s.	n. s.
DeLisi et al.	1986	26	VBR	+	Krawiecka scale	n. s.	n. s.
Johnstone et al.	1976	13	VBR	+	Krawiecka scale	n. s.	n. s.
Kemali et al.	1985	33	VBR	+	SANS	n. s.	$P<0.05$
Losonczy, Song, Mohs, Small, et al.	1986	28	VBR	−	SANS	n. s.	n. s.
Luchins, Lewine et al.	1984	45	VBR	−	SADS-C	$P<0.06$	n. s.
Naber et al.	1985	36	VBR	−	BPRS	n. s.	n. s.
Nasrallah, Kuperman, Hamara, et al.	1983	55	VBR	+	Clinical imprs.	n. s.	n. s.
Nyman, Nyback, Wiesel, Oxentierna, & Schalling	1986	23	VIX, III	−	CPRS	n. s.	n. s.
Owens et al.	1985	110	VBR	+	Krawiecka scale	n. s.	n. s.
Pandurangi et al.	1986	23	VBR	+	SANS	n. s.	n. s.
Pearlson et al.	1985	19	VBR	+	Standard ratings	n. s.	$P<0.1$
van Kammen et al.	1986	53	CA	−	BPRS, Schneider	n. s.	n. s.
Williams et al.	1985	40	VBR	+	PSE	n. s.	$P<0.02$

[a]VBR, ventricular-brain ratio (Synek & Reuben, 1976); CA, cortical atrophy ratings; VIX, a linear measure of the lateral ventricles; III, third ventricular width.
[b]Medication status: plus and minus signs pertain to on and off drugs, respectively.
[c]Instrument: SANS, Scale for the Assessment of Negative Symptoms (Andreasen et al., 1982); SADS, Schedule for Affective Disorders and Schizophrenia (Endicott & Spitzer, 1978); BPRS, Brief Psychiatric Rating Scale (Overall & Gorham, 1962); Krawiecka Scale (Krawiecka, Goldberg, & Vaughan, 1977); SADS-C (Lewine, Fogg, & Meltzer, 1983); CPRS, Comprehensive Psychological Rating Scale (Asberg, Montgomery, Perris, Schalling, & Sedval, 1978); PSE, Present State Examination (Wing, Cooper, & Sartorius, 1974).

son, Garbacz, Moberg, Ahn, and DePaulo (1985) found a nonsignificant trend, suggestive of a relationship between ventriculomegaly and negative symptoms in patients with bipolar illness.

The Study of CT Abnormalities and Diagnostic Subtypes in the Psychoses

Relatively little systematic study of diagnostic subtypes has been reported, perhaps because the alternative phenomenological classification, the positive/negative symptoms dichotomy straightforwardly linked to Crow's two-syndrome hypothesis, has had more appeal to investigators in the field.

Weinberger et al. (1982) found enlarged ventricles and cortical atrophy in 35 patients with first-episode schizophreniform psychosis. In a follow-up study of this sample, DeLisi et al. (1983) found that patients with enlarged ventricles had poorer outcome than those whose ventricular size was classified as normal. Outcome was measured on the Strauss/Carpenter Outcome Scale (Strauss & Carpenter, 1974) and the Global Assessment Scale (Endicott, Spitzer, Fleiss, & Cohen, 1976).

The results of three CT studies of patients with schizoaffective disorder (Weinberger, Torrey, Neophytides, *et al.,* 1979; Kling, Kurtz, Tachiki, & Orzeck, 1983; Williams, Reveley, Kolakowska, Ardern, & Mandelbrote, 1985) indicated that ventriculomegaly was less pronounced in these patients than in chronic schizophrenics. Rieder, Mann, Weinberger, Van Kammen, and Post (1983) could not demonstrate differences between schizoaffective and schizophrenic patients on any CT measures, but the lack of nonpsychotic controls in their study did not allow interpretation of these findings. It was unclear whether both psychotic groups had enlarged ventricles or whether the schizophrenic group was atypical, with normal ventricular size. The results of comparisons between paranoid and nonparanoid schizophrenics on CT measures are quite inconsistent: Kling *et al.* (1983) and Tachiki *et al.* (1984) reported that paranoid schizophrenics had significantly smaller lateral ventricles in comparison to undifferentiated or residual schizophrenics. Golden *et al.* (1980b) and Boronow *et al.* (1985) found no differences between paranoid and undifferentiated schizophrenics, and Jeste *et al.* (1982) found no differences between patients with and without "paranoid features" on the ventricle–brain ratio (VBR). Frangos and Athanassenas (1982) reported that paranoid schizophrenics had greater VBRs than hebephrenics, whereas Nasrallah, Jacoby, McCalley-Whitters, and Kuperman (1982), who could not demonstrate differences in VBR between these two subtypes, found the former group to have larger ventricles than those of undifferentiated schizophrenics.

Perhaps the conflicting findings in this area of investigation result in part from differences in interrater agreement between laboratories in establishing diagnoses for the paranoid and nonparanoid subtypes or, alternatively, from differences in severity of illness between the paranoid patients used in various samples. In fact, some of the studies failed to document the clinical features of their patients by subtype, whereas others (e.g., Frangos and Athanassenas, 1982; Nasrallah *et al.,* 1982) reported that their paranoid and nonparanoid patients had similar clinical characteristics when compared on such illness severity indices as age of onset, cumulative length of hospitalization, or number of hospitalizations. Thus, it is possible that studies in which paranoid schizophrenics were found to have similar or greater VBRs than those of nonparanoid schizophrenics used groups of paranoid schizophrenics with a more virulent form of disorder than usually observed in this group of patients. This could have masked the expected effect of increase in CT abnormalities in the nonparanoid subtype—a group traditionally considered to have a poor prognosis.

Three groups of investigators attempted to explore the relationships between diagnostic subtype and CT abnormalities in patients with affective disorders. Targum, Rosen, DeLisi, Weinberger, and Citrin (1983) found that patients with delusional depression were significantly more likely to be classified as having large ventricles than were patients with nondelusional depression. Luchins and Meltzer (1983a) similarly discovered that patients with psychotic depression were more likely to be classified as having dilated ventricles than were those with nonpsychotic depression.

Dolan, Calloway, and Hahn (1985), who compared patients with unipolar illness to those with bipolar illness, could not detect differences in ventricular size. Putative differences could have been masked in their study by a stepwise adjustment of the effect of subtype to the effect of sex. The latter variable explained a significant amount of the variance in the study by Dolan *et al.* (1985), although it is usually found to be unrelated to ventricular volume corrected for cranium size in normal individuals (see Pfefferbaum, Zatz, & Jernigan, 1986).

To summarize, the positive/negative-symptoms dichotomy has thus far not yielded meaningful relationships with structural brain abnormalities in the psychoses. The results of studies investigating the association between subtypes of schizophrenia and CT abnormalities are mixed. The study of patients with schizophreniform psychosis and schizoaffective disorder tends to support the view that the presence of CT abnormalities is linked to severity of illness or quality of prognosis. However, studies in which paranoid schizophrenics, a group with relatively favorable outcome (Kendler, Gruenberg, & Tsuang, 1984), were compared to nonparanoid schizophrenics have yielded conflicting results. Interestingly, in two of three CT studies of subtypes in patients with affective disorders, the CT scans of those with a more virulent form of the disorder were shown to be more likely to be classified as abnormal than were the CT scans of those with less severe illness.

Genetic and Environmental Factors and CT Abnormalities in the Psychoses

The first attempt to link morphological brain abnormalities to etiological factors in schizophrenia was a study by Campbell et al. (1979). They proposed two separate sets of etiological factors: one environmental (CNS trauma), the other genetic. They hypothesized that environmental factors would be associated with abnormal CT scans. Reveley and associates further suggested that "schizophrenics with manifest genetic predisposition (such as those with other ill family members) would not show ventricular enlargement" (Reveley, Reveley, & Murray, 1984, p. 89). If this were true, then ventricular enlargement (or, by implication, adverse environmental factors) and genetic predisposition would be dependent events. These hypotheses have guided the efforts in several recent investigations of the relation contribution of genetic and environmental factors to CT abnormalities in schizophrenia.

Genetic Factors in CT Studies of Schizophrenia and Affective Disorders

Three methods have been used to investigate the relationships between genetic factors and CT abnormalities in schizophrenia. These methods are described below.

The first method—comparing CT abnormalities among schizophrenics, their unaffected first-degree relatives, and normal controls—has been used in three studies in which schizophrenics were found to have significantly larger ventricles than their unaffected siblings (Weinberger, DeLisi, Neophytides, & Wyatt, 1981; DeLisi et al., 1986) or discordant co-twins (Reveley, Reveley, Clifford, & Murray, 1982). The schizophrenics only (DeLisi et al., 1986), or both schizophrenics and their relatives (Reveley et al., 1982), were also reported to have had perinatal complications. Reveley and her coinvestigators suggested, therefore, that "some common environmental factor, possibly perinatal damage, may have led to the increase in ventricular size in the schizophrenia-discordant pairs, with schizophrenia developing in the more severely affected twin" (Reveley et al., 1982, p. 540).

Interestingly, in two studies (Weinberger et al., 1981; Reveley et al., 1982) ventricular size in the nonschizophrenic relatives was found to be greater than that of healthy volunteers. This may be viewed as suggesting that mild ventriculomegaly may be a marker of vulnerability to the disorder in family members of schizophrenics. However, the data reported by Reveley et al. (1982) suggest that the mild ventriculomegaly observed in the sample of co-twins could be attributed to perinatal complications rather than to genetic factors. This view may be supported to some extent by a study (DeLisi et al., 1986) in which previously reported findings of a difference in ventricular size between nonschizophrenic siblings of schizophrenic probands and controls could not be replicated. None of the nonschizophrenic siblings in that sample had a history of obstetric complications.

The second method, examining CT abnormalities in affected and unaffected relatives of schizophrenic probands, has been applied in two studies (Schulsinger et al., 1984; DeLisi et al., 1986). Schulsinger and his colleagues, who studied the CT scans of schizophrenic, schizotypal, and mentally healthy offspring of schizophrenic mothers, found the schizophrenics to have larger ventricular size than the other offspring. A significant negative correlation found between ventricular size and obstetric variables (e.g., weight at birth, $r(25) = -0.60$, $P < 0.001$) led the authors

TABLE 2. Proportion of Variance in VBR Accounted for by
Genetic, Environmental, and Nonspecific Factors in 34
Members of 11 Families with Presumed Predisposition
to Schizophrenia[a]

	Frontal horns	Body of lateral ventricles
Age	0.10 ($P = 0.004$)	0.07 ($P = 0.05$)
Birth complications	0.34 ($P = 0.0001$)	0.03 ($P = 0.19$)
Head injuries	0.13 ($P = 0.0009$)	0.17 ($P = 0.003$)
Family	0.23 ($P = 0.031$)	0.45 ($P = 0.019$)
Diagnosis	0.04 ($P = 0.052$)	0.00 ($P = 0.87$)

[a]Calculated from results of analysis of variance provided by DeLisi et al. (1986).
[b]VBRs at two cuts were used as separate dependent measures in statistical analysis.

to hypothesize that "neurological insult may decompensate schizotypal individuals toward florid schizophrenia" (Schulsinger et al., 1984, p. 602).

DeLisi et al. (1986) investigated genetic and environmental factors putatively associated with CT abnormalities in 34 members of 11 families with presumed genetic vulnerability to schizophrenia: at least two members in each family were affected by the illness. The ventricles of the probands and of their ill and well siblings were measured at two cuts: through the frontal horns and through the bodies of the lateral ventricles. The proportion of variance accounted for by each variable, calculated from data provided in the article, is shown in Table 2. It should be noted that variables were entered sequentially, thereby decreasing the likelihood of the latter variables (e.g., "diagnosis") accounting for a large portion of the variance in ventricular size.

DeLisi and associates concluded from these findings that their data, in contrast to other reports, "are consistent with an association of increased ventricular size seen in schizophrenic patients with inherited vulnerability toward schizophrenia" (DeLisi et al., 1986, p. 153). This conclusion, was apparently based on evidence of a significant familial component, with family membership accounting for a large share of the variance in VBR as shown above. It should be noted, however, that this familial component stands for variance accounted for by membership in any one of 11 specific families with presumed predisposition to schizophrenia. Thus, it may be inferred from the study that ventricular size in members of "schizophrenic" families is, in a large part, genetically determined. This is consistent with the finding that ventricular volume, like other physical attributes, has a genetic basis in normal individuals (Reveley, Reveley, Chitkara, & Clifford, 1984).

To support the hypothesis that ventricular size is partly determined by diathesis to the disorder, it should be demonstrated that membership in families with presumed genetic predisposition to schizophrenia (i.e., two or more affected members including the index case), versus membership in families without known predisposition (only the index case is diagnosed as schizophrenic), provides a unique and independent contribution to ventricular size in schizophrenics and their siblings. The abovementioned study, however, was an investigation within genetically predisposed families only. Nonetheless, the results of this study did replicate a previously reported association between frequency of birth complications and ventriculomegaly in schizophrenics. Head injury was also found to be linked to ventricular size. The former factor, however, did not yield a stable effect across CT cuts.

The third method, applied in CT investigations of genetic factors, involved comparing structural brain parameters between patients with familial and sporadic schizophrenia (without known genetic predisposition) or, alternatively, comparing the prevalence of familial schizophrenia between patients whose CT scans were classified as normal and those whose CT scans were classified as abnormal.

TABLE 3. Characteristics and Results of CT studies of Familial and Sporadic Schizophrenia

Author	Year	N	Measure[a]	Method[b]	Def.[c]	Relation[d]	Result
Boronow et al.	1985	30	LV, CA, C	—	Broad	—	n. s.
Campbell et al.	1979	35	LV, CA	3, 4	Broad	2nd	n. s.
Farmer, McGuffin, Jackson, & Storey	1985	39	LV	—	Narrow	1st	n. s.
Kemali et al.	1985	33	LV	—	Narrow	—	n. s.
Nasrallah, Kuperman, Hamara, et al.	1983	55	LV	—	Narrow	1st	$P < 0.05$[a]
Nyback, Berggren, Nyman, Sedvall, & Wiesel	1984	46	LV	2, 3	Broad	2nd	n. s.
Owens et al.	1985	112	LV	—	Narrow	—	$P < 0.002$[f]
Oxentierna et al.	1984	20	LV, CA, C	1, 2	Broad	2nd	$P < 0.05$
Pandurangi et al.	1986	20	LV	3, 4, 5	Narrow	1st	n. s.
Pearlson et al.	1985	19	LV	3, 4, 5	Broad	1st	n. s.
Reveley, Reveley, & Murray	1984	21	TV	3, 4	Broad	2nd	$P = 0.016$
Romani, Zerbi, Mariotti, Calliero, & Cosi	1986	20	LV, CA	—	Broad	1st	$P < 0.05$
Tanaka, Hazama, Kawahara, & Kobayashi	1981	40	LV, CA	—	—	—	n. s.
Turner, Toone, & Brett-Jones	1986	30	LV	5	Narrow	2nd	$P < 0.05$
Weinberger et al.	1981	51	LV, CA, C	3	Narrow	1st	n. s.
Williams et al.	1985	40	LV	3, 4, 5	Broad	1st	n. s.

[a] CT measure: LV, lateral ventricular size; TV, total ventricular volume; CA, cortical atrophy; C, cerebellar atrophy.
[b] Method of ascertainment: 1, hospital records; 2, parish registers; 3, family interview; 4, interview with patient; 5, chart review.
[c] Definition of disorder in relatives: narrow, schizophrenia; broad, psychosis or psychiatric disorder.
[d] Degree of relation: the most remote relation investigated.
[e] Effect in opposite direction (more CT abnormalities in familial schizophrenia).
[f] Significant curvilinear relationship (sporadic schizophrenia associated with either large or small ventricles, familial schizophrenia associated with average ventricular size).

As Table 3 shows, only 4 of 16 studies (25%) reported results suggestive of an association between CT abnormalities and sporadic schizophrenia or, conversely, between lack of pathological findings and familial schizophrenia. The preponderance of studies (75%) did not support this hypothesis. None of the study characteristics listed above readily appeared to bear a relationship to obtained results.

There were few CT studies of patients with affective disorders in which the relationships between CSF volume and genetic factors were investigated (Nasrallah et al., 1984; Pearlson et al., 1985; Dolan et al., 1985). None of these studies reported significant findings.

Environmental Factors and CT Abnormalities

Investigators studying the link between adverse environmental influences and CT abnormalities in schizophrenia focused primarily on events occurring during gestation, labor and birth, and early development that might have been associated with or led to subsequent CNS trauma. Perhaps the greatest difficulty in this area of retrospective research has been the unavailability of

TABLE 4. Characteristics and Results of CT Studies Investigating the Relationships between Brain Morphology and Early Physical Trauma in Schizophrenia

Author	Year	N^a	Data source[b]	Instruments[c]	Covered period	Results
Boronow et al.	1985	30	—	—	Labor and birth	n. s.
DeLisi et al.	1986	34	2	Structured questionnaire	Prenatal– postnatal	$P<0.001$ $P<0.19^d$
Kemali et al.	1985	33	—	—	Perinatal	n. s.
Oxentierna et al.	1984	30	1	—	Labor and birth	n. s.
Owens et al.	1985	110	—	—	Birth	n. s.
Pearlson et al.	1985	19	1, 2, 4	Rating scale	Prenatal– early dev.	$P<0.10$
Reveley, Reveley, & Murray	1984	21	2, 4	Structured interview	Labor and birth	n. s.
Schulsinger et al.	1984	27	—	Rating scale	Prenatal– perinatal	$P<0.001$
Turner et al.	1986	30	1, 2, 3, 4	Rating scale	Prenatal– early dev.	$P<0.01^e$
Williams et al.	1985	40	1, 2, 3, 4	—	Perinatal– early dev.	n. s.

[a]N, number of schizophrenics, with the exception of Schulsinger et al. (1984) and DeLisi et al. (1986), who also included nonschizophrenic family members.
[b]Data source: 1, patient's records; 2, first-degree relatives; 3, therapist; 4, patient.
[c]Instruments used to collect data pertaining to early physical trauma.
[d]The first result pertains to a cut through the frontal horns, and the second to a cut through the body of the lateral ventricles.
[e]This result pertains only to the 22 males included in the sample. Results were nonsignificant for the small subgroup of 8 females.

birth records or medical records from early childhood. The reliability of informants and the efforts on the part of the investigator to extract information from his sources might have therefore played a major role here. Another difficulty was related to the choice of variables putatively associated with CNS trauma. As noted by Gottesman and Shields (1982), there are multiple factors that could be selected for examination. Table 4 shows that studies also varied in the choice of data sources for verifying information pertaining to prenatal, perinatal, or early development.

Interestingly, associations between putative early CNS trauma and ventriculomegaly were reported in four of five studies relying on specifically designed questionnaires and rating scales, whereas five studies in which the use of such instruments was not reported failed to document these relationships. The probability of this occurring by chance is only $P=0.024$ (Fisher's exact probability test). Perhaps the use of rating scales or questionnaires in studies reporting positive findings is indicative of the investigators' success in eliciting specific data pertaining to potential CNS trauma from informants.

To summarize, putative relationships between genetic predisposition to psychosis and brain morphology have been revealed neither in CT investigations of schizophrenia nor in investigations of patients with affective disorders. However, environmental factors such as obstetric complications or cerebral insult have been implicated in several studies of schizophrenia as factors contributing to ventriculomegaly. Though these results are equivocal, it has been noted that investigators reporting a relationship between early trauma and ventriculomegaly have used instruments designed specifically to collect, summarize, or quantify the relevant data and were, perhaps, more likely to obtain detailed information pertaining to medical history.

Cognitive Factors and CT Abnormalities in the Psychoses

In most studies in this area of investigation, psychiatric patients were divided into two groups, those with and without CT abnormalities, whose performance on tests presumed to measure cognitive impairment was compared. In other studies, correlation coefficients were computed between measures of ventricular size and measures of cognitive impairment, again within the psychiatric group. Tables 5, 6, and 7 summarize characteristics and results of studies in which the relationships between cognitive factors and structural brain abnormalities in schizophrenia were examined. Studies were listed by three broadly defined measures used to assess cognitive impairment: mental status examination, intellectual assessment, and neuropsychological tests.

As the tables show, neuropsychological performance appears to have stronger association with CT abnormalities than do the other two forms of cognitive assessment: five of six groups of investigators reported significant results, whereas in a single study with nonsignificant findings (Carr & Wedding, 1984) the population studied was characterized by a less severe illness and included only outpatients with cumulative hospitalization of less than 12 months. Intellectual performance does not appear to be linked to CT abnormalities: in one of eight studies significant findings for all WAIS scales were reported, in another study only the Performance Scale was found to be related to sulcal widening, whereas in six studies (75%) no relationships were demonstrated between intellectual performance and enlargement of CSF spaces. Interestingly, in three studies investigating both neuropsychological and intellectual performance, significant findings were documented for overall neuropsychological performance but not for overall intellectual performance.

Two groups of investigators studied the relationships between cognitive factors and morphological brain abnormalities in patients with affective disorders. Nasrallah et al. (1984), who used the Mini-Mental State Examination, found no significant differences in performance between manic patients whose ventricles were classified as enlarged and those whose ventricles were classified as normal. Kellner, Rubinow, and Post (1986), who administered the Halstead–Reitan Category Test to a group of depressed medication-free patients, found that the VBR did not predict Category Test errors following adjustment for age.

In summary, there is evidence that schizophrenic patients with CT abnormalities obtain lower

TABLE 5. Characteristics and Results of CT Studies Investigating the Relationships between Mental Status and Brain Morphology in Schizophrenia

Author	Year	N	Course	Status[a]	Test[b]	Measure[c]	Meds.[d]	Results
Andreasen, Olsen, et al.	1982	32	Chronic	Out	MMSE	VBR	+	$P<0.05$
Johnstone et al.	1976	13	Chronic	In	W&H	VA	+	$P<0.01$
Kling et al.	1983	26	Chronic	In	MSEN	VBR	+	n. s.
Nasrallah, Kuperman, Jacoby, McCalley-Whitters, & Hamara	1983	55	Chronic	In	MMSE	SR	+	$P<0.05$
Nasrallah, Kuperman, Hamara, et al.	1983	55	Chronic	In	MMSE	VBR	+	n. s.
Owens et al.	1985	110	Chronic	In	W&H	VBR	+	n. s.

[a]Status: In, inpatients; Out, outpatients.
[b]Test: MMSE, Mini-mental State Exam (Folstein & McHugh, 1975); W & H, Withers & Hinton Battery (Withers & Hinton, 1971); MSEN, Mental Status Examination in Neurology (Strub & Black, 1981).
[c]CT measure; VBR, ventricular brain ratio; VA, ventricular area; SR, sulcal ratings.
[d]Meds: +, on medication.

TABLE 6. Characteristics and Results of CT Studies Investigating the Relationships between Intellectual Performance and Brain Morphology in Schizophrenia

Author	Year	N	Course	Status[a]	Test	Measure[b]	Meds.[c]	Results
Boronow et al.	1985	16	Chronic	In	WAIS	VBR, SR	*	n. s.
Donnelly, Weinberger, Waldman, & Wyatt	1980	15	Chronic	In	WAIS	VBR, SW	+	$P<0.004$
Kemali et al.	1985	33	Chronic	In/out	WAIS	VBR	+	n. s.
Nyback et al.	1984	20	Acute	In	WAIS	VIX, III	−	n. s.
Pandurangi et al.	1986	23	Chronic	—	WAIS	VBR, HD	+	n. s.
Rieder, Donnelly, Herdt, & Waldman	1979	8	Chronic	Out	WAIS	SA	+	n. s.[d]
Weinberger, Neophytides, et al.	1979	14	Chronic	In	WAIS	VBR	+	n. s.
Williams et al.	1985	40	Chronic Acute	In/out	Ad hoc battery	VBR	+	n. s.

[a]Status: in, inpatients; out, outpatients.
[b]CT measure: VBR, ventricular brain ratio; SR, sulcal ratings; C, cerebellar atrophy; SW, sulcal width; VIX, linear ventricular index; III, third ventricle width; HD, hyperdensity measures; SA, sulcal area.
[c]Meds: +, on medication; −, off medication; *, unspecified.
[d]Not significant for full-scale IQ and for verbal IQ. Performance IQ, $P=0.052$.

scores on measures of neuropsychological performance than schizophrenics without CT abnormalities. Measures of intellectual performance have generally not yielded similar findings, perhaps because they are less sensitive to the sequelae of brain damage than neuropsychological batteries. At present, there is no evidence for an association between cognitive impairment and dilation of CSF-filled cavities in patients with affective disorders.

Drug Response and CT Abnormalities in the Psychoses

In the first CT study of drug response in schizophrenics (Weinberger et al., 1980), it was found that the presence of morphological brain abnormalities was related to poor response to neuroleptics. Several groups of investigators attempted to replicate this finding using the change

TABLE 7. Characteristics and Results of CT Studies Investigating the Relationships between Neuropsychological Performance and Brain Morphology in Schizophrenia

Author	Year	N	Course	Status[a]	Test[b]	Measure[c]	Meds.[d]	Result[e]
Carr & Wedding	1984	21	Chronic	Out	HRB	VBR	+	n. s.
Donnelly et al.	1980	15	Chronic	In	HRB	VBR, SW	+	$P<0.002$
Golden, Moses et al.	1980	42	Chronic	In/out	LNNB	VBR	+	$P<0.001$
Kemali et al.	1985	33	Chronic	In/out	LNNB	VBR	+	$P<0.05$
Pandurangi et al.	1986	23	Chronic	—	HRB	VBR, HD	+	$P<0.05$
Rieder et al.	1979	8	Chronic	Out	HRB	SA	+	$P<0.004$

[a]Status: in, inpatients; out, outpatients.
[b]Tests: HRB, Halstead–Reitan Battery; LNNB, Luria Nebraska Neuropsychological Battery.
[c]VBR, ventricular brain ratio; SW, sulcal width; HD, hyperdensity measures; SA, sulcal area.
[d]Meds, on medication.
[e]Indices of overall performance were used.

TABLE 8. Characteristics and Results of Studies Investigating the Relationships between Drug Response and the VBR in Schizophrenia

Author	Year	N	Blind	Drug-free (weeks)[a]	Drug treatment (weeks)	Dose[b]	Instrument[c]	Result
Boronow et al.	1985	25	+	6	>4	Ind.	BHRS, BPRS	n. s.
Jeste et al.	1982	20	+	8	8	Ind.	BPRS	$P < .05$
Losonczy, Song, Mohs, Small, et al.	1986	19	−	3	6	Fix.	BPRS, CGI	n. s.
Luchins et al.	1983	35	+	2.5	>5	Ind.	GAS, SADS-C	$P < .05$
Nasrallah, Kuperman, Hammara, et al.	1983	55	+	N.A.	N.A.	Ind.	Chart rev.	n. s.
Schulz, Sinicro, Kishore, & Friedee	1983	12	+	—	3.5	Ind.	BPRS	$P = .07$
Smith & Maser	1983	35	+	1.5	>3.5	Fix.	BPRS, NHSI	n. s.
Weinberger et al.	1980	20	+	8	8	Ind.	BPRS	$P < .05$
Williams et al.	1985	40	+	N.A.	N.A.	Ind.	Chart rev.	n. s.

[a] Length of drug washout period (when a range was provided, the midpoint was recorded); N.A., not applicable.
[b] Dose: ind., individualized; fix., fixed.
[c] Symptom ratings scale: BHRS, Bunney–Hamburg Ratings Scale; BPRS, Brief Psychiatric Rating Scale; CGI, Clinical Global Impression rating; GAS, Global Assessment Scale; SADS-C, Schedule for Affective Disorders and Schizophrenia; Chart. rev., chart review, NHSI, New Haven Schizophrenia Index.

scores on standardized measures of symptomatology between a drug washout-placebo period and a drug treatment period to assess drug response. As Table 8 shows, studies varied in the length of time allowed for drug washout and for the treatment period. Since studies also varied on a number of other dimensions, it was difficult to determine whether or how the time period allotted to each state affected the findings.

Four of the nine studies investigating the relationships between drug response and the VBR reported findings indicating that schizophrenics with ventriculomegaly tended to be poor responders to neuroleptics. It should be noted that two of the studies in which nonsignificant findings were reported (Nasrallah, Kuperman, Hamara, & McCalley-Whitters, 1983; Williams et al., 1985) used retrospective chart review rather than a standard method of comparing on-drug with off-drug periods in their sample.

A major problem in these drug studies is that of subject attrition. As noted by Luchins (1983), data about attrition rates and the reasons for attrition have usually been lacking in CT studies of drug response. Subjects may drop out for different reasons: Losonczy et al. (1986a), for instance, reported attrition of subjects whose condition had deteriorated apparently during the drug washout period, whereas Luchins, Lewine, and Meltzer (1983, 1984b) reported attrition of patients who had not completed the 5-week neuroleptic treatment, apparently because they failed to respond. These poor responders were later found to have larger ventricles than patients who remained in the study. Obviously, the attrition of both good and poor responders from a sample will restrict the variability in drug response and perhaps in ventricular size as well, thereby decreasing the likelihood of finding an effect in the expected direction. With this difficulty in mind, it is noteworthy that four of the seven studies in which standardized symptom ratings scales were implemented (57%) did report statistically significant results.

There were only two studies of drug response in patients with affective disorders by Luchins *et al.* (1983) and Nasrallah *et al.* (1984). In the former study, a mixed group of 22 affective spectrum patients was studied, and the change score on the GAS and SADS-C was used to estimate drug response. In the latter study, 24 manic males were studied, and retrospective chart review was used to assess drug response. In neither of these studies was an association found between drug response and ventriculomegaly.

Neurochemical Dysregulation and CT Abnormalities in the Psychoses

The investigation of the relationships between biochemical aberrations and morphological brain abnormalities has been largely inspired by Crow's (1980) hypothesis, that neurotransmitter dysregulation plays a more important role in type I schizophrenia than in schizophrenia type II. To test this hypothesis, various biochemical substances have been studied in the CSF, in the blood, and in the urine. These substances included monoamines, their metabolites, enzymes involved in the breakdown of monoamines, and hormones regulated by monoamine action.

Investigations of monoamine metabolite concentrations focused on the dopamine (DA) metabolites homovanillic acid (HVA) and dihydroxyphenylacetic acid (DOPAC); 5-hydroxyindoleacetic acid (5-HIAA), the major metabolite of serotonin (5-HT); and 3-methoxy-4-hydroxyphenylglycol (MHPG or MOPEG), norepinephrine's (NE) major metabolite. Enzyme activity studies focused mainly on monoamine oxidase (MAO) and dopamine-β-hydroxylase (DBH), a marker of central noradrenergic activity (van Kammen *et al.*, 1983). Serum concentrations of prolactin, a hormone regulated by dopamine, were investigated in several studies as a peripheral index of CNS dopaminergic activity.

As Table 9 shows, with the exception of Boronow *et al.* (1985), studies of dopamine metabolites found inverse relationships between metabolite concentrations and measures of ventricular or sulcal fluid volume. In none of the studies using prolactin levels as a peripheral measure of dopaminergic activity were significant results obtained.

With the exception of the study of Oxentierna, Bergstrand, Bjerkenstedt, Sedvall, and Wik (1984), in which patients were not withdrawn from neuroleptics, and the study by Boronow *et al.* (1985), in CSF studies of serotonin's metabolite 5-HIAA, inverse relationships were found between concentrations and measures of ventricular or sulcal fluid volume. There was a single study (DeLisi, Neckers, Weinberger, & Wyatt, 1981) in which whole-blood serotonin concentrations were measured in schizophrenics and found to be positively correlated with ventricular size. The authors noted, however, that the relationship between brain and blood serotonin levels was unclear.

In contrast with the rather consistent inverse relationships found between concentrations of dopamine and serotonin metabolites and CSF volume estimates, investigations of CSF levels of MHPG, a norepinephrine metabolite, have found no significant correlations with structural brain parameters (see Table 9). Enzyme studies also have not yielded conclusive results.

Though the studies of CSF metabolites of dopamine and serotonin have yielded fairly consistent results that appear to support Crow's hypothesis, a few methodological issues should be considered before conclusions are drawn. First, since most of the studies lacked normal controls, apparently because of ethical considerations, the interpretation of the results is ambiguous: either schizophrenics with normal CT scans have high CSF concentrations of monoamine metabolites or schizophrenics with abnormal CT scans have low CSF concentrations of monoamine metabolites or perhaps both. Secondly, since psychiatric controls were not included in those studies, it is uncertain whether the findings are specific to schizophrenia, or whether they have etiological significance. Thirdly, the inverse relationship between CSF monoamine concentrations and CT abnormalities may be attributed to dilution effect: the greater volume of an expanded CSF system may lead to lower levels of metabolites. Though van Kammen, Mann, Seppala, and Linnoila (1986) argue against this, relying on lack of statistically significant differences in peptide and protein concentrations between schizophrenics with and without CT abnormalities, the possibility

TABLE 9. Characteristics and Results of CT Studies Investigating the Relationships between Brain Morphology and Biochemical Indices of Brain Activity in Schizophrenic Patients

Authors	Year	N^a	CT measure[b]	Biochemical substance	Site[c]	Meds.[d]	Findings r	Findings P
Studies of dopamine metabolite concentrations and hormones regulated by dopamine								
Boronow et al.	1985	22	VBR, SW	HVA	CSF	28		n. s.
Jennings, Schulz, Narosimhachari, Hamara, & Friedel	1985	16	VBR	HVA	CSF	42	−0.56	<0.03
Losonczy, Song, Mohs, & Mathe	1986	17	VBR	HVA	CSF	22	−0.55	<0.05
		28		Prolactin	Plasma	22		n. s.
Luchins, Robertson, et al.	1984	23	VBR	Prolactin	Serum	17	0.05	n. s.
Nyback et al.	1984	26	VIX	HVA	CSF	−	−0.41	<0.05
Oxentierena et al.	1984	30	SW, CR	HVA	CSF	+		<0.1
van Kammen et al.	1986	50	SW	HVA	CSF	32	−0.46	<0.01
		33		DOPAC	CSF	32	−0.37	<0.01
Studies of serotonin and serotonin metabolite concentrations								
Boronow et al.	1985	22	VBR, SW	5-HIAA	CSF	28		n. s.
DeLisi et al.	1981	33	VBR	5-HT	Blood	+	0.40	<0.03
Jennings et al.	1985	15	VBR	5-HIAA	CSF	42	−0.45	<0.09
Losonczy, Song, Mohs, Mathe, et al.	1986	17	VBR	5-HIAA	CSF	22	− 0.48	<0.06
Nyback et al.	1984	26	VIX	5-HIAA	CSF	−	−0.40	<0.05
Oxentierna et al.	1984	30	SW, CR	5-HIAA	CSF	+		n s.
Potkin, Weinberger, Linnoila, & Wyatt	1983	24	VBR	5-HIAA	CSF	24	−0.6	<0.01
van Kammen, Mann, Scheinen, Van Kammen, & Linnoila	1984	56	VBR, SW	5-HIAA	CSF	32	−0.35	<0.01
Studies of norepinephrine and NE's metabolite concentrations								
Boronow et al.	1985	22	VBR, SW	MHPG	CSF	28		n. s.
Losonczy, Song, Mohs, Mathe, et al.	1986	17	VBR	MHPG	CSF, urine	22		n. s.
Nyback et al.	1984	26	VIX	MHPG	CSF	−	−0.24	n. s.
Oxentierna et al.	1984	30	SW, CR	MOPEG	CSF	+		n. s.
van Kammen et al.	1984	48	VBR, SW	NE	CSF	32	−0.29	<0.05
				MHPG		32		n. s.
Enzyme activity studies								
Jeste et al.	1982	36	VBR	MAO	Platelet	+		n. s.
Meltzer et al.	1984	46	VBR	DBH	Serum	*		n. s.
Pandurangi et al.	1986	20	VBR, HD	DBH	Plasma	+		n. s.
				MAO	Platelet	+		n. s.
Tachiki et al.	1984	25	VBR	MAOb	Blood	+	0.55	<0.01
van Kammen et al.	1983	25	VBR	DBH	CSF	33	−0.56	<0.01
		24	SW		CSF	33	−0.14	n. s.

[a]N, number of schizophrenics, except Jennings et al. (1985), who used a mixed group of psychiatric patients.
[b]CT measure: VBR, ventricular brain ratio; VIX, ventricular index, a linear measure; SW, ratings of sulcal/fissure widening; CR, cerebellar atrophy ratings; HD, hyperdensity measures.
[c]Site: blood, whole blood; CSF, cerebrospinal fluid.
[d]Meds: +, on medication; −, off medication; number, days without medication.

of a dilution effect can not be completely discarded, at least until it is established beyond doubt that an inverse relationship between CSF monoamine metabolite concentrations and presence of CT abnormalities can not be demonstrated in nonpsychotic control groups with enlarged ventricles.

There were only a few studies investigating the relationships between biochemical variables and CT abnormalities in patients with affective disorders. Kellner et al. (1983) found a positive correlation between the VBR and urinary free cortisol levels in 10 manic–depressive patients. They suggested, therefore, that ventricular enlargement in patients with affective disorder might be related to excessive production of endogenous cortisol secondary to overactivation of the hypothalamic–pituitary–adrenal axis. Meltzer, Tong, and Luchins (1984) found an inverse relationship ($r = -0.41$, $P < 0.05$) between DBH activity and the VBR in 28 patients with major depression. Because a weak trend in the same direction was also found in schizophrenics ($r = -0.15$), the authors speculated that this inverse relationship may be a manifestation of a genetic predisposition common to the major psychoses.

Standish-Barry, Bouras, Hale, Bridges, and Bartlett (1986) investigated the relationships between ventricular size and levels of precursors and metabolites of biogenic amines in patients with affective disorders. They found that plasma levels of free tryptophan (a precursor of serotonin) were inversely related to the size of the lateral ventricles measured by Evans' ratio ($r = -0.46$, $P < 0.05$), whereas CSF concentration of 5-HIAA correlated positively with the same linear measure ($r = 0.55$, $P < 0.01$). The correlation between ventricular size and HVA levels in the CSF was not significant. Unfortunately, in addition to the lack of adequate controls, the lack of replicated findings in this area does not allow one to render a summary statement on the biochemical correlates of ventriculomegaly in affective disorders.

INTEGRATION OF FINDINGS

It is proposed here that the findings from the CT literature have not, thus far, provided convincing evidence in support of a two-syndrome model of schizophrenia. In other words, the results may not be readily interpreted as indicative of discontinuity between the two presumed conditions—that with and that without CT abnormalities. Rather, the findings may be incorporated within the parsimonious view of a unitary illness expressed in various levels or grades of severity. Thus, group differences obtained on measures of neuropsychological functioning, drug response, or other clinical/biological variables may be viewed as quantitative differences in severity of illness between those with and those without morphological brain abnormalities. Though a two-subtype model cannot be completely discarded, it is suggested that until convincing evidence is found, the simpler model of unitary illness be adopted.

The findings from the CT literature allow one to speculate about the mechanism linking morphological brain abnormalities and illness severity. The convergence of measures of neuropsychological functioning and measures of brain structure in schizophrenics, indicating that poor neuropsychological performance is associated with structural brain abnormalities, supports the view that some sort of CNS damage has occurred in these patients. There is evidence to suggest that these abnormalities do not emerge during the course of the schizophrenic illness but antedate its onset (Weinberger et al., 1982). The origin of this putative brain damage might, perhaps, be traced back to the prenatal or perinatal period. It is suggested that viewing the occurrence of adverse environmental influences as an all-or-none phenomenon is probably oversimplified. A continuum of such influences, reflected in a continuum of neuropathological sequelae differing in degree of severity, should probably be postulated. Ventricular size in schizophrenics could be one index of cumulative, environmentally inflicted CNS damage.

Adverse environmental events may be hypothesized to contribute to psychotic illness either by combining with a genetic predisposition to increase liability to the disorder or by serving as etiological factors that may elicit psychotic illness independently of genetic factors. At present, there is no support for the latter view, because in populations with elevated risk for obstetric complications (e.g., males, twins—particularly MZ twins), there is no evidence for an increase in the prevalence of schizophrenia (Gottesman & Shields, 1982). Evidence for the diathesis–stress model has been cited, however, in several reports. For instance, Parnas et al. (1982), who investigated schizophrenic, borderline, and healthy offspring of schizophrenic mothers in a prospective study, found that schizophrenics had the most complicated births. They proposed that "birth complications can decompensate borderline individuals toward schizophrenic breakdown" (ibid. p. 416). Their findings also suggest that adverse environmental influences may exacerbate an already apparent illness. This view may be supported by McNeil and Kaij (1978), who reported an increased frequency of obstetric complications in process schizophrenics when compared to patients with a less virulent form of illness (e.g., atypical or schizoaffective psychosis). Thus, adverse environmental influences seem to constitute a risk-increasing factor when combined with a genetic predisposition. They may either reduce the threshold for the expression of psychotic illness or exacerbate an already apparent illness, pushing an individual toward the more virulent subtypes within the schizophrenia spectrum.

The notion of exacerbated illness in schizophrenics with evidence of birth complications is not incompatible with the notion of aggravated illness in those with enlarged CSF-filled spaces. An association between severity of illness and ventricular size has been suggested in several studies awaiting replication. Luchins and Meltzer (1986) found that lateral ventricular size was significantly increased in chronic-ward schizophrenics in comparison to either acute-ward schizophrenics or controls. The two schizophrenic groups were similar in age and length of illness. There was also a significant positive correlation between percentage of illness spent in the hospital, an index of illness severity, and VBR. DeLisi et al. (1983) found that patients with a first-episode schizophreniform psychosis whose ventricles were enlarged had a significantly worse outcome than patients with normal ventricular size in a 2-year follow-up. Kolakowska, Williams, Ardern, et al. (1985) and Kolakowska, Williams, Jambor, and Ardern (1985) reported increased ventricular size in patients with unfavorable outcome compared to patients with good outcome. Outcome was defined, in this study, on the basis of symptom severity and social functioning. Similar findings were reported by Seidman, Sokolov, McElroy, Knapp, and Sabin (1987), who used measures of social and residential outcome.

Though ventriculomegaly has been documented in several studies of RDC acute schizophrenics, this evidence should not necessarily be taken as contradicting the postulated relationship between ventricular size and illness severity, because many of these patients later develop a chronic course (American Psychiatric Association, 1980). Andreasen, Smith, Jacoby, Dennert, and Olsen (1982), who studied chronic schizophrenics with milder illness who had a remitting course and lived in the community, discovered only moderate ventriculomegaly in their patients compared to other studies.

It has been reported that patients with a more virulent form of disorder within either the schizophrenia or the affective spectrum are also characterized by increased genetic risk. Farmer, McGuffin, and Gottesman (1984), for instance, found that co-twins of schizophrenics with hebephrenialike psychosis were significantly more likely to receive a diagnosis of schizophrenia than co-twins of schizophrenics with paranoidlike psychosis, a disorder with a relatively good prognosis within the schizophrenia spectrum. Gershon et al. (1982) found that lifetime prevalence of affective-spectrum illness in adult relatives of patients with schizoaffective disorder, bipolar illness, and unipolar illness was 36%, 25%, and 20%, respectively. Thus, the amount of genetic risk directly corresponded to quality of prognosis within affective spectrum disorders.

The abovementioned findings from studies of obstetric complications suggest that CNS trauma may constitute a second important risk factor linked to severity of illness. The degree of enlargement of CSF-filled spaces, particularly the ventricular system, may indirectly reflect the amount of cumulative CNS damage sustained by an individual, allowing assessment of the role of this particular class of environmental events in the clinical phenomenology of schizophrenia. The mechanism by which these events exert their influence is, at present, unclear. Murray *et al.* (1985), however, speculating about the causes of brain pathology in schizophrenia, suggested that intraventricular and subependymal hemorrhages, observed in 40% of newborns weighing under 1.5 kg, may play an important role in increasing the liability for the disorder. It may be added to this that these intracranial abnormalities have also been reported in 6% of full-term healthy infants with uncomplicated gestation (Winchester, Brill, Cooper, Krauss, & Peterson, 1986).

At present, little is known about factors associated with ventricular enlargement in patients with affective disorders. It has been demonstrated in two studies, however, that similarly to the findings reported in schizophrenia, enlargement of CSF-filled spaces is more likely to be found in patients with a severe form of disorder (psychotic or delusional depression) than in those with milder illness. The putative contribution of genetic or environmental factors to structural brain abnormalities has been largely unexplored in patients with affective disorders. Kellner, Rubinow, Gold, and Post (1983), suggested that increased endogenous cortisol levels rather than external factors might be the cause for enlargement of CSF-filled spaces observed in patients with affective disorder. This hypothesis awaits further investigation. Another possibility is that CNS trauma during early development constitutes a risk-increasing factor in affective disorders in a similar fashion to its putative role in schizophrenia. Dalen (1965), for instance, found an association between EEG abnormalities and perinatal complications in a group of manic patients. It remains to be explored whether such relationships could be demonstrated between ventricular size and CNS damage in affective-spectrum disorders.

ACKNOWLEDGMENT. Supported by a fellowship from the American Association of University Women (Mabel Sutton Wasson and Virginia B. Sloan endowments).

REFERENCES

American Psychiatric Association, Committee on Nomenclature and Statistics (1980). *Diagnostic and Statistical Manual of Mental Disorders* (3rd ed.). Washington: American Psychiatric Association.

Andreasen, N. C., Dennert, J. W., Olsen, S. A., & Damasio, A. R. (1982). Hemispheric asymmetries and schizophrenia. *American Journal of Psychiatry, 139,* 427–430.

Andreasen, N. C., Olsen, S. A., Dennert, J. W., & Smith, M. R. (1982). Ventricular enlargement in schizophrenia: Relationship to positive and negative symptoms. *American Journal of Psychiatry, 139,* 297–302.

Andreasen, N. C., Smith, M. R., Jacoby, C. G., Dennert, J. W., & Olsen, S. A. (1982). Ventricular enlargement in schizophrenia: Definition and prevalence. *American Journal of Psychiatry, 139,* 292–296.

Asberg, M., Montgomery, S., Perris, C., Schalling, D., & Sedval, G. (1978). CPRS—the Comprehensive Psychological Rating Scale. *Acta Psychiatrica Scandinavica* (Supplement 271).

Becker, H., Schneider, E., Hacker, H., & Fischer, P. A. (1979). Cerebral atrophy in Parkinson's disease represented in CT. *Arkiv fur Psychiatrie und Nervenkrankheiten, 227,* 81–88.

Behar, D., Rapoport, J. L., Berg, C. J., Denckla, M. B., Mann, L., Cox, C., Fedio, P., Zahn, T., & Wolfman, M. G. (1984). Computerized tomography and neuropsychological test measures in adolescents with obsessive–compulsive disorder. *American Journal of Psychiatry, 141,* 363–369.

Bentson, J. R., Reza, M., Winter, J., & Wilson, G. (1978). Steroids and apparent cerebral atrophy on computed tomography scans. *Journal of Computer Assisted Tomography, 2,* 16–19.

Bilaniuk, L. T., Patel, S., & Zimmerman, R. A. (1977). Computed tomography of systemic lupus erythematosus, *Radiology, 124,* 119–121.

Bishop, R. J., Golden, C. J., Macinnes, W. D., Chu, C., Ruedrich, S. L., & Wilson, J. (1983). The BPRS in assessing symptom correlates of cerebral ventricular enlargement in acute and chronic schizophrenia. *Psychiatry Research, 9,* 225–231.

Boronow, J., Pickar, D., Ninan, P. T., Roy, A., Hommer, D., Linnoila, M., & Paul, S. M. (1985). Atrophy limited to the third ventricle in chronic schizophrenic patients: Report of a controlled series. *Archives of General Psychiatry, 42,* 266–271.

Campbell, R., Hays, P., Russell, D. B., & Zacks, D. J. (1979). CT scan variants and genetic heterogeneity in schizophrenia. *American Journal of Psychiatry, 136,* 722–723.

Carlen, P. L., Wortzman, G., Holgate, R. C., Wilkinson, D. A., & Rankin, J. G. (1978). Reversible cerebral atrophy in recently abstinent chronic alcoholics measured by computed tomography scans. *Science, 200,* 1076–1078.

Carr, E. G., & Wedding, D. (1984). Neuropsychological assessment of cerebral ventricular size in chronic schizophrenics. *The International Journal of Clinical Neuropsychology, 6,* 106–111.

Coffman, J. A., & Nasrallah, H. A. (1984). Brain density patterns in schizophrenia and mania. *Journal of Affective Disorders, 6,* 307–315.

Coffman, J. A., Mefferd, J., Golden, C. J., Bloch, S., & Graber, B. (1981). Cerebellar atrophy in chronic schizophrenia. *Lancet, 1,* 666.

Crow, T. J. (1980). Molecular pathology of schizophrenia: More than one disease? *British Medical Journal, 28,* 66–68.

Crow, T. J. (1983). Is schizophrenia an infectious disease? *Lancet, 1,* 173–175.

Crow, T. J. (1984). A reevaluation of the viral hypothesis: Is psychosis the result of retroviral integration at a site close to the cerebral dominance gene? *British Journal of Psychiatry, 145,* 243–253.

Cullum, C. M., & Bigler, E. D. (1986). Ventricle size, cortical atrophy and the relationship with neuropsychological status in closed head injury: A quantitative analysis. *Journal of Clinical and Experimental Neuropsychology, 8,* 437–452.

Dalen, P. (1965). Family history, the electroencephalograms and perinatal factors in manic conditions. *Acta Psychiatrica Scandinavica, 41,* 527–563.

Datlof, S., Coleman, P. D., Forbes, G. B., & Kreipe, R. E. (1986). Ventricular dilation on CAT scans of patients with anorexia nervosa. *American Journal of Psychiatry, 143,* 96–98.

DeLisi, L. E., Neckers, L. M., Weinberger, D. R., & Wyatt, R. J. (1981). Increased whole blood serotonin concentrations in chronic schizophrenic patients. *Archives of General Psychiatry, 38,* 647–655.

DeLisi, L. E., Schwartz, C. C., Targum, S. D., Byrnes, S. M., Cannon-Spoor, E., Weinberger, D. R., & Wyatt, R. J. (1983). Ventricular brain enlargement and outcome of acute schizophreniform disorder. *Psychiatry Research, 9,* 169–171.

DeLisi, L. E., Goldin, L. R., Hamovit, J. R., Maxwell, M. E., Kurtz, D., & Gershon, E. S. (1986). A family study of the association of increased ventricular size with schizophrenia. *Archives of General Psychiatry, 43,* 148–153.

Dewan, M. J., Pandurangi, A. K., Lee, S. H., Ramachandran, T., Levy, B. F., Boucher, M., Yozawitz, A., & Major, L. (1983). Cerebellar morphology in chronic schizophrenic patients: A controlled computed tomography study. *Psychiatry Research, 10,* 97–103.

Dolan, R. J., Calloway, S. P., & Hann, A. H. (1985). Cerebral ventricular size in depressed subjects. *Psychological Medicine, 15,* 873–878.

Donnelly, E. F., Weinberger, D. R. Waldman, I. N., & Wyatt, R. J. (1980). Cognitive impairment associated with morphological brain abnormalities on computed tomography in chronic schizophrenic patients. *Journal of Nervous and Mental Disease, 168,* 305–308.

El-Tatawy, S., Badrawi, N., & El-Bishlawy, A. (1983). Cerebral atrophy in infants with protein energy malnutrition. *American Journal of Neuroradiology, 4,* 434–436.

Endicott, J., & Spitzer, R. L. (1978). A diagnostic interview: The Schedule for Affective Disorders and Schizophrenia (SADS). *Archives of General Psychiatry, 35,* 837–844.

Endicott, J., Spitzer, R. L., Fleiss, J. L., & Cohen, J. (1976). The Global Assessment Scale: A procedure for measuring overall severity of psychiatric disturbance. *Archives of Psychiatry, 33,* 766–781.

Farmer, A., McGuffin, P., & Gottesman, I. I. (1984). Searching for the split in schizophrenia: A twin study perspective. *Psychiatry Research, 13,* 109–118.

Farmer, A., McGuffin, P., Jackson, R., & Storey, P. (1985). Classifying schizophrenia. *Lancet, 1,* 1333.

Farmer, A., Jackson, R., McGuffin, P., & Storey, P. (1987). Cerebral ventricular enlargement in chronic schizophrenia: Consistencies and contradictions. *British Journal of Psychiatry, 150,* 324–330.

Folstein, M. F., & McHugh, P. R. (1975). Mini-Mental State. *Journal of Psychiatric Research, 12,* 189–198.

Fox. J. H., Topel, J. L., & Huckman, M. S. (1975). Use of computerized tomography in senile dementia. *Journal of Neurology, Neurosurgery, and Psychiatry, 38,* 948–953.

Frangos, E., & Athanassenas, G. (1982). Differences in lateral brain ventricular size among various types of chronic schizophrenics. *Acta Psychiatrica Scandinavica, 66,* 459–463.

Gershon, E. S., Hamovit, J., Guroff, J. J., Dibble, E., Leckman, J. F., Sceery, W., Targum, S. D., Nurnberger, J. I., Goldin, L. R., & Bunney, W. E. (1982). A family study of schizoaffective, bipolar I, bipolar II, unipolar, and normal control probands. *Archives of General Psychiatry, 39,* 1157–1167.

Goldberg, S. C. (1985). Negative and deficit symptoms in schizophrenia do respond to neuroleptics. *Schizophrenia Bulletin, 11,* 453–456.

Golden, C. J., Graber, B., Coffman, J., Berg, R., Bloch, S., & Brogan, D. (1980). Brain density deficits in chronic schizophrenia. *Psychiatry Research, 3,* 179–184.

Golden, C. J., Moses, J. A., Zelazowski, R., Graber, B., Zatz, L. M., Horvath, T. B., & Berger, P. A. (1980). Cerebral ventricular size and neuropsychological impairment in young chronic schizophrenics. *Archives of General Psychiatry, 37,* 619–623.

Gonzales-Scarano, F., Lisak, R. P., Bilaniuk, L. T., Zimmerman, R. A., Atkins, P. C., & Zweiman, B. (1979). Cranial computed tomography in the diagnosis of systemic lupus erythematosus. *Annals of Neurology, 5,* 158–165.

Gottesman, I., & Shields, J. (1982). *Schizophrenia: The epigenetic puzzle.* Cambridge: Cambridge University Press.

Haug, J. O. (1962). Pneumoencephalographic studies in mental disease. *Acta Psychiatrica Scandinavica [Suppl.] 165,* 1–114.

Heinz, E. R., Martinez, J., & Haenggeli, A. (1977). Reversibility of cerebral atrophy in anorexia nervosa and Cushing's syndrome. *Journal of Computer Assisted Tomography, 1,* 415–417.

Hier, D. B., LeMay, M., & Rosenberger, P. B. (1978a). Autism: Association with reversed cerebral asymmetry. *Neurology, 28,* 348.

Hier, D. B., LeMay, M., & Rosenberger, P. B. (1978b). Developmental dyslexia: Evidence for a subgroup with reversal of cerebral asymmetry. *Archives of Neurology, 35,* 90–92.

Hungerford, G. D., du Boulay, G. H., & Zilkah, K. J. (1976). Computerised axial tomography in patients with severe migraine: A preliminary report. *Journal of Neurology, Neurosurgery, and Psychiatry, 39,* 990–994.

Jacobs, L., Kinkel, W. R., Painter, F., Murawski, J., & Heffner, R. R. (1978). Computerized tomography in dementia with special reference to changes in size of normal ventricles during aging and normal pressure hydrocephalus. In R. Katzman, R. D. Terry, K. L. Bick (eds.). *Alzheimer's disease: Senile dementia and related disorders* (pp. 241–260). New York: Raven Press.

Jennings, W., Schulz, S. C., Narasimhachari, N., Hamer, R. M., & Friedel, R. O. (1985). Brain ventricular size and CSF monoamine metabolites in an adolescent inpatient population. *Psychiatry Research, 16,* 87–94.

Jernigan, T. L., Zatz, L. M., Moses, J. A., & Berger, P. A. (1982). Computed tomography in schizophrenics and normal volunteers. I: Fluid volume. *Archives of General Psychiatry, 39,* 765–770.

Jernigan, T. L., Zatz, L. M., Moses, J. A., & Cardellino, J. P. (1982). Computed tomography in schizophrenics and normal volunteers. II: Cranial asymmetry. *Archives of General Psychiatry, 39,* 771–773.

Jeste, D. V., Kleinman, J. E., Potkin, S. G., Luchins, D. J., & Weinberger, D. R. (1982). *Ex uno multi:* Subtyping the schizophrenic syndrome. *Biological Psychiatry, 17,* 199–222.

Johnstone, E. C., Crow, T. J., Frith, C. D., Stevens, J., & Kreel, L. (1976). Cerebral ventricular size and cognitive impairment in chronic schizophrenia. *Lancet, 2,* 924–926.

Kanaba, S., Shima, S., Tsukumo, D., Masuda, Y., & Asai, M. (1984). Brain CT density in chronic schizophrenia. *Biological Psychiatry, 19,* 273–274.

Kellner, C. H., Rubinow, D. R., Gold, P. W., & Post, R. M. (1983). Relationship of cortisol hypersecretion to brain CT scan alterations in depressed patients. *Psychiatry Research, 8,* 191–197.

Kellner, C. H., Rubinow, D. R., & Post, R. M. (1986). Cerebral ventricular size and cognitive impairment in depression. *Journal of Affective Disorders, 10,* 215–219.

Kemali, D., Maj, M., Galderisi, S., Ariano, M. G., Cesarelli, M., Milici, N., Salvati, A., Valente, A., & Volpe, M. (1985). Clinical and neuropsychological correlates of cerebral ventricular enlargement in schizophrenia. *Journal of Psychiatric Research, 19,* 587–596.

Kendler, K. S., Gruenberg, A. M., & Tsuang, M. (1984). Outcome of schizophrenic subtypes defined by four diagnostic systems. *Archives of General Psychiatry, 41,* 149–154.

Kling, A. S., Kurtz, N., Tachiki, K., & Orzeck, A. (1983). CT scans in sub-groups of chronic schizophrenics. *Journal of Psychiatric Research, 17*, 375–384.

Kolakowska, T., Williams, A. O., Ardern, M., Reveley, M. A., Jambor, K., Gelder, M. G., & Mandelbrote, B. M. (1985). Schizophrenia with good and poor outcome. I: Early clinical features, response to neuroleptics and signs of organic dysfunction. *British Journal of Psychiatry, 146*, 229–239.

Kolakowska, T., Williams, A. O., Jambor, K., & Ardern, M. (1985). Schizophrenia with good and poor outcome. III: Neurological 'soft' signs, cognitive impairment and their clinical significance. *British Journal of Psychiatry, 146*, 348–357.

Krawiecka, M., Goldberg, D., & Vaughan, M. (1977). A standardized psychiatric assessment scale for rating chronic psychotic patients. *Acta Psychiatrica Scandinavica, 55*, 299–308.

Laing, R. D. (1960). *The divided self,* London: Tavistock Publications.

Levin, H. S., Meyers, C. A., Grossman, R. G., & Sarwar, M. (1981). Ventricular enlargement after closed head injury. *Archives of Neurology, 38*, 623–629.

Lewine, R., Fogg, L., & Meltzer, H. Y. (1983). The development of scales for the assessment of negative and positive symptoms in schizophrenia. *Schizophrenia Bulletin, 9*, 368–376.

Lippmann, S., Manshadi, M., Baldwin, H., Drasin, G., Rice, J., & Alrajeh, S. (1982). Cerebellar vermis dimensions on computerized tomographic scans of schizophrenic and bipolar patients. *American Journal of Psychiatry, 139*, 667–668.

Losonczy, M. F., Song, I. S., Mohs, R. C., Small, N. A., Davidson, M., Johns, C. A., & Davis, K. (1986). Correlates of lateral ventricular size in chronic schizophrenia, I: Behavioral and treatment response measures. *American Journal of Psychiatry, 143*, 976–981.

Losonczy, M. F., Song, I. S., Mohs, R. C., Mathe, A. A., Davidson, M., Davis, B., & Davis, K. (1986). Correlates of lateral ventricular size in chronic schizophrenia, II: Biological measures. *American Journal of Psychiatry, 143*, 1113–1118.

Luchins, D. J., & Meltzer, H. Y. (1983a). Ventricular size and psychosis in affective disorder. *Biological Psychiatry, 18*, 1197–1198.

Luchins, D. J., & Meltzer, H. Y. (1983b). A blind controlled study of occipital cerebral asymmetry in schizophrenia. *Psychiatry Research, 10*, 87–95.

Luchins, D. J., & Meltzer, H. Y. (1986). A comparison of CT findings in acute and chronic ward schizophrenics. *Psychiatry Research, 17*, 7–14.

Luchins, D. J., Weinberger, D. R., & Wyatt, R. J. (1982). Schizophrenia and cerebral asymmetry detected by computed tomography. *American Journal of Psychiatry, 139*, 753–757.

Luchins, D. J., Lewine, R. R. J., & Melzer, H. Y. (1983). Lateral ventricular size in the psychoses: Relations to psychopathology and therapeutic adverse response to medications. *Psychopharmacology Bulletin, 19*, 518–522.

Luchins, D. J., Lewine, R. R. J., & Meltzer, H. Y. (1984). Lateral ventricular size, psychopathology, and medication response in the psychoses. *Biological Psychiatry, 19*, 29–44.

Luchins, D. J., Robertson, A. G., & Meltzer, H. Y. (1984). Serum prolactin, psychopathology, and ventricular size in chronic schizophrenia. *Psychiatry Research, 12*, 149–153.

Mackay, A. V. P. (1980). Positive and negative schizophrenic symptoms and the role of dopamine. *British Journal of Psychiatry, 137*, 379–386.

McNeil, T. F., & Kaij, L. (1978). Obstetric factors in the development of schizophrenia: Complications in the births of preschizophrenics and in reproduction by schizophrenic parents. In L. C. Wynne, R. L. Cromwell, & S. Mathysse (Eds.). *The Nature of Schizophrenia: New Approaches to Research and Treatment* (pp. 401–429). New York: John Wiley & Sons.

Meltzer, H. Y., Tong, C., & Luchins, D. J. (1984). Serum dopamine-β-hydroxylase activity and lateral ventricular size in affective disorders and schizophrenia. *Biological Psychiatry, 19*, 1395–1401.

Meltzer, H. Y., Sommers, A. A., & Luchins, D. J. (1986). The effect of neuroleptics and other psychotropic drugs on negative symptoms in schizophrenia. *Journal of Clinical Psychopharmacology, 6*, 329–338.

Murray, R. M., Lewis, S. W., & Reveley, A. M. (1985). Towards an aetiological classification of schizophrenia. *Lancet, 1*, 1023–1026.

Naber, D., Albus, M., Burke, H., Muller-Spahn, F., Munch, U., Reinertshofer, T., Wissmann, J., & Ackenheil, M. (1985). Neuroleptic withdrawal in chronic schizophrenia: CT and endocrine variables relating to psychopathology. *Psychiatry Research, 16*, 207–219.

Naeser, M. A., Gebhardt, C., & Levine, H. L. (1980). Decreased computerized tomography numbers in patients with pre-senile dementia. *Archives of Neurology, 37*, 401–409.

Nasrallah, H. A., Jacoby, C. G., McCalley-Whitters, M., & Kuperman, S. (1982). Cerebral ventricular enlargement in subtypes of chronic schizophrenia. *Archives of General Psychiatry, 39,* 774–777.

Nasrallah, H. A., Kuperman, S., Hamara, B. J., & McCalley-Whitters, M. (1983). Clinical differences between schizophrenic patients with and without large cerebral ventricles. *Journal of Clinical Psychiatry, 44,* 407–409.

Nasrallah, H. A., Kuperman, S., Jacoby, C. G., McCalley-Whitters, M., & Hamara, B. (1983). Clinical correlates of sulcal widening in schizophrenia. *Psychiatry Research, 10,* 237–242.

Nasrallah, H. A., McCalley-Whitters, M., & Pfohl, B. (1984). Clinical significance of large cerebral ventricles in manic males. *Psychiatry Research, 13,* 151–156.

Nyback, H., Wiesel, F.-A., Bergren, B.-M., & Hindmarsh, T. (1982). Computed tomography of the brain in patients with acute psychosis and in healthy volunteers. *Acta Psychiatrica Scandinavica, 65,* 403–414.

Nyback, H., Bergren, B.-M., Nynan, H., Sedvall, G., & Wiesel, F.-A. (1984). Cerebroventricular volume, cerebrospinal fluid monoamine metabolites, and intellectual performance in schizophrenic patients. In E. Usdin, A. Carlson, A. Dahlstrom, & J. Engel (Eds.), *Catecholamines: Neuropharmacology and central nervous system—therapeutic aspects* (pp. 161–165). New York: Alan R. Liss.

Nyman, H., Nyback, H., Wiesel, F. A., Oxentierna, G., & Schalling, D. (1986). Neuropsychological test performance, brain morphological measures and CSF monoamine metabolites in schizophrenic patients. *Acta Psychiatrica Scandinavica, 74,* 292–301.

Okuno, T., Ito, M., Konishi, Y., Yoshioka, M., & Nakano, Y. (1980). Cerebral atrophy following ACTH therapy. *Journal of Computer Assisted Tomography, 4,* 20–23.

Overall, J. E., & Gorham, D. R. (1962). The Brief Psychiatric Rating Scale. *Psychological Reports, 10,* 789–812.

Owens, D. G. C., Johnstone, E. C., Crow, T. J., Frith, C. D., Jagoe, J. R., & Kreel, L (1985). Lateral ventricular size in schizophrenia: Relationship to the disease process and its clinical manifestations. *Psychological Medicine, 15,* 27–41.

Oxentierna, G., Bergstrand, G., Bjerkenstedt, L., Sedvall, G., & Wik, G., (1984). Evidence of disturbed CSF circulation and brain atrophy in cases of schizophrenic psychosis. *British Journal of Psychiatry, 144,* 654–661.

Pandurangi, A. K., Dewan, M. J., Lee, S. H., Ramachandran, T., Levy, B. F., Boucher, M., Yozawitz, A., & Major, L. (1984). The ventricular system in chronic schizophrenic patients: A controlled computed tomography study. *British Journal of Psychiatry, 144,* 172–176.

Pandurangi, A. K., Dewan, M. J., Boucher, M., Levy, B., Ramachandran, T., Bartell, K., Bick, P. A., Phelps, B. H., & Major, L. (1986). A comprehensive study of chronic schizophrenic patients. II: Biological, neuropsychological, and clinical correlates of CT abnormality. *Acta Psychiatrica Scandinavica, 73,* 161–171.

Parnas, J., Schulsinger, F., Teasdale, T. W., Schuljsinger, H., Feldman, P. M., & Mednick, S. A. (1982). Perinatal complications and clinical outcome within the schizophrenia spectrum. *British Journal of Psychiatry, 140,* 416–420.

Pearlson, G. D., & Veroff, A. E. (1981). Computerised tomographic scan changes in manic–depressive illness. *Lancet, 2,* 470.

Pearlson, G. D., Garbacz, D. J., Breakey, W. R., Ahn, H. S., & DePaulo, J. R. (1984). Lateral ventricular enlargement associated with persistent unemployment and negative symptoms in both schizophrenia and bipolar disorder. *Psychiatry Research, 12,* 1–9.

Pfefferbaum, A. , Zatz, L. M. , & Jernigan, T. L. (1986). Computer-interactive method for quantifying cerebrospinal fluid and tissue in brain CT scans: Effects of aging. *Journal of Computer Assisted Tomography, 10,* 571–578.

Potkin, S. G., Weinberger, D. R., Linnoila, M., & Wyatt, R. J. (1983). Low CSF 5-hydroxyindoleacetic acid in schizophrenic patients with enlarged cerebral ventricles. *American Journal of Psychiatry, 140,* 21–25.

Probst, F. P. (1973). Gas distension of the lateral ventricles at encephalography. *Acta Radiologica, 14,* 1–4

Raz, S., Raz, N., & Bigler, E.D. (1988). Ventriculomegaly in schizophrenia: Is the choice of controls important? *Psychiatry Research, 24,* 71–77.

Reveley, A. M., & Reveley, M. A. (1983). Aqueduct stenosis and schizophrenia. *Journal of Neurology, Neurosurgery and Psychiatry, 46,* 18–22.

Reveley, A. M., Reveley, M. A., Clifford, C. A., & Murray, R. M. (1982). Cerebral ventricular size in twins discordant for schizophrenia. *Lancet, 1,* 540–541.

Reveley, A. M., Reveley, M. A., Chitkara, B., & Clifford, C. (1984). The genetic basis of cerebral ventricular volume. *Psychiatry Research, 13,* 261–266.

Reveley, A. M., Reveley, M. A., & Murray, R. M. (1984). Cerebral ventricular enlargement in non-genetic schizophrenia: A controlled twin study. *British Journal of Psychiatry, 144,* 89–93.

Rieder, R. O., Donnelly, E. F., Herdt, J. R., & Waldman, I. N. (1979). Sulcal prominence in young chronic schizophrenic patients. CT scan findings associated with impairment on neuropsychological tests. *Psychiatry Research, 1,* 1–8.

Romani, A., Zerbi, F., Mariotti, G., Callieco, R., & Cosi, V. (1986). Computed tomography and pattern reversal visual evoked potentials in chronic schizophrenic patients. *Acta Psychiatrica Scandinavica, 73,* 566–573.

Schulsinger, F., Parnas, J., Petersen, E. T., Schulsinger, H., Teasdale, T. W., Mednick, S. A., Moller, L., & Silverton, L. (1984). Cerebral ventricular size in the offspring of schizophrenic mothers: A preliminary study. *Archives of General Psychiatry, 41,* 602–606.

Schulz, S. C., Sinicrope, P., Kishore, P., & Friedel, R. O. (1983). Treatment response and ventricular enlargement in young schizophrenic patients. *Psychopharmacology Bulletin, 19,* 510–512.

Scott, M. L., Golden, C. J., Ruedrich, S. L., & Bishop, R. J. (1983). Ventricular enlargement in major depression. *Psychiatry Research, 8, c. 91–93.*

Scidman, L. J., Sokolov, R. L., McElroy, C., Knapp, P., & Sabin, T. (1987). Lateral ventricular size and social network differentiation in young nonchronic schizophrenic patients. *American Journal of Psychiatry, 144,* 512–514.

Shelton, R. C., Weinberger, D. R., Doran, A., & Pickar, D. (1985). Cerebral structural pathology in schizophrenia. Presented at the Fourth World Congress of Biological Psychiatry, Philadelphia.

Smith, R. C., & Maser, J. (1983). Morphological and neuropsychological abnormalities as predictors of clinical response to psychotropic drugs. *Psychopharmacology Bulletin, 19,* 505–509.

Sommers, A. A. (1985). "Negative symptoms": Conceptual and methodological problems. *Schizophrenia Bulletin, 11,* 365–379.

Standish-Barry, H. M. A. S., Bouras, N., Hale, A. S., Bridges, P. K., & Bartlett, J. R. (1986). Ventricular size and CSF transmitter metabolite concentrations in severe endogenous depression. *British Journal of Psychiatry, 148,* 386–392.

Strauss, J. S., & Carpenter, W. T. (1974). The prediction of outcome in schizophrenia: II. Relationship between predictor and outcome variables. *Archives of General Psychiatry, 31,* 37–42.

Strub, R. L., & Black, W. F. (1981). *The Mental Status Examination in Neurology.* Philadelphia: F.A. Davis.

Synek, V., & Reuben, J. R. (1976). The ventricular-brain ratio using planimetric measurement of EMI scans. *British Journal of Radiology, 49,* 233–237.

Tachiki, K. H., Kurtz, N., Kling, A. S., & Hullett, F. J. (1984). Blood monoamine oxidases and CT scans in subgroups of chronic schizophrenics. *Journal of Psychiatric Research, 18,* 233–243.

Tanaka, Y., Hazama, H., Kawahara, R., & Kobayashi, K. (1981). Computerized tomography of the brain in schizophrenic patients. *Acta Psychiatrica Scandinavica, 63,* 191–197.

Targum, S. D., Rosen, L. N., DeLisi, L. E., Weinberger, D. R., & Citrin, C. M. (1983). Cerebral ventricular size in major depressive disorder: Association with delusional symptoms. *Biological Psychiatry, 18,* 329–336.

Turner, S. W., Toone, B. K., & Brett-Jones, J. (1986). Computerized tomographic scan changes in early schizophrenia—preliminary findings. *Psychological Medicine, 16,* 219–225.

van Kammen, D. P., Mann, L. S., Sternberg, D. E., Scheinin, M., Ninan, P. T., Marder, S., van Kammen, W. B. Rieder, R. O., & Linnoila, M. (1983). Dopamine beta-hydroxylase activity and homovanillic acid in spinal fluid of schizophrenics with brain atrophy. *Science, 220,* 974–977.

van Kammen, D. P., Mann, L. S., Scheinin, M., van Kammen, W. B., & Linnoila, M. (1984). Spinal fluid monoamine metabolites and anticytomegalovirus antibodies and brain scan evaluation in schizophrenia. *Psychopharmacology Bulletin, 20,* 519–522.

van Kammen, D. P., van Kammen, W. B., Mann, L. S., Seppala, T., & Linnoila, M. (1986). Dopamine metabolism in the cerebrospinal fluid of drug-free schizophrenic patients with and without cortical atrophy. *Archives of General Psychiatry, 43,* 978–983.

Weinberger, D. R. (1984). Computed tomography (CT) findings in schizophrenia: Speculation on the meaning of it all. *Journal of Psychiatric Research, 18,* 477–490.

Weinberger, D. R., Torrey, E. F., Neophytides, A., & Wyatt, R. J. (1979). Lateral cerebral ventricular enlargement in chronic schizophrenia. *Archives of General Psychiatry, 36,* 735–739.

Weinberger, D. R., Torrey, E. F., & Wyatt, R. J. (1979). Cerebellar atrophy in chronic schizophrenia. *Lancet, 1,* 718–719.

Weinberger, D. R., Bigelow, L. B., Kleinman, J. E., Klein, S. T., Rosenblatt, J. E., & Wyatt, R. J. (1980). Cerebral ventricular enlargement in chronic schizophrenia, an association with poor response to treatment. *Archives of General Psychiatry, 37,* 11–13.

Weinberger, D. R., DeLisi, L. E., Neophytides, A. N., & Wyatt, R. J. (1981). Familial aspects of CT scans abnormalities in chronic schizophrenic patients. *Psychiatry Research, 4,* 65–71.

Weinberger, D. R., DeLisi, L. E., Perman, G. P., Targum, S., & Wyatt, R. J. (1982). Computed tomography in schizophreniform disorder and other acute psychiatric disorders. *Archives of General Psychiatry, 39,* 778–783.

Weinberger, D. R., Wagner, R. L., & Wyatt, R. J. (1983). Neuropathological studies of schizophrenia: A selective review. *Schizophrenia Bulletin, 9,* 193–212.

Wilkinson, D. A. (1982). Examination of alcoholics by computed tomographic (CT) scans: A critical review. *Alcoholism: Clinical and Experimental Research, 6,* 31–45.

Williams, A. O., Reveley, M. A., Kolakowska, T., Ardern, M., & Mandelbrote, B. M. (1985). Schizophrenia with good and poor outcome, II: Cerebral ventricular size and its clinical significance. *British Journal of Psychiatry, 146,* 239–246.

Winchester, P., Brill, P. W., Cooper, R., Krauss, A. N., & Peterson, H. deC. (1986). Prevalence of ''compressed'' and asymmetric lateral ventricles in healthy full term neonates: Sonographic study. *American Journal of Neuroradiology, 7,* 149–153.

Wing, J. K., Cooper, J. E., & Sartorius, N. (1974). *The measurement and classification of psychiatric symptoms.* Cambridge: Cambridge University Press.

Withers, E., & Hinton, J. (1971). Three forms of the clinical tests of the sensorium and their reliability. *British Journal of Psychiatry, 119,* 1–8.

Yates, W. R., Jacoby, C. G., & Andreasen, N. C. (1987). Cerebellar atrophy in schizophrenia and affective disorder. *American Journal of Psychiatry, 144,* 465–467.

10

Cerebral Imaging and Emotional Correlates

C. MUNRO CULLUM

INTRODUCTION

Emotional changes associated with cerebral damage are common and may range from depression to hypomania or complete denial of illness. Affective changes or reactions following a neurological event may vary depending on the nature, severity, and chronicity of the disorder and also may vary as a result of intraindividual factors such as age at onset, general health, additional physical sequelae, social supports, and premorbid personality. In addition to the psychological reactions to perceived alteration or loss of function that might be expected, emotional and psychological changes also may result as a direct consequence of damage to the brain (i.e., a neuroaffective disorder; see Figure 1).

Figure 1 depicts a highly simplified schematic overview of some of the primary components involved in emotional behavior. The solid lines represent the typical situation, wherein environmental events interact with the preexisting genetic and personality components of the CNS. The next sequence of events involves an individual's perception/interpretation of environmental and intrapsychic events, which then leads to an emotional response. It should be noted that these interrelationships are not unidirectional, and, for example, a person's emotional response may affect his/her perception and interpretation of external as well as internal events. In the case of a cerebral lesion (represented by the dotted lines), the lesion itself may directly influence CNS functions at essentially all levels, thereby causing any of a number of alterations in the emotional functioning of an individual. Of particular note is the potential direct effect of the lesion on emotional behavior (i.e, the neuroaffective effect); this may not only represent an interaction with the person's preexisting CNS status, personality, and experiences, but may also result in stereotyped affective output that may not be responsive to feedback via the typical cerebrocortical routes.

Clinically, the emotional and characterological changes that may occur as a result of cerebral damage may prove more devastating to patients and their families than the physical and cognitive sequelae that may be present (e.g., see Lezak, 1978). Consider, for example, the neurological patient who has undergone characterological changes as a result of injury: outwardly, in a physical

C. MUNRO CULLUM ● Department of Psychiatry, University of California at San Diego, La Jolla, California 92103; *present address:* Department of Psychiatry, University of Colorado Health Sciences Center, Denver, Colorado 80262.

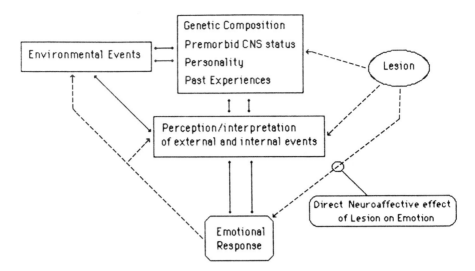

FIGURE 1. Schematic depiction of some of the interrelationships between acquired cerebral lesions and emotional functioning.

sense, he or she may be unchanged, yet interpersonally and behaviorally, the patient might appear a total stranger. The psychological impact of such a situation can be devastating to both the patient and the family, and the multifaceted effects of neurological injury pose critical issues to the assessment, management, and treatment of such patients.

Various studies have examined the cognitive effects of cerebral insult, and there has long been an interest in localization-of-function issues in neuropsychology. Investigations into the cerebral mechanisms determining mood, however, have been few in comparison with those concerned with cognitive processes. Still fewer have pursued the question of the possible localization of emotional or psychological functions (for general reviews, see Campbell, 1982; Tucker, 1981).

Even without regard to functional laterality differences, the association between cognition and emotion is widely acknowledged on experimental, clinical, and intuitive grounds (e.g., see Pribram, 1981; Schacter, 1964). Given the different cognitive processing propensities of the cerebral hemispheres (e.g., see Benson & Zaidel, 1985), it futhermore seems logical that damage to a cerebral system processing in a verbal, more sequential, analytic style might result in affective sequelae different from damage to a more nonverbal. holistically processing system.

Whereas the left hemisphere's general predominance in terms of language functions has been well established, the right hemisphere has been noted to play a relatively larger role in visuospatial functions and various aspects of emotion (Heilman, Bowers, Valenstein & Watson, 1986). Ross (1981, 1984), for example, has postulated the right hemisphere's predominant role in prosody, which includes the pauses, melody, intonation, and accents that are used during speech. These aspects of communication are responsible for the nonverbal, "emotional" components of speech by which we are able to convey our feelings regardless of the specific words used. A dysprosodic patient, for example, may speak in a low, monotone voice when expressing highly emotional feelings such as sadness, anger, or extreme happiness, which is in contrast to the intact individual, whose tone of voice quite often denotes his/her underlying mood. Ross has extended this idea and presented evidence to suggest that there are several types of aprosodia that may be associated with regional right hemispheric damage and that these parallel the aphasias (e.g., motor, sensory, transcortical) that are associated with dysfunction of homologous speech areas of the left hemisphere. It also has been suggested that the organization of affective language in the right hemisphere mirrors the organization of propositional speech in the left hemisphere (Ross, 1984; Ross & Mesulam, 1979).

RESEARCH CONSIDERATIONS

Studies of patients with relatively localized cerebral lesions have provided much of the information available with respect to the question of localization of emotional processes. In research with neurological subjects, however, there are a plethora of variables to consider, even though control of all of the potentially relevant biopsychosocial factors is not possible, given the inherent complexity of the problem. For one, the etiology of the disorder (e.g., cerebrovascular accident, neoplasm, head injury) should be considered, since different pathophysiological processes may produce different clinical manifestations (i.e., neurobehavioral as well as emotional), have different courses of recovery, and differentially affect various cerebral systems.

Another critical yet often overlooked variable is the extent of damage or lesion size. In view of the various cortical and subcortical regions that can be affected by lesions and the individual differences in cerebral organization that exist (Kertesz, 1983; also see Chapter 11), the assessment of lesion size and location is of particular importance (especially in lateralization or localization studies). Without specific data regarding the size and location of cerebral lesions, appropriate attributes of functional localization cannot be proposed.

Measurement of brain and lesion parameters have posed significant difficulties in the past, although techniques are available that permit relatively accurate estimates of lesion size and location using computerized tomography (CT) or magnetic resonance imaging (MRI). These techniques vary in methodology, complexity, and accuracy (see Chapter 3 for a review). Briefly, the more basic and common methods use linear measures to estimate various brain parameters such as lesion size. Such methods may inaccurately estimate actual lesion size, however, since the extent of damage may vary depending on the CT or MRI level from which the measurements are taken. More advanced computer-assisted techniques have moved from linear to volumetric estimates of brain and lesion size in an attempt to depict these parameters more accurately and reliably.

Some of the techniques such as those of Jernigan and co-workers (Jernigan, Zatz, & Naeser, 1979; Pfefferbaum, Zatz, & Jernigan, 1986) utilize digital information directly from the CT or MRI scanner, which allows for rather precise estimation of the volume of selected cerebral structures and areas of different density gradients. Whereas such sophisticated measures can provide relatively accurate estimates of lesion size, unless the digital information is retained from the scanning session (and many clinical CT and MRI labs do not routinely do so), these estimated brain and lesion parameters cannot be obtained. Other computerized volumetric methods of lesion size estimation allow for the calculation of volumetric cerebral and lesion parameters from the familiar "hard copy" of CT or MRI films (e.g., see Cullum & Bigler, 1986; Turkheimer, Yeo, & Bigler, 1983). These techniques have been shown to be reliable (Yeo, Turkheimer, & Bigler, 1983), and although they may be less precise and potentially more subject to measurement error than some of the methods that use information directly from CT or MRI scanners, their relative simplicity may make them more widely applicable. Furthermore, since all of these techniques involve various user-defined guidelines in delineating intracranial structures, it should be noted that some degree of subjectivity and measurement error is inherent in each of the procedures mentioned.

In addition to consideration of lesion size, the stage of recovery from cerebral damage is an important factor in research with neurological patients and certainly impacts on emotional sequelae as well. Depending on the nature and severity of insult, the time of evaluation post-onset can be a critical variable in terms of clinical status. Rates of recovery vary among individuals, and in localization studies in particular, the patients utilized need to be similar in terms of recovery stage.

This can be especially important in investigations of emotional processes, too, since the immediate postinjury affective state may differ from more long-term, residual changes. Thus, there are a number of relevant factors that merit consideration in examining the affective sequelae of cerebral lesions.

FOCAL AND ASYMMETRIC EMOTIONAL PROCESSING EVIDENCE

Epilepsy Studies

In an often-cited investigation, Bear and Fedio (1977) evaluated self-report interictal personality attributes of temporal lobe epileptics with right and left hemispheric epileptogenic foci. The right-temporal-lobe subjects reportedly exhibited greater mood lability and symptoms of denial, and left-temporal-lobe subjects tended toward self-deprecation and "catastrophic" overemphasis of negative behavioral qualities (observer ratings of the same patients, however, revealed opposite and equally significant findings). A number of investigations in corroboration with Bear and Fedio's results have subsequently been published and are reviewed by Flor-Henry (1983). Recently, Mendez, Cummings, and Benson (1986) found a significantly increased incidence of depression among epileptic patients when compared with matched control subjects. Their data further suggested an association between left hemisphere seizure foci and depressive symptomatology, although these authors pointed out that "any relationship of a unilateral seizure focus to affective disorders remains controversial" (p. 769).

The majority of studies of personality attributes associated with epilepsy have involved the assessment of interictal behaviors and self-perceptions. Interictal assessment of subjects may further complicate conclusions, however, as the specific contribution of epileptogenic lesions per se to the observed personality/emotional results cannot be ascertained. To address this issue, Sackheim et al. (1982) reviewed the literature with regard to the incidence of emotional outbursts of laughing and crying as ictal phenomena. As in pathological affect, such outbursts reportedly are unrelated to external circumstances and may or may not involve concomitant subjective mood changes (Sackheim et al., 1982). In reviewing the reported cases, it was found that 88% presented largely with laughing outbursts (gelastic epilepsy), 6% with crying (dacrystic epilepsy), and the remaining cases with a combination of the two. Laughing outbursts were significantly associated with left hemisphere seizure foci, and although the number of dacrystic epilepsy cases was small, there was a tendency (4 out of 6 cases) for right-sided foci to be associated with crying. In a most interesting case study, Hurwitz, Wada, Kosaka, and Strauss (1985) recently reported on a patient with independent left- and right-sided temporal lobe seizure foci who displayed differential affective features depending on the side of epileptogenic discharge. Following left-side discharges, the patient reportedly became aphasic and depressed, yet after right-sided events, she evidenced laughing and postictal hypomania.

Based on the pathological affect and epilepsy literature, Sackheim et al. (1982) concluded that the left hemisphere typically plays a greater role in regulating positive emotion, and the right hemisphere generally subserves negative emotion to a greater extent. Whereas these conclusions may appear to be in contrast with the interictal and postictal findings of some other studies (e.g., Bear & Fedio, 1977), it should be noted that the literature reviewed by Sackheim and associates pertained only to ictal displays of affect and thus may represent somewhat different (and possibly more "pure") neurobehavioral mechanisms. Hemispheric differences in terms of the regulation of positive (left hemisphere) and negative (right hemisphere) emotions also have been noted in studies of normal subjects using various procedures (Ahern & Schwartz, 1985; Dimond, Farrington, & Johnson, 1976), although there nevertheless exists controversy regarding such findings. In a recent review of the gelastic epilepsy literature, Myslobodsky (1983) concluded that although outbursts of laughter as ictal phenomena do tend to be associated with left hemisphere foci, the mechanisms involved are not sufficiently understood to warrant firm conclusions.

Ross and Stewart (1987) reported on two patients with a history of depression who, with the onset of right inferior frontal lesions, began to evidence displays of pathological crying. These authors speculated that an interaction between structural and physiopharmacological brain disturbances may lead to abnormal behaviors. Specifically, disorders such as depression may result in state changes in the temporal limbic system that may lower the threshold for extreme, mood-congruent affective behaviors (p. 170). They go on to suggest that the display of pathological affect following right frontal opercular lesions may be a clinical sign of an underlying depressive disorder, thereby further attesting to some of the complex interactions that may occur in studies of the neurological basis of emotion.

Psychiatric Disorders

Although the topic of neuroimaging and psychiatric disorders is dealt with in detail in Chapter 12, some brief mention of some of the popular findings as they pertain to emotional functions is in order. When cerebral structural abnormalities have been reported in psychiatric patients, the findings typically have been generalized (i.e., ventricular dilation, cortical atrophy), without specific focal phenomena. As an example, ventricular dilation has been reported in some schizophrenic patients (Luchins, Lewine, & Meltzer, 1984; Nasrallah & Coffman, 1985; Weinberger, 1984), although such findings are not specific to schizophrenia and are not observed in all schizophrenics. Some studies, in fact, have found no evidence of ventricular enlargement among schizophrenics (Jernigan, Zatz, Moses, & Cardellino, 1982; Largen et al., 1984; Shima et al., 1985).

In terms of emotional/mood disorders, similar findings in patients with bipolar affective disorder have been reported (Nasrallah, McCalley-Whitters, & Jacoby, 1982; also see Nasrallah & Coffman, 1985), although here, too, the relationship of such findings to clinical symptomatology remains unclear (aside from the implication of some cerebrostructural abnormality, i.e., a generalized reduction of cerebral tissue). Ventricular enlargement in patients with a diagnosis of major depression also has been reported (Shima et al., 1984; Standish-Barry, Bouras, Hale, Bridges, & Bartlett, 1986), even in the absence of neuropsychological and neurological findings (see Figure 2).

Affective Asymmetry and Localized Cerebral Damage

In terms of laterality differences in emotional behavior following cerebral damage, a number of studies and clinical anecdotes have noted differential emotional sequelae depending on the hemispheric side of the lesion. One of the earliest systematic observations was that of Goldstein (1948), who noted that patients with left hemisphere damage tended to exhibit depressive, "catastrophic reactions" (i.e., an exaggerated negative response far beyond what would normally be expected). Also around that time, patients with damage to the right hemisphere were reported to demonstrate a tendency toward rather euphoric reactions, indifference, or denial of illness altogether (Denny-Brown, Meyer, & Horenstein, 1952; Hecaen, 1962). These observations have since been supported by a number of authors (e.g., Heilman, Watson, & Bowers, 1983; Lishman, 1973).

In an often-cited study of 160 patients with lateralized brain damage, Gainotti (1972) observed that left-hemisphere patients exhibited anxious–depressive reactions, swearing, tearful outbursts, and depressive self-statements in response to failure on a battery of neuropsychological measures. Conversely, patients with right hemisphere damage tended to exhibit indifference reactions, which include minimization, joking, and anosognosia. It should be noted that the "catastrophic" reactions observed with left hemisphere damage occurred primarily among those subjects with aphasia, and the indifference reactions of the right-hemisphere-damaged patients were associated with some degree of contralateral neglect.

With unilateral carotid artery injections of sodium amobarbital used to temporarily anesthetize one hemisphere, some similar results have been obtained, with sad or depressive reactions

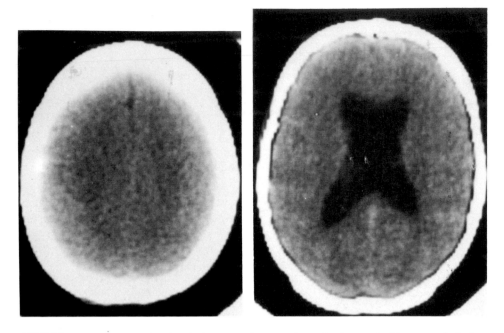

FIGURE 2. This 34-year-old patient had severe endogenous depression but no significant neurological or neuropsychological findings. Note the mild-to-moderate ventricular enlargement on CT.

reportedly appearing when the left hemisphere is anesthetized and more indifferent or euphoric reactions when the right hemisphere is inactivated (Bryden, 1982; Heilman *et al.*, 1983; Nebes, 1978). Although the majority of published studies using these techniques generally have tended to lend some support to these observations, there have been others who have indicated no differential effects for emotional response during hemispheric anesthetization (e.g., see Milner, 1974). It is likely that various methodological factors contribute to the mixed findings across studies, although an important point to be noted is that attempting to study emotional processes using the amobarbital technique is a most difficult (if validly possible) task.

Further support for the notion of some degree of asymmetric affective organization/processing derives from a variety of techniques and observations in normal as well as brain-injured individuals. The recognition and perception of emotion in tachistoscopically presented faces, for example, consistently has demonstrated a left visual field/right hemisphere advantage among normal subjects, even when the effects of facial recognition per se are controlled for (Bryden & Ley, 1983; Safer, 1981). Similar results of a right hemisphere advantage for recognition of emotional stimuli have come from dichotic listening studies (Bryden, 1982; Ley & Bryden, 1982). Furthermore, there is evidence to suggest that the facial expression of emotion may be asymmetric, with the left side of the face exhibiting emotions more strongly, thereby possibly suggesting a greater role for the right hemisphere (Borod & Caron, 1980; Etcoff, 1986; Moscovitch & Olds, 1982; Sackheim, Gur, & Saucy, 1978). Among brain-injured patients, right-hemisphere-damaged individuals have been found to be more dysfunctional than their left-hemisphere-damaged counterparts on a variety of affect-related tasks (Borod, Koff, Lorch, & Nicholas, 1985; Buck & Duffy, 1980; Gainotti, 1979; Heilman, Scholes, & Watson, 1975; Tucker, Watson, & Heilman, 1976).

In terms of the nature of the affective changes that may be seen in patients with cerebrovascular accidents (CVA), depressive symptoms reportedly occur in approximately 30 to 60% of poststroke patients (Folstein, Maiberger, & McHugh, 1977; Robinson, Starr, Kubos, & Price,

1983). Furthermore, the association between depression and cerebral damage appears to represent more than a functional reaction to loss of ability and has been demonstrated to be relatively independent of degree of impairment (Finklestein *et al.*, 1982; Folstein *et al.*, 1977; Robinson & Szetela, 1981).

In a series of studies of depressive symptomatology in poststroke patients, Robinson and colleagues have reported the highest incidence of depression among patients with left hemisphere lesions (Lipsey, Spencer, Rabins, & Robinson, 1986; Robinson & Price, 1982; Robinson, Lipsey, & Price, 1985). In addition, lesion location has been found to be associated with depressive symptomatology as measured on a variety of clinical instruments in the majority of these studies. Specifically, the distance of the lesion from the frontal pole (as visualized on CT films) among left-hemisphere-stroke patients was found to be significantly (negatively) associated with depression severity. That is, patients with more anterior left hemisphere lesions demonstrated greater symptoms of depression. A significant positive correlation between lesion size and depressive symptomatology also has been reported, although this effect may be mediated by lesion duration effects (Robinson, Starr, Lipsey, Rao, & Price, 1985). Among right-hemisphere-stroke patients, a significant positive correlation between lesion distance from the frontal pole and depression severity was found (Robinson, Kubos, Starr, Rao, & Price, 1984), suggesting both inter- and intrahemispheric effects of lesions with respect to depressive symptomatology. It was also reported that right frontal lesions were associated with inappropriate cheerfulness and hypomania.

To help explain such findings, Robinson and co-workers (e.g., see Robinson *et al.*, 1984) postulated that the ''lateralized frontal affective syndrome'' observed with left frontal lesions may be ''.   a behavioral manifestation of asymmetrical depletion in the cortical biogenic amine pathways with more anterior lesions producing greater depletion of these amines'' (p. 91). Some related support for this hypothesis is derived from animal studies, as greater disruption of noradrenergic fibers has been reported to be associated with anterior cortical lesions (Morrison, Molliver, & Grzanna, 1979), and a global but asymmetric depletion of catecholamine concentrations has been reported, depending on the laterality of the lesion (Robinson, 1979). Such a hypothesis would appear to be highly plausible, particularly in terms of helping to explain the relationship between brain-lesion-induced and so-called endogenous depression, which have been reported to be similar in clinical symptomatology (Berrios & Samuel, 1987; Lipsey, Spencer, Rabins, & Robinson, 1986).

Despite the relative consistency of the popularly reported findings of Robinson and colleagues, however, other investigators have found that lesion location along left–right dimensions is not reliably associated with depressive symptomatology (Feibel & Springer, 1982; Finklestein *et al.*, 1982). Still others have found greater depression among stroke patients with right as opposed to left hemisphere pathology (Folstein *et al.*, 1977). Similarly, in a recent study of patients with penetrating brain wounds, Grafman, Vance, Weingartner, Salazar, and Amin (1986) reported that right orbitofrontal lesions were associated with increased anxiety and depression, whereas left dorsofrontal lesions were more associated with feelings of anger and hostility. Thus, the specific intrahemispheric location of a lesion may also be an important variable and may contribute to the some of the inconsistent results across studies.

In a recent attempt to replicate some of the findings of Robinson and co-workers, Sinyor *et al.* (1986) obtained CT and depression measures in patients with focal left and right cerebral lesions. The relationship between proximity of the lesion to the frontal pole was statistically significant among right-hemisphere-damage patients and approached a significant level among the left-hemisphere-damage patients. However, no laterality differences in terms of depressive symptomatology as measured on a number of standard clinical scales were observed. Various methodological and sample differences (e.g., time since onset, lesion size, sample characteristics, specific procedures) may account for the differential findings across investigations, although it appears that the relationship between depressive symptoms and lesion parameters is highly complex (i.e., nonlinear) and may be influenced by a number of factors. To further illustrate this point, whereas Robinson and co-workers have reported significant positive correlations between left frontal lesion size and depressive symptoms (Robinson et al., 1984, 1985), a recent study by their group reported significantly smaller lesion volumes among depressed stroke patients when compared with nondepressed stroke patients matched for global cognitive ability (Robinson, Bolla-Wilson, Kaplan, Lipsey, & Price, 1986).

In addition to the depressed patients with right inferior frontal lesions reported by Ross and Stewart (1987), Grafman et al. (1987) found that patients with right orbitofrontal lesions tended to display increased anxiety and depression, whereas left dorsofrontal lesions tended to be associated with increased anger and hostility. Thus, from the various results of the published studies in this area, it appears that the association between cerebral lesions and their emotional sequelae is highly complex, and clear–cut brain-behavior relationships may not be attainable given our present state of knowledge and technology.

PERSONALITY ASSESSMENT WITH NEUROLOGICAL PATIENTS

The consideration of personality and mood factors in the evaluation of neurological patients is a critical yet too often underemphasized practice. The presence of brain damage per se does not, for example, eliminate the effects of preexisting intraindividual dynamics, abilities, and emotional attributes. Neurobehavioral and neuroaffective sequelae of circumscribed cerebral lesions may vary among individuals depending on a host of factors, including premorbid personality. Moreover, the effects of brain damage cannot be viewed in strict isolation from these other factors (see Figure 1). As Maloney and Ward (1976) stated, "The lesion exists in a unique person whose age, premorbid level of functioning, personality variables, supports and demands of his environment play a significant role in determining the effects of lesions" (p. 264).

The patient's attitude, level of cooperation, and motivation are known to be critical factors in the neuropsychological evaluation of patients with neurological disorders. In some cases, the changes in emotional and social behavior patterns that may be associated with particular types of cerebral dysfunction also may play a critical role in the diagnostic process (e.g., the disinhibited, inappropriate behavior seen in some frontal lobe patients). Thus, since changes in affective responsivity and personality are not uncommonly associated with damage to the brain, these factors merit investigation and should be routinely addressed in the neuropsychological and neurobehavioral evaluation.

The importance of affective and personality assessment in neurological patients is also indicated in situations dealing with recovery of function and rehabilitation issues. Neglect of such emotional factors may result in greater frustration in the patient as well as in those working with him or her. For example, whereas some degree of depressive symptomatology naturally may be seen following any loss of cognitive or physical abilities, significant biological or endogenous (neuroaffective) depression also may result from the brain injury and, without proper assessment and treatment, actually may hinder the recovery and rehabilitation process (Robinson et al., 1986; Ross & Rush, 1981; Sinyor et al., 1986). Furthermore, it has been noted that patients with either "functional" or "psychogenic" poststroke depressive disorders tend to demonstrate similar symptoms (Berrios & Samuel, 1987; Lipsey et al., 1986), and there is some evidence to suggest that depressive symptoms in stroke patients also may respond to psychopharmacological intervention (Lipsey, Robinson, Pearlson, Rao, & Price, 1984).

Some of the inherent difficulties in the assessment of personality factors in patients with acquired brain damage include the effects of premorbid personality traits, the patient's psychological reaction to the disorder, and the personality changes that may occur as a direct result of cerebral damage. Since premorbid personality data aside from qualitative information gathered from the patient and others familiar with him or her is not available in most cases, normative parameters (and comparisons with depressed groups of patients) generally are relied on. This also is the case, albeit to a lesser extent, with cognitive/intellectual abilities, and the limitations of estimating premorbid attributes in individual patients are obvious. Despite these limitations, however, such issues are of relatively less importance when the goal is the comparison of patients with lateralized brain damage. If patients with similar lateralized lesions are found to differ in terms of certain personality variables, for example, the differences may be more confidently attributed to lateralized affective responses, provided the effects of various background variables (e.g., age, sex, education, intellectual status) are adequately considered.

In terms of personality/emotional assessment instruments, the Minnesota Multiphasic Personality Inventory (MMPI) is the most widely used (Lubin, Larsen, & Matarazzo, 1984). The standard form consists of 550 empirically derived true–false items in self-report format that are associated with test-taking attitudes and various facets of psychopathology. One major concern regarding the MMPI is that traditional interpretation of the standard validity and clinical scale profiles may not be applicable to patients with brain damage. In fact, although some authors question the usefulness of the MMPI in evaluating personality dimensions in patients with neuropsychological disabilities (e.g., Lezak, 1983, p. 607), others such as Dikmen and Reitan (1977) have suggested that a careful selection of the MMPI items that are sensitive to emotional issues in brain-damaged patients might provide more accurate information with respect to the emotional status of these patients. They and others have cautioned that abnormal MMPI profiles among neurological patients may represent feelings and symptoms naturally associated with the disorder and may not have traditional psychopathological implications.

In terms of its clinical use in neurological patients, the data obtained from the MMPI (or any other psychometric instrument) should be examined in the context of all available historical and neuromedical information, behavioral observations, and neuropsychological assessment data. One asset of the MMPI with certain neurological subjects is that the self-report nature of the instrument may be particularly useful, since the overt affective appearance of some patients may not be consistent with their subjective feelings. In a comprehensive review of MMPI research in patients with brain damage, Mack (1979) concluded that although the findings must be interpreted carefully, the MMPI can provide useful information with regard to the personality adjustment of neurological patients. Mack aptly added, however, that many methodological and theoretical issues remain in the area that need to be addressed (p. 68).

One of the major limitations of the MMPI is that the standard 550-item form is quite lengthy, particularly for patients with compromised cerebral functioning and attentional deficits. Motivation, concentration, and potential fatigue effects are critical factors in obtaining valid responses on essentially all psychometric measures, and significant deficits in these areas may particularly confound the accuracy of patients' self-reports. To help circumvent some of these difficulties and enhance practicality, a number of short forms of the MMPI have been developed but have not been widely used in published studies with neurological populations. Such abbreviated versions are not without limitations, however (e.g., see Butcher & Tellegen, 1978; Green, 1980), although some such as the 166-item Faschingbauer Abbreviated MMPI (Faschingbauer, 1974) have demonstrated relatively good reliability with respect to the standard long-form MMPI (Faschingbauer, 1974; Willcockson, Bolton & Dana, 1983). The Faschingbauer MMPI was developed using cluster analysis techniques with the original MMPI items. Any Faschingbauer MMPI scales demonstrating less than a 0.85 correlation with the original scales were altered by the addition or deletion of items until this criterion was met. The resultant scales contain between 46% and 67% of the full-scale items, and the correlations with original MMPI scales range from 0.85 to 0.93 (Faschingbauer, 1974). It should be noted, however, that although none of the short forms of the MMPI can be considered equivalent to the standard instrument (Butcher & Tellegen, 1978) and thus must be interpreted with caution, the Faschingbauer MMPI appears to offer a reasonably good estimate of the overall presence or absence of psychological maladjustment (Skenazy & Bigler, 1985) and is easily administered.

The MMPI and Lateralized Cerebral Lesions

A number of studies have examined MMPI profiles among patients with lateralized brain damage, and the overall results have been mixed (although more recent studies have tended to be more consistent). The earliest published reports of the effects of unilateral brain damage on MMPI profiles indicated no relationship between lesion laterality and differential personality disturbance (Meier & French, 1965; Vogel, 1962). However, comparisons of groups of patients with lateralized brain damage mandate a careful examination of a host of potentially confounding variables, including the presence and extent of cognitive and aphasic deficits.

An early study by Doehring and Reitan (1960) indicated essentially no differences between aphasic and nonaphasic brain-damaged patients on the MMPI, although a later series of studies by Dikmen and Reitan (1974a, 1977) found that MMPI variables were related to general adaptive abilities and the presence of aphasia. Aside from such relationships, these authors reported that differential MMPI performance was not, however, related to lesion laterality per se (Dikmen & Reitan, 1974b). Although often ignored, consideration of the extent of aphasic symptoms is of critical importance in studies of this type, since the MMPI requires intact reading (unless orally administered) and comprehension abilities. Along these lines, Robinson and Benson (1981) found that type of aphasia was specifically related to depression as assessed on several depression-rating scales. Furthermore, patients with anterior lesions and nonfluent aphasia were found to evidence greater depressive symptoms than fluent aphasic patients with more posterior lesions. Such findings were supported by the recent findings of Sinyor et al. (1986) and Robinson et al. (1984, 1985) wherein anterior left hemisphere lesion location was found to be associated with increased depressive symptomatology.

Using the MMPI and Reitan–Indiana Aphasia Screening Test scores, Dikmen and Reitan (1974b) found that the presence of aphasia was significantly related to depressive symptomatology as assessed by MMPI scale 2 (D).* At the same time, lesion laterality per se (aside from associated aphasic and cognitive deficits) was not related to MMPI variables in any of the Dikmen and Reitan studies. A more recent investigation by Black and Black (1982) likewise revealed no relationship between the side of lesion and differential MMPI performance despite the inclusion of some patients with aphasic deficits.

Gasparrini, Satz, Heilman, and Coolidge (1978) compared the MMPI profiles from LHD and RHD patients who did not differ in terms of general cognitive or motor abilities and who exhibited "no detectable signs of dysphasia." A multivariate analysis of variance indicated that the groups were different in terms of overall MMPI profiles. Further analysis revealed that only scale 2 significantly differentiated the groups, with the LHD patients exhibiting significantly higher scores. The authors concluded that hemispheric differences are related to MMPI performance (particularly on scale 2) and that such differences are not attributable to cognitive or expressive deficits. The findings of this investigation must be interpreted with caution, however, in view of the small sample size ($N = 24$). Furthermore, the indication by these authors of similar verbal and performance IQ scores and similar motoric performances in their patients with lateralized brain damage cells to question sample composition issues in view of the commonly reported association between differential effects of right versus left brain damage on verbal and performance IQ measures (Russell, 1979), not to mention lateralized motor deficits.

To address more carefully the relationship between the degree of aphasia and MMPI status, Gass and Russell (1986) examined the MMPI profiles of patients with left hemisphere damage who demonstrated signs of no, mild, and moderate aphasic deficits as reflected by Aphasia Screening Test scores. Results indicated no significant relationship between degree of aphasic deficits and any MMPI scales when the effects of education were statistically controlled. Cullum and Bigler (1988), using similar procedures, recently replicated these findings in a large group of neurologic patients.

---

*See Appendix for a key to MMPI scale names and numbers.

Despite the apparent consistency of some of the previous findings indicating no laterality effects on MMPI results, a number of investigators have found significant lateralization effects. Furthermore, when differences have emerged, it generally has been found that patients with left hemisphere damage (LHD) tend to exhibit greater psychopathology on the MMPI. As noted previously, Gasparrini et al. (1978) found significantly higher scale 2 scores among their left-hemisphere-damaged patients. In a study of patients with lateralized penetrating missile wounds matched for age, education, and time since injury, Black (1975) found the MMPI profiles of those with right hemisphere damage (RHD) to be within normal limits. The LHD subjects, however, exhibited significant elevations on clinical scales 2 (D), 8 (Sc), and 1 (Hs). This was interpreted to reflect the depressive, catastrophic reaction previously mentioned in relation to left hemispheric damage. In a study of 15 pairs of patients with lateralized brain damage matched for age and neuropsychological status, Louks, Calsyn, and Lindsay (1976) found that subjects with RHD demonstrated generally more "neurotic" MMPI profiles (67%), whereas the LHD subjects' profiles appeared more "psychotic" (80%) as reflected by the Goldberg (1965) Neurotic–Psychotic Index.

The specific effects of cognitive deficits on MMPI performance have been carefully addressed in only a few studies, with early results indicating significant relationships between verbal intellectual deficits and increased MMPI psychopathology (Dikmen & Reitan, 1977; Osmon & Golden, 1978). A more recent study by Gass and Russell (1985), however, found no significant association between WAIS Verbal IQ and MMPI performance across groups of patients with below-average, average, and above-average Verbal IQ scores when the effects of education were statistically controlled. Support for these latter findings was recently provided by an MMPI analysis of neurological patients with WAIS-R Full-Scale IQ scores above and below 100, wherein similar overall profiles were obtained (Cullum, 1986).

In terms of overall MMPI differences between LHD and RHD patients, Woodward, Bisbee, and Bennett (1984) found significant MMPI profile differences between 21 LHD and 19 RHD patients using multivariate statistical techniques. In contrast to the majority of previous studies, however, the left-hemisphere patients in their sample had MMPI profiles within normal limits, and right-hemisphere patients demonstrated a number of significant elevations on several of the clinical scales [8, 2, 3 (Hy), and 4 (Pd)]. The authors presented several methodological points of contrast between their study and previous works, citing duration of disorder as a potentially crucial variable. They noted, for example, that the subjects in Black's (1975) study were evaluated within 4 months of penetrating head injuries. Although data regarding the time since onset of the disorder in the heterogeneous (i.e., vascular lesion, tumor, focal trauma, surgery, and "other") sample of Woodward et al. (1984) were not available, they estimated that most of the conditions had been in existence for longer than 4 months.

Lezak (1983) discusses the differences in emotional reactions initially following insult and the more enduring affective changes that may appear over time. She noted that patients with left hemisphere language deficits initially react with the depression but over time begin to adapt and compensate for their deficits, with a concomitant decrease in depressive feelings. On the other hand, patients with right hemisphere lesions tend to appear more indifferent, but over time may become depressed as their lack of awareness and poor perception of emotional cues (and thus greater likelihood of interpersonal difficulties and alienation) and unmet, unrealistic goals may lead to increasing dysphoria. Thus, according to Lezak (1983), depression in patients with RHD may be more delayed, but when it does occur it is more likely to be more chronic, debilitating, and resistant to intervention (p. 62).

The importance of lesion duration on measures of depression was highlighted in a longitudinal study of CVA patients wherein Robinson et al. (1985) found that the relationship between lesion location and severity of depression in RHD patients changed or reversed by 3 or 6 months post-stroke. In the Cullum and Bigler (1987) study, examination of the composite MMPI profiles of LHD and RHD patients with short (less than or equal to 12 months) and long (greater than 12 months) durations revealed generally higher scores overall among patients with longer lesion durations, particularly among the RHD patients. This finding of lower average MMPI scores across RHD and, to a lesser extent, LHD subjects with shorter lesion durations may lend some tentative support to the notion of greater indifference reactions (as reflected in lower MMPI scores) among short-duration (particularly RHD) patients.

The finding of greater depressive symptomatology among RHD as opposed to LHD patients in the Woodward *et al.* (1984) MMPI study also was interpreted as consistent with such a notion, since the average time since onset of disorder in their study was estimated to be greater (''considerably more than 4 months,'' p. 967) than in some previous investigations. Whereas the aforementioned studies were cross-sectional in nature, such findings do provide support for the idea of increased depression and general psychological distress among RHD (and, to a lesser extent, LHD) patients over time. A longitudinal and more multidimensional follow-up of LHD and RHD patients would be needed to better address this issue.

Recently, Gass and Russell (1986), based on the work of Gass (1985), analyzed the MMPI profiles of a relatively large sample ($N = 90$) of patients with lateralized cerebral lesions. By multivariate statistical techniques, comparison of the composite MMPI profiles of LHD and RHD groups failed to reveal any significant overall differences. Both groups produced highly similar profiles that generally were within normal limits, with the highest scores in both groups on scale 2.When examined in terms of the number of patients within each group obtaining MMPI T scores above 70 across all scales, similar patterns again were observed, with most differences (expressed as the percentage of MMPI T scores above 70) between groups being less than 6%. The overall results suggested that, in general, lesion laterality had little effect on MMPI performance. Information regarding lesion size, precise location, and duration was unavailable, however, and the authors suggested that these variables be examined in future studies of the MMPI in patients with lateralized cerebral lesions.

Although the number of MMPI studies of individuals with lateralized cerebral damage is relatively small, the overall results have been inconsistent. This is underscored by the negative laterality results of the Black and Black (1982) study, which used some of the same subjects that were used in an earlier investigation (Black, 1975), wherein significant MMPI laterality effects were found. Similar inconsistencies are evident in those MMPI studies examining the effects of anterior versus posterior lesion location. Whereas some positive results have been found, generally indicating elevated MMPI scores among patients with posterior lesions (Anderson & Hanvik, 1950; Black & Black, 1982; Williams, 1952), other investigations have produced negative results, indicating no effects of anterior–posterior lesion location on the MMPI (Dikmen & Reitan, 1974a,b, 1977; Woodward *et al.*, 1984).

The inconsistency in the literature with respect to lateralized cerebral lesion effects on MMPI profiles may be the result of a host of methodological issues. First, the sample sizes in the majority of published studies have been relatively small (typically 10–20 per group), thereby significantly limiting the power of the statistical analyses and the generalizability of results. Furthermore, several investigations employed multiple *t* tests (without adjusting significance levels) in comparing MMPI scale scores, thereby capitalizing on finding positive but spuriously significant results. Multivariate techniques, on the other hand, may be too global and potentially insensitive to the more subtle and specific differences (e.g., depressive symptomatology) that may exist between LHD and RHD patients in terms of psychological status.

Yet another potentially critical methodological factor is that the overwhelming majority of investigations using the MMPI have included patients of various etiologies in their samples. Whereas unilateral cerebral damage may have similar sequelae across diagnostic categories, it is nevertheless known that different neuropathological processes can and do produce differential behavioral effects. To illustrate, patients with cerebral tumors typically have been included in MMPI studies of lateralized brain damage along with HI and CVA patients grouped together. Whether these patients represent similar cases from the neuropathological standpoint is an obvious question (i.e., considering lesion size, growth rate, pressure effects, duration, pre- versus postoperative status, etc.), with further implications for psychological effects per se.

Although there reportedly is little evidence to suggest that different neurological disorders produce discernably different personality adjustment patterns (Mack, 1979), a more rigorous investigative strategy is to compare different etiological groups with similar lesion parameters. Along with this, it should be noted that whereas lesions restricted exclusively to one hemisphere are not uncommon, the effects generally will be more widespread and involve other regions to some degree. This is particularly true with traumatic brain injury and disconnection effects of focal lesions, as these may have additional nonspecific effects throughout the brain (see Chapter 5). Other potentially relevant variables that have not been thoroughly addressed in many of the previous studies using the MMPI include the effects of lesion size and location, severity of accompanying (i.e., motor or sensory) deficits, and duration of the disorder.

Recent MMPI Findings in Patients with Lateralized Cerebral Lesions Identified by CT

Cullum and Bigler (1987, 1988) conducted a series of studies based on the work of Cullum (1986) to attempt to address many of the aforementioned issues, replicate some previous findings, and help lend further clarification to the literature in the area of MMPI findings in patients with predominantly lateralized cerebral lesions as visualized on CT. The overall results of this series of investigations using the Faschingbauer abbreviated MMPI (Faschingbauer, 1974) suggested that predominantly unilateral cerebral damage (LHD, $N = 47$; RHD, $N = 47$) tends to be associated with similar levels of psychological disturbance as reflected on overall short-form MMPI profile configurations. The similarity of MMPI profiles of LHD and RHD patients has been reported in the majority of previous studies using the standard MMPI in patients with lateralized brain damage (Black & Black, 1982; Dikmen & Reitan, 1974b; Gass & Russell, 1986; Meier & French, 1965; Reitan, 1976; Vogel, 1962). Such findings, using a variety of statistical techniques for comparison, have lent support to the hypothesis that cerebral damage, regardless of laterality, tends to result in similar overall patterns of psychological distress as assessed by the MMPI, although a number of potentially relevant factors have been relatively unexplored in previous studies.

Patients with evidence of primarily diffuse cerebral damage ($N = 59$) were included by Cullum and Bigler as a "control" group in order to help ascertain whether MMPI findings might be a result of lateralized cerebral dysfunction or simply a result of brain damage per se. The resultant overall profile of the diffuse-damage group proved to be similar to both lateralized groups, although it more closely resembled the RHD profile (see Figure 3). The majority of short-form MMPI scores across all three groups were within normal limits, with the highest scores on scales 8 and 2. Similar 8–2 or 2–8 profiles have been noted previously in LHD (Black, 1975; Gasparrini et al., 1978; Gass, 1985; Woodward et al., 1984) and RHD patients (Woodward et al., 1978). In a study that included a sample of patients with traumatic head injury, similar elevations on MMPI scales 8 and 2 were observed by Heaton, Smith, Lehman, & Vogt (1978). Such findings suggest moderate to high levels of anxiety, depression, and confusion in such groups of brain-injured patients, and this appears to characterize these groups largely irrespective of the primary location of damage.

With regard to the often-reported association between LHD and depressive symptomatology, the Cullum (1986) and Cullum and Bigler (1988) results suggested that, although statistically nonsignificant, a trend toward significance could be inferred from the LHD–RHD difference between mean scale 2 scores, and this trend remained when the effects of age were statistically controlled for. Furthermore, and probably more importantly, the LHD sample was the only group to evidence a significant (above $T = 70$) elevation on scale 2. The tendency for LHD patients to endorse scale 2 items more frequently also was highlighted in the comparison with the diffuse-damage group, wherein scale 2 scores were not significantly elevated.

Further tentative support for the association between LHD and MMPI evidence of depression was gained from the examination of the frequency of significant scale 2 elevations among LHD,

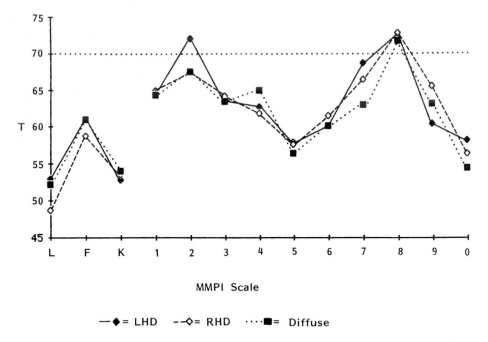

FIGURE 3. Composite short-form MMPI profiles of LHD, RHD, and diffuse-damage patients. See Appendix, MMPI Scales, for key to scale names. (From Cullum, 1986.)

RHD, and diffuse-damage groups. It was found that approximately 62% of LHD patients, as compared to 42% of RHD and 40% of diffuse damage patients, demonstrated significant scale 2 elevations (see Table 1). This represents the largest difference between groups on any of the MMPI scores, and comparison of the distributions between LHD and RHD groups on scale 2 approached statistical significance at the $P = 0.05$ level [X^2 (1) = 3.45, $P = 0.06$].

In a mixed group of patients with lateralized brain lesions, Gasparrini et al. (1978) found significant scale 2 elevations among 44% of LHD and 0% of RHD patients who were similar in terms of overall cognitive and motor status. Similarly, in a particularly well-done study of the MMPI and lateralized cerebral damage, Black (1975) examined the MMPI profiles of 35 relatively homogeneous patients with lateralized penetrating brain injuries. Whereas the overall profile configurations for LHD and RHD groups were somewhat similar, the LHD sample demonstrated significant elevations on scales 1, 2, and 8, with 2 and 8 as the high-point pair. In contrast, all scales of the RHD group were within normal limits.

In studies of lateralized cerebral damage that have utilized indices of depression other than the MMPI, the association between depressive symptoms and left hemisphere damage generally has received greater support. From the observations of Hecaen (1962) to the often-cited study of Gainotti (1972) and the series of investigations by Robinson and co-workers (1982, 1983, 1984, 1985, 1986), depressive symptomatology has rather consistently been observed to be more commonly associated with damage to the left hemisphere. As noted, other convergent lines of evidence include studies of epilepsy (Bear, 1983; Mendez et al., 1986), pathological affect (Sackheim et al., 1982), and observations during sodium amobarbital testing (Bryden, 1982; Heilman et al., 1983). Differential emotional sequelae of lateralized cerebral damage have also been supported by the "indifference" or "euphoric" reactions that are not uncommonly observed in patients with RHD but are much less frequent in cases of LHD (Hecaen, 1962; Heilman et al., 1983).

TABLE 1. Percentage of Short-Form MMPI T Scores over 70 by Lesion Group

MMPI scale[a]	Left (N=47) Percentage	N	Right (N=47) Percentage	N	Diffuse (N=59) Percentage	N
L	8.51	4	0	0	0	0
F	19.15	9	17.02	8	11.86	7
K	6.38	3	6.38	3	3.39	2
1 (Hs)	31.91	15	34.04	16	32.20	19
2 (D)	61.70	29	42.55	20	40.68	24
3 (Hy)	14.89	7	27.66	13	25.42	15
4 (Pd)	21.28	10	31.91	15	35.60	21
5 (Mf)	8.51	4	19.15	9	10.17	6
6 (Pa)	27.66	13	19.15	9	23.73	14
7 (Pt)	42.55	20	36.17	17	28.81	17
8 (Sc)	53.19	25	42.55	20	50.85	30
9 (Ma)	21.28	10	38.30	18	27.12	16
0 (Si)	10.64	5	8.51	4	6.78	4

[a]See Appendix for key to scale names.

In interpreting the results of any of the aforementioned studies of groups of patients with predominantly localized brain damage, it should be noted that although some group trends have been noted, there is relatively wide interindividual variability, and the predictive ability regarding emotional changes in any specific case is limited. Two illustrative cases of patients with lateralized cerebral lesions are presented in Figure 4 along with their respective CTs and MMPI profiles. Although patient 1's MMPI is technically invalid, the results (consistent with the patient's clinical presentation) nonetheless are suggestive of rather marked psychological disturbance with prominent scale elevations. Similar profiles are not uncommon in HI and CVA patients, and there appears to be some tendency for such significant elevations to be more frequently associated with LHD. Clinically, this patient displayed generalized cognitive deficits and affective disturbance characterized primarily by emotional lability, acting out, depression, and alienation from others. In contrast, patient 2's overall profile was generally within normal limits despite a left hemiplegia and expressive dysprosodia. Such findings, which appear to be more frequent in RHD patients (Cullum, 1986), appear to be consistent with the notion of greater indifference reactions associated with RHD.

Lateralized Cerebral Lesions and the Effects of Etiology

As noted, most previous MMPI studies of patients with lateralized lesions have included various etiological groups, and none except Cullum (1986) and Cullum and Bigler (1988) have specifically addressed the potential effects of etiology. In a series of head injury (HI) and cerebrovascular accident (CVA) patients, we found that the LHD subjects tended to demonstrate similar overall profiles, with the HI group having somewhat higher scores on the majority of MMPI scales. When examined in terms of the average T scores above 70, LHD HI patients demonstrated significant elevations on scales 2, 7, and 8, whereas LHD CVA patients averaged above 70 only on scale 8, with scale 2 approaching that level.

Whereas those patients with LHD produced overall MMPI profiles across HI and CVA groups that were highly similar, RHD HI patients tended to demonstrate lower scores on several of the clinical scales in comparison with the RHD CVA group. The RHD HI patients produced an

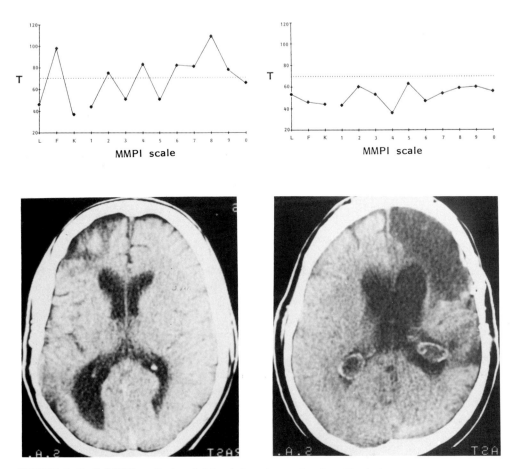

FIGURE 4. (Left) MMPI profile in a left-hemisphere-damaged patient. The brain injury was secondary to cerebral trauma. (Top) MMPI profile. (Bottom) CT scan depicting left frontal and parietal–occipital damage. Although the MMPI profile is technically invalid (significantly elevated F), the remainder of the profile is one often seen with left-hemisphere damage, that is, prominent elevations on subscales 2 and 8. (Right) MMPI profile (Top) in a right-hemisphere-damaged patient with the brain injury secondary to cerebral trauma. (Bottom) CT scan, which indicates prominent right frontal damage. Despite the marked organic structural damage present, the patient's MMPI profile demonstrates no significant elevations. See Appendix, MMPI Scales, for key to scale names.

overall profile entirely within normal limits; RHD CVA patients, on the other hand, demonstrated higher scores on scales 1, 2, 7, and 8, averaging above 70 on scales 2 and 8.

The finding of significant depression scores among both lateralized groups of CVA patients and LHD HI patients in the Cullum (1986) sample and decreased MMPI scale 2 scores among the RHD HI group suggests that etiological factors may, in fact, play a significant role in the mixed findings across previous studies. In the Cullum and Bigler (1988) study, significant elevations on scale 2 were more common among CVA in comparison with HI subjects overall. A relationship between age and higher scale 2 scores across all the sample as a whole was noted, and the CVA group was older than their HI counterparts. However, since the LHD HI patients, who were similar in age to the RHD HI subjects, did not differ from either CVA group on scale 2, age per se did not seem to be the critical factor. In a study of LHD HI and CVA patients of similar age, education, and lesion size, clinically significant depression (using measures other than the MMPI) was observed in more than 60% of CVA patients compared to approximately 20% among the HI sample (Robinson and Szetela, 1981), thereby further suggesting the influence of etiological effects on depressive symptoms in these patients.

Lesion Size Effects

It should be noted that comparisons between LHD and RHD patient groups in terms of lesion size and location in the various published reports that include such data are not absolute, since greater restrictions on patients with LHD are necessary in such studies. To illustrate, RHD patients with relatively large lesions may be included in such investigations, yet LHD patients with similar areas of damage may have to be excluded because of severe aphasia and comprehension deficits, which may preclude valid completion of the MMPI or similar measures. Furthermore, since an unknown proportion of those LHD patients who are excluded may have been depressed, it is likely that the level of depression among LHD patients has been underestimated in the majority of studies. This is a particularly important point that has not been given sufficient attention in the literature. Until more appropriate measures of depression are developed that are validly applicable in patients with major language deficits, however, this potentially significant underestimation of depression associated with LHD will continue.

Because of the reported association between left frontal lesion parameters and depressive symptomatology (e.g., Robinson et al., 1984, 1985), Cullum and Bigler (1988) examined the relationship between MMPI scale 2 scores and estimated left frontal lesion volume. Relationships involving left frontal lesion size did demonstrate some significant findings, but these were not necessarily in the expected direction, as increased scale 2 scores were associated with smaller left frontal lesions. The finding of a negative association between left frontal lesion size and depressive symptomatology was inconsistent not only with the authors' hypothesis but with previous reports as well (Robinson et al., 1985). In their sample of 10 LHD patients with anterior lesions, Robinson et al. (1984) reported a positive correlation of 0.72 between lesion volume and severity of depressive symptoms. A significant, albeit weaker, association between left frontal lesion size also was found in the Cullum and Bigler study ($r = -0.56$, $P = 0.038$), but as noted, the correlation was in the negative direction. It should be noted that the Robinson et al. findings were based on patients with lesions reportedly confined exclusively to the left anterior frontal lobe, however, whereas this was not the case in the Cullum and Bigler study because of the small sample size that would have resulted from such specificity.

It may be that the smaller left frontal lesions as observed in our sample allow for greater awareness of perceived changes and deficits (and thus, presumably, greater complaints of depression and anxiety) and that with larger lesions, patients' awareness may be diminished. A similar explanation may help account for the recent findings of Robinson et al. (1986) wherein groups of depressed and nondepressed stroke patients were compared. Interestingly, these authors found that the depressed group had significantly smaller lesions than their nondepressed counterparts.

Such findings also may be associated with some degree of the denial phenomenon that is most often associated with right hemisphere pathology but has been reported among patients with LHD as well (Heilman, Watson, Valenstein, & Damasio, 1983). It should be noted that the Cullum and Bigler analyses utilized rather gross lesion location measures (i.e., the proportion of brain quadrant damaged) and that more specific measures (e.g., more anatomically based or delineated along cortical–subcortical dimensions) might be more illuminating with respect to lesion location issues.

Another and highly compelling factor in interpreting the somewhat discrepant results regarding lesion location and depressive symptomatology across studies is the inherently complex nature of the problem. Recently, Sinyor et al. (1986) examined depressive symptomatology in 35 patients with lateralized CVAs and found that although there does appear to be a relationship between depression and lesion location, the relationship is not a simple linear one and is mediated by laterality effects. In their study, significant correlations between lesion size and indices of depression were observed, although these findings were limited to their RHD subjects. To underscore further the complexity of the lesion size–depression relationship, Robinson and co-workers (Robinson et al., 1986) recently found that stroke patients with depression had significantly smaller lesions than nondepressed patients matched for gross cognitive integrity. Thus, the question of the relationship among lesion size, specific location, and depressive symptoms appears to be highly complex and to require a multifaceted investigative approach.

To illustrate further the complex nature of relating lesion size to depressive symptomatology, relationships involving lesions in areas other than the left frontal region have yielded different results in some studies. In contrast to the significant negative relationships involving depressive symptomatology and left frontal lesion size, Cullum and Bigler (1988) found positive associations between right posterior lesion size and indices of depression (Faschingbauer MMPI scale 2 and a modified version of the MMPI Depression factor of Johnson, Butcher, Null, & Johnson, 1984) that approached or reached significant levels. In addition, right frontal and left posterior lesion size were not related to depression scores. These findings supported the results of Robinson et al. (1984), wherein RHD patients with posterior lesions were found to exhibit greater symptoms of depression than RHD patients with anterior lesions. Thus, it appears that inter- and intrahemispheric lesion location does seem to play an important role in the behavioral expression of psychological symptoms following cerebral damage.

The right hemisphere's predominant role in emotional behavior has been postulated and was discussed earlier, although despite a number of lines of converging information, the precise underlying neural mechanisms remain largely unknown. Such findings may suggest that specific critical regions within the left and/or right hemisphere must be damaged in order to produce greater psychological disturbance. Along these lines, Ross and co-workers (Ross, 1985; Ross & Stewart, 1987) have postulated a number of interesting and provocative cerebral mechanisms and interrelationships in an attempt to reconcile clinical observations of affective behaviors in brain-damaged patients with their possible neuroanatomic and neurophysiological substrates. Figure 5 (from Ross, 1985) presents a schematicized diagram of the neocortical and limbic motor systems, including the lateralized linguistic and affective processing mechanisms of the left and right cerebral hemispheres, respectively.

It is hypothesized that unilateral damage to these systems will have different clinical sequelae, although the specific behavioral changes that might be seen will be a result of the particular mechanisms involved (i.e., emphasizing the importance of lesion location). Regarding the notion of the effects of lesion location, the critical variable may be the specific pathways affected rather than the ostensible location of the lesion per se. To complicate matters further, it is well known that the effects of focal lesions may be more widespread and involve other functional areas as well. It also should be noted that although lateralized cortical functions have received substantial support (although this is less true with regard to emotional processes per se), whether there exists an asymmetry of emotional function within the limbic system itself remains unknown.

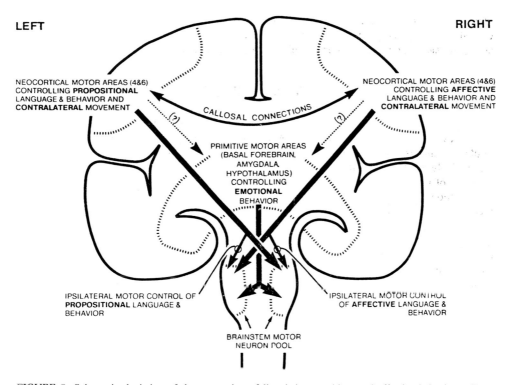

FIGURE 5. Schematic depiction of the processing of linguistic, cognitive, and affective behaviors. (From Ross, 1985, with permission.)

Another point that should be mentioned is that neurobehavioral disorders typically represent conglomerations of symptoms and are not isolated, unidimensional phenomena. The syndrome of depression, for example, represents a variety of symptoms, some of which are primarily affective and neurovegetative and others more cognitive (e.g., negative self-statements and thoughts of hopelessness). Patients may differ in the nature and severity of their symptoms, and, likewise, psychometric instruments to assess depression also vary in the types of symptoms addressed. With respect to cerebral laterality issues, as Ross and Rush discussed in their insightful 1981 article, it is most likely that both hemispheres play a role in depression, and it may be that the right hemisphere is more associated with the primary affective and neurovegetative symptoms, whereas the left hemisphere may be more involved in the more "cognitive" aspects of the disorder. The predominance of symptoms in either sphere may have important implications for treatment approaches as well.

As suggested by the foregoing discussion of depression, the interactive nature of hemispheric functions should not be ignored. It seems clear that the hemispheres influence each other in various ways, some facilitative and others inhibitory. As an example, the left hemisphere (typically) may play a more modulating role on affect (assuming affective processes are predominantly right-hemisphere phenomenon) through reciprocal inhibitory processes. In the case of LHD, such modulation may be disrupted, thereby possibly releasing the more "pure" emotional reactions from right-hemisphere mechanisms. With damage to the more affective (right-hemisphere) system, the affective component of the patient's reaction may be disrupted, yet the more cognitive–linguistic style of the intact left hemisphere may be relatively preserved. Such a model would predict less of an emotional reaction in such patients, which, as noted, has been reported in a number of studies. The inter- and intrahemispheric communication of information may be the critical variable in the pursuit of the neural substrate of emotion, and the multifaceted and inter-connected nature of cerebral systems likely is where the answers await discovery.

CONCLUSIONS

Acquired cerebral lesions (i.e., at least in terms of lesion size and location) as visualized by CT and MRI do appear to bear some relationship to psychological status, although the relationship is highly complex and is mediated by a host of factors. Whereas depressive symptoms seem to be relatively common sequelae of cerebral damage (particularly left anterior lesions), the neural substrate of these and other emotions remains largely a mystery and warrants further systematic investigation. Future studies incorporating multifactorial assessment procedures, including more precise quantitative structural and functional neuroimaging techniques, may be particularly valu-able along these lines.

Using any measure of psychological status in neurological patients requires a readjustment of the traditional interpretive framework, lest potentially misleading interpretations be made. In terms of future studies, factor-analytic and individual-item analyses of psychometric measures such as the MMPI in large groups of neurological patients would be useful in order for a more detailed assessment to be made of the specific types of items endorsed by patients with various types of cerebral damage. From such analyses, a revised personality inventory that may be more applicable to neurological patients might be derived, with further information pertaining to the differential psychological effects of localized cerebral lesions. Additional ratings by the patient, family members, and/or staff who are in frequent contact with the patient also would be valuable sources of information in future investigations of the emotional correlates of cerebral lesions. In addition, such data should address both the neurovegetative as well as the cognitive symptoms of depression to address more qualitatively inter- and intrahemispheric lesion effects when examined in conjunction with neuroimaging findings.

It should also be noted that the use of any single measure no doubt represents an inadequate assessment of the dimensions of depression, and it is argued that a more multidimensional assess-ment approach be taken in future investigations. This is further underscored by the evidence to suggest that the psychological sequelae of cerebral lesions may additionally be associated with factors such as etiology, time since onset, the degree of lesion laterality, and other relevant lesion parameters (e.g., absolute size and inter- and intrahemispheric location).

In view of the variety and number of factors that appear to play mediating and interactive roles in the effects of localized cerebral damage on psychological/emotional status following ce-rebral damage, the mixed overall results of some of the previous studies using various indices of emotional state are not surprising. Whereas studies using the MMPI or other psychological mea-sures in patients with predominantly focal cerebral lesions have addressed a number of relevant issues to some degree, they generally have examined different components of the highly complex brain–behavior picture and often have neglected to utilize specific neuroimaging data. The com-bined and systematic application of modern psychological, neuropsychological, and neuroimaging techniques may provide the vehicle for obtaining new insights into the intricate and multifaceted interrelationships involving cerebral lesions and emotional processes.

APPENDIX: MMPI SCALES

Scale   Traditional name

L   "Lie scale"
F   "Frequency scale"

K	"K scale"
1	(Hs) "Hypochondriasis scale"
2	(D) "Depression scale"
3	(Hy) "Hysteria scale"
4	(Pd) "Psychopathic deviate scale"
5	(Mf) "Masculinity–femininity scale"
6	(Pa) "Paranoia scale"
7	(Pt) "Psychasthenia scale"
8	(Sc) "Schizophrenia scale"
9	(Ma) "Hypomania scale"
0	(Si) "Social introversion scale"

REFERENCES

Ahern, G. A., & Schwartz, G. E. (1985). Differential lateralization for positive and negative emotion in the human brain: EEG spectral analysis. *Neuropsychologia, 23,* 745–755.

Andersen, A. L., & Hanvik, L. J. (1950). The psychometric localization of brain lesions: The differential effect of frontal and parietal lesions on MMPI profiles. *Journal of Clinical Psychology, 6,* 177–180.

Bear, D. M. (1983). Hemispheric specialization and the neurology of emotion. *Archives of Neurology, 40,* 195–202.

Bear, D. M., & Fedio, P. (1977). Quantitative analysis of interictal behavior in temporal lobe epilepsy. *Archives of Neurology, 34,* 454–467.

Benson, D. F., & Zaidel, E. (Eds.). (1985). *The dual brain.* New York: Guilford Press.

Berrios, G. E., & Samuel, C. (1987). Affective disorder in the neurological patient. *Journal of Nervous and Mental Disease, 175,* 173–176.

Black, F. W. (1975). Unilateral brain lesions and MMPI performance: A preliminary study. *Perceptual and Motor Skills, 40,* 87–93.

Black, F. W., & Black, I. L. (1982). Anterior–posterior locus of lesion and personality: Support for the caudality hypothesis. *Journal of Clinical Psychology, 38,* 468–477.

Borod, J. C., & Caron, H. S. (1980). Facedness and emotion related to lateral dominance, sex and expression type. *Neuropsychologia, 18,* 237–241.

Borod, J. C., Koff, E., Lorch, M. P., & Nicholas, M. (1985). Channels of emotional expression in patients with unilateral brain damage. *Archives of Neurology, 42,* 345–348.

Bryden, M. P. (1982). *Laterality—functional asymmetry in the intact brain.* New York: Academic Press.

Bryden, M. P., & Ley, R. G. (1983). Right-hemispheric involvement in the perception and expression of emotion in normal humans. In K. M. Heilman & P. Satz (Eds.), *Neuropsychology of human emotion* (pp. 6–44). New York: Guilford Press.

Buck, R., & Duffy, R. J. (1980). Nonverbal communication of affect in brain-damaged patients. *Cortex, 16,* 351–362.

Butcher, J. N., & Tellegen, A. (1978). Common methodological problems in MMPI Research. *Journal of Consulting and Clinical Psychology, 46,* 620–628.

Campbell, R. (1982). The lateralisation of emotion: A critical review. *International Journal of Psychology, 17,* 211–229.

Cullum, C. M. (1986). *Localized cerebral damage and the MMPI: Neuropsychological and CT-derived parameters.* (Doctoral dissertation, The University of Texas at Austin.) *Dissertation Abstracts International 47,* 5049-B.

Cullum, C. M., & Bigler, E. D. (1986). Ventricular enlargement, cortical atrophy and the relationship with neuropsychological status in closed head injury: A quantitative analysis. *Journal of Clinical and Experimental Neuropsychology, 8,* 437–452.

Cullum, C. M., & Bigler, E. D. (1987). Lateralized cerebral dysfunction and the MMPI revisited. Paper presented at the 15th annual meeting of the International Neuropsychological Society, Washington.

Cullum, C. M., & Bigler, E. D. (1988). Short form MMPI findings in patients with predominantly lateralized cerebral dysfunction: Neuropsychological and CT-derived parameters. *Journal of Nervous and Mental Disease, 176,* 332–342.

Denny-Brown, D., Meyer, J. S., & Horenstein, S. (1952). The significance of perceptual rivalry resulting from parietal lesions. *Brain, 75,* 433–471.

Dikmen, S., & Reitan, R. M. (1974a). MMPI correlates of localized cerebral lesions. *Perceptual and Motor Skills, 39,* 831–840.

Dikmen, S., & Reitan, R. M. (1974b). Minnesota Multiphasic Personality Inventory correlates of dysphasic language disturbances. *Journal of Abnormal Psychology, 83,* 675–679.

Dikmen, S., & Reitan, R. M. (1977). MMPI correlates of adaptive ability deficits in patients with brain lesions. *Journal of Nervous and Mental Disease, 165,* 247–254.

Dimond, S. J., Farrington, L., & Johnson, P. (1976). Differing emotional response from right and left hemispheres. *Nature, 261,* 690–692.

Doehring, D. G., & Reitan, R. M. (1960). MMPI performance of aphasic and nonaphasic brain-damaged patients. *Journal of Clinical Psychology, 16,* 307–309.

Etcoff, N. L. (1986). The neuropsychology of emotional expression. In G. Goldstein & R. E. Tarter (Eds.), *Advances in clinical neuropsychology* (pp. 127–179). New York: Plenum Press.

Faschingbauer, T. R. (1974). A 166-item written short form of the group MMPI: The FAM. *Journal of Consulting and Clinical Psychology, 42,* 645–655.

Feibel, J. H., & Springer, C. J. (1982). Depression and failure to resume social activities after stroke. *Archives of Physical Medicine and Rehabilitation, 63,* 276–278.

Finklestein, S., Benowitz, L. I., Baldessarini, R. J., Arana, G., Levine, D., Woo, E., Bear, D., Moya, K., & Stoll, A. L. (1982). Mood, vegetative disturbance, and dexamethasone suppression test after stroke. *Annals of Neurology, 12,* 463–468.

Flor-Henry, P. (1983). *Cerebral basis of psychopathology.* Cambridge, MA: John Wright.

Folstein, M. F., Maiberger, R., & McHugh, P. R. (1977). Mood disorder as a specific complication of stroke. *Journal of Neurology, Neurosurgery and Psychiatry, 40,* 1018–1020.

Gainotti, G. (1972). Emotional behavior and hemispheric side of the lesion. *Cortex, 8,* 41–55.

Gainotti, G. (1979). The relationship between emotions and cerebral dominance: A review of clinical and experimental evidence. In J. Gruzelier & P. Flor-Henry (Eds.), *Hemispheric asymmetries of function in psychopathology.* (pp. 21–34). Amsterdam: Elsevier/North-Holland Biomedical Press.

Gasparrini, W. G., Satz, P., Heilman, K. M., & Coolidge, F. L. (1978). Hemispheric asymmetries of affective processing as determined by the Minnesota Multiphasic Personality Inventory. *Journal of Neurology, Neurosurgery, and Psychiatry, 41,* 470–473.

Gass, C. S. (1985). *Minnesota Multiphasic Personality Inventory correlates of lateralized cerebral lesions and cognitive-behavioral deficits. Dissertation Abstracts International, 46,* 2805-B.

Gass, C. S., & Russell, E. W. (1985). MMPI correlates of verbal-intellectual deficits in patients with left hemisphere lesions. *Journal of Clinical Psychology, 41,* 664–670.

Gass, C. S., & Russell, E. W. (1986). Minnesota Multiphasic Personality Inventory correlates of lateralized cerebral lesions and aphasic deficits. *Journal of Consulting and Clinical Psychology, 3,* 359–363.

Goldberg, L. R. (1965). Diagnosticians vs. diagnostic signs: The diagnosis of psychosis vs. neurosis from the MMPI. *Psychological Monographs, 79* (whole No. 602).

Goldstein, K. (1948). *Language and language disturbances.* New York: Grune & Stratton.

Grafman, J., Vance, S. C., Weingartner, H., Salazar, A. M., & Amin, D. (1986). The effects of lateralized lesions on mood regulation. *Brain, 109,* 1127–1148.

Greene, R. L. (1980). *The MMPI: An interpretative manual.* New York: Grune & Stratton.

Heaton, R. K., Smith, H. H., Lehman, A. W., & Vogt, A. T. (1978). Prospects for faking believable deficits on neuropsychological testing. *Journal of Consulting and Clinical Psychology, 46,* 892–900.

Hecaen, H. (1962). Clinical symptomatology in right and left hemispheric lesions. In V. B. Mountcastle (Ed.), *Interhemispheric relations and cerebral dominance.* (pp. 215–243). Baltimore: Johns Hopkins Press.

Heilman, K. M., Scholes, R., & Watson, R. (1975). Auditory affective agnosia. *Journal of Neurology, Neurosurgery and Psychiatry, 38,* 69–72.

Heilman, K. M., Watson, R. T., & Bowers, D. (1983). Affective disorders associated with hemispheric disease. In K. M. Heilman & P. Satz, *Neuropsychology of human emotion.* (pp. 45–64). New York: Guilford Press.

Heilman, K. M., Bowers, D., Valenstein, E., & Watson, R. T. (1986). The right hemisphere: Neuropsychological functions. *Journal of Neurosurgery, 64,* 693–704.

Hurwitz, T. A., Wada, J. A., Kosaka, B. A., & Strauss, E. H. (1985). Cerebral organization of affect suggested by temporal lobe seizures. *Neurology, 35,* 1335–1337.

Jernigan, T. L., Zatz, L. M., & Naeser, M. A. (1979). Semiautomated methods for quantitating CSF volume on cranial computed tomography. *Radiology, 32,* 463–466.

Jernigan, T. L., Zatz, L. M., Moses, J. A., & Cardellino, J. P. (1982). Computed tomography in schizophrenics and normal volunteers. *Archives of General Psychiatry, 39,* 771–773.

Johnson, J. H., Butcher, J. N., Null, C., & Johnson, K. N. (1984). Replicated item factor analysis of the full MMPI. *Journal of Personality and Social Psychology, 47,* 105–114.

Kertesz, A. (Ed.). (1983). *Localization in neuropsychology.* New York: Academic Press.

Ley, R. G., & Bryden, M. P. (1982). A dissociation of right and left hemispheric effects for recognizing emotional tone and verbal content. *Brain and Cognition, 1,* 3–9.

Lezak, M. D. (1978). Living with the characterologically altered brain injured patient. *Journal of Clinical Psychiatry, 39,* 592–598.

Lezak, M. D. (1983). *Neuropsychological assessment* (2nd ed.). New York: Oxford University Press.

Lipsey, J. R., Robinson, R. G., Pearlson, G. D., Rao, K., & Price, T. R. (1984). Nortriptyline treatment of post-stroke depression: A double-blind study. *Lancet, 1,* 297–300.

Lipsey, J. R., Spencer, W. C., Rabins, P. V., & Robinson, R. G. (1986). Phenomenological comparison of poststroke depression and functional depression. *American Journal of Psychiatry, 143,* 527–529.

Louks, J., Calsyn, D., & Lindsay, F. (1976). Personality dysfunction and lateralized deficits in cerebral functions as measured by the MMPI and Reitan–Halstead battery. *Perceptual and Motor Skills, 43,* 655–659.

Lubin, B., Larsen, R. M., & Matarazzo, J. D. (1984). Patterns of psychological test usage in the United States: 1935–1982. *American Psychologist, 39,* 451–454.

Luchins, D. J., Lewine, R. J., & Meltzer, H. Y. (1984). Lateral ventricle size, psychopathology, and medication response in the psychoses. *Biological Psychiatry, 19,* 29–44.

Lishman, W. A. (1973). The psychiatric sequelae of head injury. A review. *Psychological Medicine, 3,* 304–318.

Mack, J. L. (1979). The MMPI and neurological dysfunction. In C. S. Newmark (Ed.), *MMPI: Clinical and research trends.* (pp. 53–79). New York: Praeger.

Maloney, M. P., & Ward, M. P. (1976) *Psychological assessment: A conceptual approach.* New York: Oxford University Press.

Meier, M. J., & French, L. A. (1965). Some personality correlates of unilateral and bilateral EEG abnormalities in psychomotor epileptics. *Journal of Clinical Psychology, 31,* 3–9.

Mendez, M. F., Cummings, J. L., & Benson, D. F. (1986). Depression in epilepsy: Significance and phenomenology. *Archives of Neurology, 43,* 766–770.

Milner, B. (1974). Hemispheric specialization: Scope and limits. In F. O. Schmitt & F. G. Worden (Eds.), *The neurosciences: Third study program.* (pp. 75–88). Cambridge, MA: MIT Press.

Morrison, J. H., Molliver, M. E., & Grzanna, R. (1979). Noradrenergic innervation of cerebral cortex: Widespread effects of local cortical lesions. *Science, 205,* 313–316.

Moscovitch, M., & Olds, J. (1982). Asymmetries in spontaneous facial expressions and their possible relation to hemispheric specialization. *Neuropsychologia, 20,* 71–81.

Myslobodsky, M. S. (1983). Epileptic laughter. In M. S. Myslobodsky (Ed.), *Hemisyndromes: Psychobiology, neurology, psychiatry* (pp. 239–263). New York: Academic Press.

Nasrallah, H. A., & Coffman, J. A. (1985). Computer tomography in psychiatry. *Psychiatric Annals, 14,* 239–249.

Nasrallah, H. A., McCalley-Whitters, M., & Jacoby, C. G. (1982). Cerebral ventricular enlargement in young manic males. *Journal of Affective Disorders, 4,* 15–19.

Nebes, R. D. (1978). Direct examination of cognitive function in the right and left hemispheres. In M. Kinsbourne (Ed.), *Asymmetrical function of the brain* (p. 128). Cambridge, MA: Cambridge University Press.

Osmon, O. C., & Golden, C. J. (1978). Minnesota Multiphasic Personality Inventory correlates of neuropsychological deficits. *International Journal of Neuroscience, 8,* 113–122.

Pfefferbaum, A., Zatz, L. M., & Jernigan, T. L. (1986). Computer-interactive method for quantifying cerebrospinal fluid and tissue in brain CT scans: Effects of aging. *Journal of Computer Assisted Tomography, 10,* 571–578.

Pribram, K. H. (1981). Emotions. In S. B. Filskov & T. J. Boll (Eds.), *Handbook of clinical neuropsychology* (pp. 102–134). New York: John Wiley & Sons.

Reitan, R. M. (1976). Neurological and physiological bases of psychopathology. *Annual Review of Psychology, 27,* 189–216.

Robinson, R. G. (1979). Differential behavioral and biochemical effects of right and left hemispheric cerebral infarction in the rat. *Science, 205,* 707–710.

Robinson, R. G., & Benson, D. F. (1981). Depression in aphasic patients: Frequency, severity, and clinical–pathological correlations. *Brain and Language, 14,* 282–291.

Robinson, R. G., & Price, T. R. (1982). Post-stroke depressive disorders: A follow-up study of 103 patients. *Stroke, 13,* 635–641.

Robinson, R. G., & Szetela, B. (1981). Mood change following left hemispheric brain injury. *Annals of Neurology, 9,* 447–453.

Robinson, R. G., Starr, L. B., Kubos, K. L., & Price, T. R. (1983). A two-year longitudinal study of post-stroke mood disorders: Findings during the initial evaluation. *Stroke, 14,* 736–741.

Robinson, R. G., Kubos, K. L, Starr, L. B., Rao, K., & Price, T. R. (1984). Mood disorders in stroke patients: Importance of location of lesion. *Brain, 107,* 81–93.

Robinson, R. G., Starr, L. B., Lipsey, J. R., Rao, K., & Price, T. R. (1985). A two-year longitudinal study of poststroke mood disorders. *The Journal of Nervous and Mental Disease, 173,* 221–226.

Robinson, R. G., Bolla-Wilson, K., Kaplan, E., Lipsey, J. R., & Price, T. R. (1986). Depression influences intellectual impairment in stroke patients. *British Journal of Psychiatry, 148,* 541–547.

Ross, E. D. (1981). The aprosodias. Functional–anatomical organization of the affective components of language in the right hemisphere. *Archives of Neurology, 38,* 561–569.

Ross, E. D. (1984). Disturbances of emotional language with right hemisphere lesions. In A. Ardila & F. Ostrosky-Solis (Eds.), *The right hemisphere: Neurology & neuropsychology* (pp. 109-123). New York: Gordon & Breach.

Ross, E. D. (1985). Modulation of affect and nonverbal communication by the right hemisphere. In M. M. Mesulam (Ed.), *Principles of behavioral neurology* (pp. 239–257). Philadelphia: F. A. Davis.

Ross, E. D., & Mesulam, M. M. (1979). Dominant language functions of the right hemisphere? Prosody and emotional gesturing. *Archives of Neurology, 36,* 144–148.

Ross, E. D., & Rush, A. J. (1981). Diagnosis and neuroanatomical correlates of depression in brain-damaged patients. *Archives of General Psychiatry, 38,* 1344–1354.

Ross, E. D., & Stewart, R. S. (1987). Pathological display of affect in patients with depression and right frontal brain damage. *Journal of Nervous and Mental Disease, 175,* 165–172.

Russell, E. W. (1979). Three patterns of brain damage on the WAIS. *Journal of Clinical Psychology, 35,* 611–620.

Sackheim, H. A., Gur, R. C., & Saucy, M. C. (1978). Emotions are expressed more intensely on the left side of the face. *Annals of the New York Academy of Sciences, 202,* 424–435.

Sackheim, H. A., Greenberg, M. S., Weiman, A. L., Gur, R. C., Hungerbuhler, J. P., & Geschwind, N. (1982). Hemispheric asymmetry in the expression of positive and negative emotions: Neurologic evidence. *Archives of Neurology, 39,* 210–218.

Safer, M. A. (1981). Sex and hemisphere differences in access to codes for processing emotional expressions and faces. *Journal of Experimental Psychology: General, 110,* 86–100.

Schacter, S. (1964). The interaction of cognitive and physiological determinants of emotional state. In L. Berkowitz (Ed.), *Advances in experimental social psychology, Vol. 1* (pp. 49–80). New York: Academic Press.

Shima, S., Shikano, T., Kitamura, T., Masuda, Y, Tsukumo, T., Kanba, S., & Asai, M. (1984). Depression and ventricular enlargement. *Acta Psychiatrica Scandinavica, 70,* 275–277.

Shima, S., Kanba, S., Masuda, Y., Tsukumo, T., Kitamura, T., & Asai, M. (1985). Normal ventricles in chronic schizophrenics. *Acta Psychiatrica Scandinavica, 71,* 25–29.

Sinyor, D., Jacques, P., Kaloupek, D. G., Becker, R., Goldenberg, M., & Coopersmith, H. (1986). Post-stroke depression and lesion location: An attempted replication. *Brain, 109,* 537–546.

Skenazy, J. A., & Bigler, E. D. (1985). Psychological adjustment and neuropsychological performance in diabetic patients. *Journal of Clinical Psychology, 41,* 391–396.

Standish-Barry, H. M. A. S., Bouras, N., Hale, A. S., Bridges, P. K., & Bartlett, J. R. (1986). Ventricle size and CSF transmitter metabolite concentrations in severe endogenous depression. *British Journal of Psychiatry, 148,* 386–392.

Tucker, D. M. (1981). Lateral brain function, emotion, and conceptualization. *Psychological Bulletin, 89,* 19–46.

Tucker, D. M., Watson, R. T., & Heilman, K. M. (1976). Affective discrimination and evocation in patients with right parietal disease. *Neurology, 27,* 947–950.

Turkheimer, E., Yeo, R., & Bigler, E. D. (1983). Digital planimetry in APLSF. *Behavior Research Methods and Instrumentation, 15,* 471–473.

Vogel, W. (1962). Some effects of brain lesions on MMPI profiles. *Journal of Consulting Psychology, 26,* 412–415.

Willcockson, J. C., Bolton, B., & Dana, R. H. (1983). A comparison of six MMPI short forms: Code type correspondence and indices of psychopathology. *Journal of Clinical Psychology, 39,* 968–969.

Williams, H. L. (1952). The development of a caudality scale for the MMPI. *Journal of Clinical Psychology, 8,* 293–297.

Woodward, J. A., Bisbee, C. T., & Bennett, J. E. (1984). MMPI correlates of relatively localized brain damage. *Journal of Clinical Psychology, 40,* 961–969.

Yeo, R., Turkheimer, E., & Bigler, E. D. (1983). Computer analysis of lesion volume: Reliability, utility and neuropsychological applications. *Clinical Neuropsychology, 44,* 683.

11

Individual Differences

RONALD A. YEO

INTRODUCTION

Human brains differ in a variety of ways. Language skills, for example, are typically represented in the left hemisphere but in some individuals may be represented in the right hemisphere or bilaterally (Hecaen & Sauget, 1971). Naming ability is typically a function represented in the posterior portion of the left perisylvian region but in some individuals may be more widespread within the left hemisphere (Ojemann, 1983). To what extent are these two types of differences (between and within hemispheres) linked to particular patterns of cognitive performances? Most broadly, neuropsychological approaches to individual differences aim to understand the strengths and weaknesses of different neural designs and thereby to elucidate basic principles of brain–behavior relationships. This approach complements traditional functional neuroanatomy: rather than asking "What function does a given structure perform?" one can ask "What structural or dynamic characteristics confer optimal (or suboptimal) function?"

Still in their infancy, neuropsychological approaches to individual differences have been unable to answer this broadly important question. Investigators have proceeded along two major paths. One approach involves clarification of cognitive differences between groups that putatively differ in brain organization. Thus, the cognitive skills of left- and right-handers have been compared, or those of men and women, or younger versus older people. Another approach aims to clarify the nature of the brain differences among individuals presumed to show different patterns of cognitive skills (again, typically, men versus women or younger versus older people). With careful analysis of both brain and behavioral differences, it may eventually be possible to answer the question posed above and understand some of the mechanisms underlying individual differences in behavior.

The goal of this chapter is to explore selected aspects of individual differences in brain organization in groups that differ in patterns of cognitive skills. Differences among brains are investigated using the focal lesion method. The logic of this approach is straightforward. If similar lesions produce different sequelae in two groups of people, one can infer that the brains differ between groups. There are, however, a number of methodological issues that must be addressed before the focal lesion method can be optimally applied and the nature of brain differences appropriately inferred.

We focus on three basic dimensions of difference among individuals, each of which has been

RONALD A. YEO • Department of Psychology, University of New Mexico, Albuquerque, New Mexico 87131.

linked to patterns of cognitive ability. The first two dimensions of difference to be explored are sex and age, perhaps the most basic ways in which people differ. Sex differences in cognitive abilities have been widely reported (e.g., Garai & Scheinfeld, 1968; Maccoby & Jacklin, 1974; Yeo & Cohen, 1983) but appear to be more restricted to specific cognitive functions and of smaller magnitude than is widely believed (Caplan, MacPherson, & Tobin, 1985). Certain verbal skills, such as verbal fluency, tend to be greater in females, whereas certain visuospatial skills, such as mental rotation ability, tend to be greater in males. With respect to the variable of age, this review focuses on age differences within the adult range. Cognitive differences associated with age are most frequently seen in visuospatial function (e.g., Berg, Hertzog, & Hunt, 1982) and memory (Craik, 1977).

The third aspect of individual differences to be explored are those associated with anatomic asymmetries of the cerebral hemispheres. Cognitive differences associated with anatomic asymmetries have been reported only recently (Yeo, Turkheimer, Raz, & Bigler, 1987) and thus must be considered much more speculative than either sex or age differences. In this investigation, we noted that among neurologically normal subjects, left hemisphere volume exceeded right hemisphere volume by approximately 2%. Also, the volumetric asymmetry of the cerebral hemispheres predicted the pattern of verbal versus nonverbal intellectual superiority. Specifically, subjects who showed greater than average anatomic asymmetry had a Verbal IQ (VIQ) that exceeded Performance IQ (PIQ). Subjects who showed less than average asymmetry had PIQs that exceeded VIQs. The correlation between volumetric anatomic asymmetry and VIQ–PIQ difference scores was $r = 0.57$. Other investigations have noted the same pattern of results in schizophrenic (Luchins, Weinberger, & Wyatt, 1982) and learning-disabled (Rosenberger & Hier, 1980) subjects, though different assessments of anatomic asymmetry were performed.

Individual differences among brains associated with handedness are not explored in this chapter, as handedness has not been reliably associated with patterns of cognitive skills (cf. Hardyck & Petrinovich, 1977). Handedness does appear, however, to interact with other individual difference variables such as sex and familial sinistrality (Harshman, Hampson, & Berenbaum, 1983). Investigations of possible neural differences associated with different combinations of handedness, sex, and familial sinistrality are exceedingly difficult to do with the focal lesion method (though see Hecaen, DeAgostini, & Monzon-Montes, 1981) because of difficulties in obtaining sufficient numbers of unilateral lesion patients in each category and thus are not considered in this chapter.

There are numerous possible dimensions of difference among normal brains, and neuropsychological methods differ in the types of differences to which they are most sensitive. For heuristic purposes, let us consider some of the major ways in which brains could differ and then the ability of specific neuropsychological investigative procedures to assess such differences. One way to classify possible differences among brains is along four dimensions: structural, representational, organizational, and activational. For this discussion, structural differences among brains refer to very basic and quantitative differences, for example, in overall brain size and the size of specific processing centers or pathways. Although brain size certainly differs among people, both older research relating head circumference or length to intellectual parameters (see Gould, 1981) and recent studies relating brain volume as determined from CT scans to IQ measures (Yeo et al., 1987) have found no important cognitive correlates of overall brain size or of hemispheric size.

The size of various specific processing centers may, however be linked to certain behavioral traits. The asymmetry in the sizes of the left and right planum temporale differs among left- and right-handers (Geschwind & Levitsky, 1968). As left- and right-handers do not appear to differ in language comprehension ability, the relevance of planum temporale asymmetry for cognitive differences among people remains to be demonstrated. Pathways, too, may vary in asymmetry. Among left-handers there is a relatively smaller percentage of pyramidal tract neurons decussating at the medullary pyramids (Kertesz & Geschwind, 1971), but, again, the relevance of such differences for cognition seems to be minimal, as left- and right-handers do not reliably differ. As discussed above, the one structural difference that has been linked to cognitive performance is the volumetric asymmetry of the cerebral hemispheres; the nature of this effect is explored with the focal lesion method later in this chapter.

Representational differences refer to variability in regional cognitive specialization. As with structural differences among brains, representational differences may be explored at a variety of levels of the nervous system. At the highest level are hemispheric differences, for example, the common observation of greater right hemisphere representation of verbal skills in left-handedness (e.g., Gloning, Gloning, Haub, & Quatember, 1969). Representational differences can also be explored within a given hemisphere, as, for example, with electrical stimulation studies (Ojemann, 1983), in which it has been observed that the left perisylvian area subserving naming is less widespread in individuals with high IQs than in individuals with low IQs.

Different patterns of neuronal organization may confer regional differences in processing ability within and between brains. Scheibel (1984) reported hemispheric differences in patterns of dendritic development in the frontal opercula that may be responsible for the differing functional abilities of these regions in each hemisphere. Semmes (1968) proposed that hemispheric differences in the representation of verbal versus spatial skills may be related to the more "focal" organization of the left hemisphere. Gur *et al.* (1982) noted a relatively higher proportion of white to gray matter in the right cerebral hemisphere, and we (Yeo, Melka, & Haaland, 1988) have recently observed hemispheric differences in the representation of elementary motor skills, consistent with the Semmes hypothesis. These two types of anatomic differences, which may underlie hemispheric differences in the representation of abilities, may also help us understand differences in the representation of abilities across people.

Organizational differences refer to variability in the interaction of components of a complex system. Denenberg (1980) has detailed the manner in which organizational differences can be inferred from the focal lesion method if data are available on performance on both patient groups that differ in lesion locus and control groups. Table 1 shows how data on group performance levels can indicate different types of organization at the hemispheric level. In model 1, where the left-lesion group (L) and the right-lesion group (R) each perform worse than controls (C), there is no evidence of laterality, but one can conclude that the hemispheres are acting synergistically. "The resultant output (of the intact, linked hemispheres) is more than one would expect from a knowledge of each individually" (Denenberg, 1980, p. 233). There is, of course, the logical possibility that the performance of all three groups would be equivalent, in which case one would

TABLE 1. Inferring Models of Hemispheric Function from Clinical Data[a]

Statistical findings	Interpretation
A. Given that L=R	
1. L<C	1. No evidence of laterality,
R<C	hemispheres interact synergistically
B. Given that L<R	
2. R=C	2. Left hemisphere dominance
L<C	
3. R<C	3. Left dominant, synergistic
L<C	
C. Given that R<L	
4. L=C	4. Right hemisphere dominance
R<C	
5. L<C	5. Right dominant, synergistic
R<C	

[a] See text for explanation.

conclude that the behavioral measure used was not sensitive to brain damage; this pattern would thus not suggest a specific model of brain organization. Model 2 may be termed "left hemisphere dominance" or "unilateral left hemisphere competence." The left hemisphere, and the left hemisphere only, controls the behavior examined. In model 3, where L < C, R < C, and R < L, both dominance and synergy are found, suggesting that in the intact organism the left hemisphere plays the larger role but that the hemispheres interact. Models 4 and 5 are analogous to models 2 and 3 except that L < R rather than R < L. Similar descriptions of organizational differences could theoretically be obtained for intrahemispheric regions, e.g., the left parietal and frontal lobes.

The biological mechanisms leading to variability in interactions of systems components are not known. It seems reasonable to suggest, however, that organizational and representational differences may not be independent. Consider two individuals who differ in the representation of a given skill across hemispheres. In individual A, with bilateral representation of given skill, there may be greater need for active inhibition of one hemisphere than in an individual with unilateral representation of that skill. Such a mechanism has been proposed to describe the differences between the brains of some stutterers and normals (Helm-Estabrooks, Yeo, Geschwind, Freedman, & Weinstein, 1986).

"Activational differences" refer to regional differences in the amount of neuronal activity on a given task, such as might be revealed by electrophysiological techniques, region cerebral blood flow (rCBF), position emission tomography (PET), or magnetoencephalography (MEG). Activational differences can also be studied at a variety of anatomic levels, from relatively small populations of neurons to hemispheres. One factor that may be related to activational differences is individual differences in neurotransmitter levels (e.g., Zuckerman, 1983).

It is important to note that the four dimensions of difference described above are neither independent nor exhaustive descriptions of ways human brains differ. Establishing links between brain differences and cognitive differences, however, will probably require an understanding of the variety of ways in which two groups of people may differ. Studies of sex differences in the sequelae of focal lesions have frequently concluded that females show more bilateral distribution of cognitive skills, a putative "representational difference." In contrast, studies of activational asymmetries have frequently revealed greater activational asymmetry in females (Gur *et al.*, 1982). Are these data contradictory, or do they suggest that sex differences in brains are multifaceted? Assuming the reliability of these patterns, it seems reasonable to presume that one would need to understand the interaction of representational and activational differences in order to understand the mechanisms underlying cognitive differences.

As specific investigative procedures typically reflect only a small number of these four dimensions of difference, a comprehensive picture of the neuropsychology of individual differences is unlikely to emerge from research within a given methodology. Laterality techniques, such as dichotic listening or tachistoscopic procedures, are limited in the specificity of the data for types of brain differences and are severely limited in terms of the possible "level of analysis." An observed group difference in asymmetry scores probably reflects a combination of all types of brain differences, though the importance of each individually is impossible to determine and probably depends on specific task parameters. Further, laterality studies can only examine differences at the hemispheric level, e.g., the degree of right hemisphere language skill.

The focal lesion method of investigating individual differences has its own strengths and weaknesses. Although it can potentially examine differences at a variety of levels of the neuraxis, interpretation of observed differences in lesion sequelae is quite difficult. First, there are a host of methodological issues that must be investigated prior to concluding that brains differ. Second, even if one might conclude that brains differ across groups, it is exceedingly difficult to determine along which dimension(s) the brains differ.

Let us briefly consider some of the methodological issues that beset focal lesion investigations of individual differences. As neuropsychological assessments are always performed at some

TABLE 2. Possible Mechanisms in the Recovery of Function

Reduction in diaschisis (see Isaacson, 1975)
  Less disruption of neural activity in nearby regions because of physical or metabolic distortions produced
    by the lesion
  Reduced astrocytic reactions at the edge of the lesion
  Reduced "irritative" reactions at the edge of the lesion
  Reduced edema
  Reduced pressure on nearby tissues caused by enclosed area of bleeding (hematomas)
  Reduced effects on cerebrospinal fluid (composition and pressure)
  Neurotransmitters return to original levels
Takeover of function
  Proximal
    Denervation supersensitivity
    Remyelination
    Synaptogenesis
      Collateral sprouting (undamaged neuronal processes in the vicinity of the damaged area may invade
        the damaged area and make contact with neurons beyond the lesion)
      Axon regeneration
      Regenerative sprouting (the damaged neuron may develop new processes)
  Distal
    Derepression of unused pathways (Geschwind, 1974)
    Takeover by an "uncommitted" region (Goldman, 1974)
Adaptation
  Development of new ways to perform tasks (i.e., the use of iconic codes when verbal codes are disrupted)
  Use of mechanical or electronic prostheses

time post-trauma, one must dissociate individual differences in recovery of function from pre-morbid brain differences. Table 2 lists the most frequently postulated mechanisms for recovery of function: reduction in diaschisis (or remote functional depression), takeover of function (by either adjacent or distal regions), and adaptation (finding new ways to perform tasks). Individual differences in lesion sequelae may reflect differences in one or several of these processes. There are two ways one might attempt to dissociate recovery-of-function differences from premorbid brain differences. The best solution is obviously to follow patients over time, as for example in the investigation of sex differences in recovery from aphasia (Kertesz & McCabe, 1977), though such follow-up data have not been reported frequently. Alternatively, one might examine the relationship between time post-lesion and neuropsychological performance. If time post-lesion is plotted along the abscissa and performance along the ordinate separately for each of the two groups under investigation, differences in the slope of the resulting regression line could be interpreted to reflect recovery differences. Group differences in the intercepts could be interpreted as reflecting pre-morbid differences.

Another obvious methodological issue concerns evaluation of possible group differences in lesion parameters. It is clearly premature to interpret individual differences in lesion sequelae as reflecting premorbid differences in brains if the investigator cannot demonstrate that the groups have equivalent lesions. Among those lesion parameters to be examined, etiology, time post-onset, location, and size are most important. Control of etiology and time post-onset presents no special difficulties for the investigator, but evaluation of lesion size and locus is more problematic. Individual differences in lesion size have been determined most accurately in studies of temporal lobectomies (Lansdell, 1968a,b, 1973) in which the amount of excised tissue may be actually measured. Unfortunately, in these studies the groups differed in the amount of tissue excised, and lesion size was not treated as a covariate. Another strategy has been to equate the groups for lesion size through the use of clinical signs of brain damage (McGlone, 1977). In this investigation, the groups did not differ in the incidence of certain clinical signs (visual field defects and contralateral weakness), so parity of lesion size was assumed. In effect, this approach assumes no sex differences in the manner in which brain damage produces some symptoms (the clinical signs) while attempting to investigate sex differences in how brain damage produces other symptoms (verbal impairment).

Neuroimaging techniques allow quantification of lesion parameters for a precise evaluation of the possibility of group differences in the lesions. Kertesz and Sheppard (1981), in an investigation of possible sex differences in brain function, used planimetry to quantify lesion size. On each slice, the area of the lesion involvement was calculated and the largest area (across all slices) was used as an index of lesion size. No significant difference was observed in lesion size for either left hemisphere lesions ($P = 0.1$) or right hemisphere lesions ($P = 0.2$). These quantification efforts were, however, applied to only a subset of the entire sample (51% of subjects), and although the differences in lesion size were not statistically significant, they may have been large enough to influence results. Such techniques, though, hold great promise in attempts to grapple with this important methodological issue. Volumetric expressions of lesion size will more accurately reflect the true configuration of the lesion than one- (linear) or two-dimensional (area) expressions.

One would also wish to be certain that the two groups under study do not differ in the locus of lesion, though such analyses have only rarely been undertaken (Yeo et al., 1984; Naeser & Borod, 1986). In the area of sex differences this would seem to be especially relevant, as epidemiologic studies suggest sex differences in the sites of cerebral arteries typically occluded (Sinderman, Bechinger, & Dichgans, 1970; Kaste & Waltimo, 1976).

A final methodological concern is the thorny issue of "strategy effects." Bryden (1979) has expressed concern that sex differences in cognitive strategies may be an important factor in determining group performance differences sometimes observed after a cerebral lesion. For example, if women are observed to show a greater deficit in spatial processing following a left-hemisphere lesion, this could reflect either more left-hemisphere representation of spatial skills or a greater tendency of women to approach spatial tasks with a verbal strategy. Cognitive strategy differences, however, may also reflect premorbid brain differences. Analysis of the role of strategy effects depends in part on what the research literature tells us is plausible. It does seem that women more often approach spatial problem solving with verbal strategies. The converse does not appear to be true; that is, women do not seem to approach verbal problems with more spatial strategies. Thus, it seems unlikely that any observed sex differences in verbal processing reflects a sex difference in cognitive strategy. One obvious approach to dealing with cognitive strategy issues is to examine performance on simple neuropsychological tests not amenable to strategy effects, such as simple sensory and motor tasks (Yeo, Turkheimer, & Bigler, 1984). Concern about strategy effects is not limited to investigations of sex differences. Age differences in cognitive strategy have been reported (Linn, 1986), as well as age differences in flexibility in using various cognitive strategies (Stern & Baldinger, 1983).

Assuming that these methodological issues (differences in recovery of function rather than premorbid brain differences, group differences in lesion parameters, strategy effects) have been effectively addressed, the focal lesion method is still ambiguous about the nature of brain differences revealed. For example, let us assume that group A shows greater verbal deficits after a left-hemisphere lesion than does group B, but verbal performance is equal in groups A and B after a right-hemisphere lesion. Several possible differences among brains could account for such an observation. Left-hemisphere verbal skills may be differently represented in the two groups. Alternatively, group difference in activation could perhaps account for these results. For example, if nonlesioned subjects from group A show greater activational asymmetry than those from group B, they may have a greater ability, after a left lesion, to activate intact parts of the left hemisphere, rendering group A less susceptible to the widespread reduction in cerebral activity seen after a significant lesion (Endo, Larsen, & Lassen, 1977). Pharmacological approaches to enhancing recovery of function often rely on such a mechanism (Feeney, Gonzalez, & Law, 1982), and it seems at least plausible that the premorbid individual differences that may exist in activational asymmetries could contribute to group differences in lesion sequelae. Analysis of patterns of cerebral activation in different groups post-lesion could help clarify these issues.

Let us now turn to the examination of differences in lesion sequelae associated with sex, age, and anatomic cerebral asymmetries. The methodological and neuropsychological concerns expressed above suggest that the focal lesion method, by itself, will not produce an integrated picture of individual differences among brains. It is, however, perhaps the single most informative approach to these questions, capable of examining differences at a variety of levels of the central nervous system (i.e., not just at the hemispheric level) as well as helping to untangle the various dimensions along which brains may differ.

SEX DIFFERENCES

In the area of sex differences, McGlone's (1980) thorough and provocative review stands as a landmark, setting forth the evidence for the bold hypothesis that there exist important differences in cerebral organization in men and women and focusing methodological and substantive criticism regarding this hypothesis. Though several recent investigations have examined sex differences, a clear consensus has yet to emerge. Methodological differences across studies, ambiguous data, and concern for the social implications of such research have all contributed to the current controversy. In this review we only briefly discuss studies of sex differences published prior to McGlone's (1980) review and then examine in more detail the results of recent investigations.

Beginning in the early 1960s, Herbert Lansdell and colleagues offered a series of reports describing sex differences in the cognitive sequelae of temporal lobectomy (Lansdell, 1961, 1962, 1968a,b, 1973; Lansdell & Urbach, 1965). When sex differences were observed, males were described as more impaired. For example, following left temporal lobectomy, males were more impaired on a proverb interpretation test (Lansdell, 1961), on verbal subtests of the Wechsler Adult Intelligence Scale (WAIS) (Lansdell, 1968b), and on a word association test (Lansdell, 1973) but not on a vocabulary test (Lansdell, 1968a). Among patients with right temporal lobectomy, males were described as more impaired on the Graves Design Judgment Test (Lansdell, 1962) and the nonverbal subtests of the WAIS (Lansdell, 1968b) but not on Mooney's Closure Faces Test (Lansdell, 1968a).

Though important in terms of alerting researchers to the possibilities of sex differences in lesion sequelae, Lansdell's investigations are complicated by methodological difficulties that present formidable problems to interpretation. First, performance of the sexes was not always compared directly (Lansdell & Urbach, 1965). Rather, each group was independently compared to a control group. Although the male lobectomy patients differed from controls and the females did not, such an outcome does not necessarily mean that the male and female lobectomy patients significantly differed. Second, all patients were suffering from longstanding epilepsy. It is possible that a chronic epileptic focus may influence brain organization, so results may not easily generalize. Third, important sex differences in lesion parameters were often noted. Females had longer intervals between surgery and testing (Lansdell, 1968a,b) as well as smaller excisions (Lansdell 1968b; Lansdell & Urbach, 1965).

A large number of recent investigations have taken up the challenge of the evaluation of sex differences in lesion sequelae. Because this literature has been reviewed (Bryden, 1979; McGlone, 1980, 1984) and metaanalyzed (Inglis & Lawson, 1981, 1982; Bornstein & Matarazzo, 1982), the reader will be spared yet another study-by-study analysis of the older literature. Rather, let us ask a few critical questions. First, are sex differences in lesion sequelae reliable? Metaanalyses are perhaps best able to provide a straightforward answer to this question. Inglis and Lawson

(1981, 1982) reviewed 16 studies examining the impact of a unilateral cerebral lesion on VIQ and PIQ scores from the various Weschsler intelligence scales. Many of these studies did not report their data separately for the sexes, so the authors examined the relationship between the magnitude of VIQ–PIQ discrepancy and the proportion of males in the study. The greater the proportion of males, the greater was the magnitude of VIQ–PIQ discrepancy (VIQ greater the PIQ after right-brain damage and greater PIQ than VIQ after left-brain damage). The rank-order correlation between the VIQ–PIQ difference and the proportion of males in the study was $r = 0.51$. Bornstein and Matarazzo's (1982) review also provided support for the hypothesis of a sex difference in lesion sequelae, as have the other major literature reviews (McGlone, 1980, 1984). Let us turn to an evaluation of recent investigations not included in these reviews and metaanalyses. Unless more recent studies provide contrary data, it seems warranted to conclude that the observed sex differences in lesion sequelae reflect more than sampling variance.

Since 1984 (the time of the last major literature review), a total of 10 studies have investigated sex differences in lesion sequelae. Of these, six report significant sex differences (Yeo *et al.*, 1984; McGlone, 1985; Blanton & Gouvier, 1986; Sundet, 1986; Friendland & Kershner, 1986; Lewis & Kamptner, 1987). The nature and interpretation of the differences observed in these investigations are discussed below. The four-remaining recent studies offered generally negative conclusions regarding sex differences. Bornstein (1984) examined the performance of 63 patients on the WAIS-R. Many of their patients had etiologies atypical for focal lesion studies; 27 (42.8%) had either trauma, focal atrophy, arteriovenous malformations, or focal epilepsy. This raises the question of whether the brain damage in this sample may not have been either sufficiently severe or lateralized to produce the pattern noted by other investigations. Indirect support for this notion comes from McGlone's (1985) investigation in which it was noted that among subjects whose lesions extended across each side of the Rolandic fissure (and presumably were thus larger), significant sex differences were noted. Among more "focal" patients, those with lesions restricted to either anterior or posterior regions, significant sex differences were not noted.

Herring and Reitan (1986) reported Wechsler–Bellevue Scale performance in 98 patients, 62 (63%) of whom had neoplastic lesions and 34 (35%) of whom had vascular lesions. As in other investigations (e.g., Inglis, Ruckman, Lawson, MacLean, & Monga, 1982; McGlone, 1978; Yeo *et al.*, 1984) and metaanalyses (Inglis & Lawson, 1981, 1982), men were noted to have greater VIQ–PIQ discrepancies after brain damage, but in this investigation this difference was not significant. The difference between the Verbal Weighted Score and Performance Weighted Score (each derived from subtest scores) was, however, significantly different in males and females. The authors also noted a significant group (left versus right lesions)-by-sex interaction for VIQ ($P = .04$) but not for PIQ; adjustment of the error rate for multiple comparisons led the authors to conclude that the VIQ difference was not statistically reliable. Herring and Reitan (1986) noted a similar pattern to that reported by others, but the effect size was smaller. This may reflect sampling variance, subject differences, or measurement differences.

Warrington and colleagues (Warrington, James, & Maciejewski, 1986) recently reported WAIS data on a large number of men and women with unilateral lesions. Results were analyzed with a group (left versus right)-by-sex-by-IQ (PIQ versus VIQ) ANOVA. The critical interaction, a hemisphere-by-sex-by-IQ interaction, did not reach conventional levels of significance ($P = 0.06$), though a main effect for sex was noted, with females of both lesion groups tending to obtain lower scores on both IQ measures. This main effect for sex, not noted by any other investigations of sex differences, suggests some manner of sampling bias. Although the "critical" interaction did not reach significance, it was in the same direction as that noted by other investigators; i.e., the VIQ–PIQ difference score was 9.1 in males and 5.5 in females. It should also be noted that Warrington *et al.* did not administer the entire WAIS; two verbal and two performance subtests were omitted for all subjects. The attendant increase in psychometric reliability with the addition of these four subtests might have been sufficient to push the "critical" interaction term into the range of statistical significance.

Snow and Sheese (1985) observed no sex differences in the performance of stroke patients on the WAIS or the Wechsler Memory Scale. Further, the nonsignificant trends observed were contrary to those observed by most investigators. Among both male and female patients with left hemisphere injuries, VIQ exceeded PIQ, by 14 points in the males and 8 points in the females. This study was also somewhat unusual in the rather brief interval between lesion onset and assessment, 19 days. Also of note is the rather small sample size, ($N = 45$), with only 5 subjects in one cell (left-damaged females).

Considered in total, the 10 most recent studies appearing in the literature seem to support the conclusion emerging from both literature reviews and metaanalyses. Although not uniformly observed, sex differences in lesion sequelae occur more often than chance. Why are such results not always reported? The possible sources of variability across studies are numerous. McGlone (1984) suggested that both severity and etiology of the lesion influence the manifestation of post-lesion sex differences. Perhaps the foremost sources of error variance are in lesion size and locus, which have been systematically examined in only a single study (Yeo et al., 1984). Subject selection criteria (e.g., aphasic severity) and the length of time between lesion onset and testing are also likely sources of differences across studies. Recent animal research suggests another interesting possibility, that there may be sex differences in neuronal plasticity. Juraska (1984) reviews evidence that manipulations of rearing conditions have greater effects on neuronal development in male than female rats. It is possible, though clearly speculative, that some of the variability in human sex difference investigations reflects true variability in the samples. As early environments may have differed across samples, and a similar environment may affect the sexes differently (Juraska, 1984), the sexes could truly differ in one study but not another.

In any case, there does seem to be a sex effect in the sequelae of cerebral lesions. What is the nature of this effect? Let us first consider possible methodological possibilities. Are there sex differences in the lesions themselves? Etiology and time post-lesion have been controlled or examined in most investigations and so seem unlikely to be causally related to sex differences. Lesion size and location have been examined in only a single study (Yeo et al., 1984). In this investigation CT data were available for a subset (63 of 78, or 81%) of focal lesion patients. The CT scans were analyzed with a computerized procedure designed to quantify important brain parameters (Yeo, Turkheimer, & Bigler, 1983; Turkheimer, Yeo, & Bigler, 1983). First, the structures visible on the CT film were traced. The tracing was then digitized with a Summagraphics Bit Pad. Areas for brain structures on each slice were computed using an interactive program written in APLSF on a DEC-20 computer. The trapezoidal rule was used to determine volumes of all structures. Lesion size was expressed as a ratio of lesion volume to brain volume. Location of the lesion was expressed as the centerpoint of the lesion, in reference to a three-dimensional coordinate system with the origin of all axes at the midpoint of the interhemispheric fissure line on the middle slice. Three location parameters were produced: values on anterior–posterior, left–right, and superior–inferior axes. A sex-by-group (left versus right) MANOVA with lesion size (a ratio of lesion volume to brain volume) and the three location variables as dependent variables revealed no significant main effects or interaction. The only variable on which the sexes differed even slightly was in lesion size, with males having slightly larger lesions. When lesion size was treated as a covariate, however, significant sex differences in lesion sequelae remained. Although one cannot be certain that sex differences in lesion size or locus did not influence the results of other investigations, there seem to be no striking differences in subject selection procedures between Yeo et al. (1984) and other studies noting sex differences. At present, there seems to be no reason to hypothesize systematic sex differences in lesion parameters that could account for observed post-lesion cognitive differences.

It also seems unlikely that strategy effects could account for these sex differences. Strategy differences may plausibly account for observations of more impaired spatial performance in females after left-brain injury and less impaired spatial performance in females after right-brain injury. Because females may more often adopt a verbal strategy in solving spatial problems, left hemisphere damage may impair performance on spatial tasks more often. Similarly, right hemisphere injury may produce less impairment in spatial skills in females, as the verbal processes they may use in problem solving remain largely intact. Strategy differences cannot, however, plausibly account for sex differences on verbal tasks, which have been noted in many studies (e.g., McGlone, 1977, 1980, 1984; Kimura, 1983; Yeo et al., 1984). Strategy differences also cannot account for sex differences on noncognitive tasks. Yeo et al. (1984) reported sex differences in lesion sequelae on both elementary motor and sensory tasks. Among left- but not right-lesioned patients, males were more impaired than females. Sex differences in cognitive strategy thus seem insufficient to account for sex differences in lesion sequelae.

Could different rates of recovery of function account for these sex differences? Unfortunately, no investigation of intellectual performance has serial data on the performance of the two sexes. Two studies, however, have reported separate recovery rates from aphasia. Kertesz and McCabe (1977) observed no sex difference of rate of recovery from aphasia in stroke patients. Pizzamiglio, Mammucari, and Razzano (1985) reported equivalent recovery rates by sex for fluent and nonfluent aphasics but a greater recovery rate in female global aphasics. Inglis et al. (1982) examined the issue of chronicity in their investigation of sex differences in intellectual performance after a unilateral lesion and concluded that sex differences in lesion sequelae were apparent in both acute and chronic stroke patients, indirectly supporting the notion of no sex differences in rate of recovery. Data for differential rates of recovery are not only sparse but also unimpressive. Long-term follow-up of intellectual performance in males and females with lateralized lesions would represent an important contribution to the field.

If sex differences in lesion sequelae are unlikely to reflect methodological artifacts, they may reflect differences among brains. One has to consider this difference to be a small effect size. Is this effect so small as to be trivial? In this regard it is important to note how limited neuropsychological investigations of sex differences with the focal lesion method have been. This research is by and large limited to a specific level of the nervous system, the hemispheric level, and to a specific set of behavioral measures, the Wechsler scales. To the extent that sex differences among brains are not limited to hemispheric differences, a research strategy comparing only "left-lesioned" and "right-lesioned" subjects of each sex will result in an incomplete analysis of possible brain differences. Reliance on the Wechsler scales seems to be a matter of convenience rather than theory. These tests were designed to eliminate any premorbid sex differences (Matarazzo, 1972) and thus may not be optimal for investigations of sex differences in lesion sequelae. Further, the verbal and performance scales of the Wechsler tests were not designed to assess left and right hemisphere damage, respectively, and thus may not be maximally sensitive to unilateral hemispheric pathology. Given these considerations, it is perhaps remarkable that postlesion sex differences have been observed as often as they have been.

Structural differences seem unlikely to contribute to the observed effects. The only reliably reported structural brain difference between the sexes (aside from structures that are related to sexual and reproductive function) is in brain size (e.g., Mettler, 1956), which does not appear to be relevant for cognitive level or pattern (Yeo et al., 1987). Sex differences in the size of the corpus callosum have been reported (deLacoste-Utamsing & Holloway, 1982) but not replicated in a larger series of subjects (Oppenheim, Lee, Nass, & Gazzaniga, 1987). Most studies of asymmetries visible on CT suggest that the sexes do not appear to differ in anatomic asymmetries (Koff, Naeser, Pieniadz, Foundas, & Levine, 1986; Chui & Damasio, 1980; Yeo et al., 1987), although one recent report noted reduced asymmetry in females (Bear, Schiff, Sauer, Greenberg, & Freeman, 1986), and asymmetry of the planum temporale may be less in women (Wada, Clarke, & Hamm, 1975).

Perhaps sex differences in lesion sequelae reflect differences in representation of function. One prominent hypothesis has been that females show less lateralization of function (e.g., Mc-Glone, 1977). Females are hypothesized to have relatively greater right hemisphere representation of verbal skills. Although not a necessary corollary of this hypothesis, it is also most often

TABLE 3. Intellectual Performance of Males and Females with
Unilateral Lesions

Group	N	VIQ	PIQ	VIQ–PIQ
Left males	465	94.4	98.8	−4.4
Left females	335	93.2	93.2	0.8
Right males	424	93.0	93.0	12.4
Right females	297	91.0	91.0	9.6

hypothesized that females have more left hemisphere representation of spatial skills. In order to evaluate this hypothesis, it is necessary to examine the specific patterns of intellectual and linguistic deficits noted across studies.

Table 3 presents summary data on intellectual performance for 1521 male and female patients with unilateral lesions, as compiled from nine studies (Inglis *et al.*, 1982; Bornstein, 1984; Kimura & Harshman, 1984; Yeo *et al.*, 1984; Snow & Sheese, 1985; McGlone, 1984; Herring & Reitan, 1986; Sundet, 1986; Warrington *et al.*, 1986). Data for McGlone's subjects were taken from her 1985 review, in which performance of patients from previous studies (McGlone, 1978) was supplemented by more recent observations. The values shown in Table 3 were determined by weighting reported group values by the number of subjects in a given study and thus reflect mean performance levels of individuals, not mean performance levels by study.

After a left hemisphere lesion, men and women are equally impaired in VIQ. Women, however, show greater PIQ deficits after left lesions. After a right hemisphere lesion, men are very slightly superior in PIQ and clearly superior in VIQ. Overall, the magnitude of these differences is rather small. Left hemisphere representation of verbal intellectual skills and right hemisphere representation of nonverbal intellectual skills do not appear to differ across the sexes. Women seem to have more left hemisphere representation of nonverbal intellectual skills and more right hemisphere representation of verbal skills. Note, however, that this pattern is not entirely consistent with the notion of sex differences in "laterality." This hypothesis would predict, in addition to the pattern of differences described above, that men show greater verbal deficits after left damage and greater nonverbal deficits after right damage.

The pattern of results shown in Table 3 thus suggests that sex differences at the hemispheric level are not easily subsumed under the rubric of representational differences. How can it be that men and women have equally impaired verbal intellectual skills after a left lesion but that women are more impaired after a right lesion? If women have more right hemisphere representation of verbal intellectual skills, why are they not superior to men in VIQ after a left hemisphere lesion?

Perhaps sex differences at the hemispheric level are better captured by "organization" and/ or "activational" differences. To evaluate organizational differences, performance of the lesioned group must be compared with control groups. Unfortunately, control groups were not available for all studies summarized in Table 3. When those studies for which control data are available are summarized (Inglis *et al.*, 1982; Yeo *et al.*, 1984; Kimura & Harshman, 1984), the results are as shown in Table 4. For the purposes of fitting these data into Denenberg's models, let us assume that a 6-point difference between groups represents a significant difference. A 40% group difference in standard deviation units (assumed to be 15, as for normative WAIS or WAIS-R data) will be significant for a total sample size of 100 on 88% of analyses, an acceptable level of statistical power (Cohen, 1977).

For males, the following differences are seen in VIQ: left less than controls, right equal to controls, left less than right. This is the pattern Denenberg refers to as "left dominance." For females, the following differences are seen in VIQ: left less than controls, right less than controls, and right less than left, a pattern of "left dominance and synergism." At the risk of simplifica-

TABLE 4. Intellectual Performance of Males and Females with
Unilateral Lesions in Those Studies Providing Data on
Control Subjects

Group	N	VIQ	PIQ	VIQ–PIQ
Left males	110	91.3	95.7	−4.4
Left females	102	95.5	94.2	1.3
Right males	112	105.0	89.2	15.8
Right females	104	97.9	91.3	6.6
Control males	63	110.5	107.4	3.1
Control females	68	104.5	105.6	1.1

tion, verbal intellectual skills are controlled by the left hemisphere in males. In females the situation is more complex, with the left hemisphere playing a dominant role, but the hemispheres also interact. With respect to PIQ, the following differences are seen in males: left less than controls, right equal to controls, and right less than left. For females, this is the pattern; left less than controls, right less than controls, and right equal to left. Again, the male pattern is of "right dominance," and the female pattern is of "right dominance and synergism." The nature of the additional hemispheric interaction seen for females on verbal and nonverbal intellectual tasks is not known; exploration of this issue may provide important data on sex differences in the human brain.

Consideration of sex differences has thus far focused only on differences in intellectual performance at the level of the cerebral hemispheres. Let us now turn toward more specific analyses. First we examine sex differences in linguistic performance and then sex differences that may appear within, rather than across, hemispheres. With respect to possible sex differences in linguistic performance, let us ask several questions of the aphasic literature.

Is there a greater incidence of aphasia in males after left hemisphere damage? McGlone (1977) reported that the incidence of aphasia following left hemisphere vascular accidents was 48% (14/29) in males and 13% (2/16) for females ($\chi^2 = 5.76$, $P < 0.02$). Kimura (1983) examined sex differences in the incidence of aphasia in a sample that included all of McGlone's subjects (McGlone's sample compromised 20% of Kimura's subjects). Of 143 male patients with left hemisphere damage, 73 (51%) were aphasic. Of 73 females, only 22 (31%) were aphasic. In contrast, Miceli et al. (1981) found no sex difference in the incidence of aphasia. Ninety-eight of 260 males (38%) and 51 of 130 (39%) of females were aphasic. Kertesz and Sheppard (1981) reported similar results.

Among aphasics, are males more severely affected? The question of a sex difference in incidence resolves to the question of differences in the severity of aphasic symptoms in an unselected group with unilateral left hemisphere lesions. When studies examine sex differences in the severity of aphasic symptoms "among aphasics," they are simply limiting their analysis to the more severely impaired, truncating the distribution, and reducing their power to detect group differences. What is not revealed by such studies is whether there are more males or females among the "nonaphasics," i.e., those patients with left hemisphere damage and fewer aphasic symptoms. McGlone's two studies (McGlone & Kertesz, 1973; McGlone, 1977) and Kimura's study (1983) are the only studies to look at severity of aphasic symptoms among left-hemisphere-damaged patients rather than patients diagnosed as aphasic. In each of these investigations there were no significant sex differences in aphasic severity. Among aphasics, Edwards, Ellams, and Thompson (1976) found males to obtain significantly poorer scores on four of the six scales of the Schuell Aphasia Battery. Sasanuma (1980) reported a relatively greater incidence of "severe" aphasias among males in a sample of Japanese aphasics. In contrast, several other investigations have not observed a sex difference in the severity of aphasia among aphasics. Kertesz and Sheppard (1981) administered the Western Aphasia Battery (an adaptation of the Boston Diagnostic Aphasia Examination) to all patients. No sex difference was observed on any aphasia subscale or total score. Another large study has produced similar results (Basso, Capitani, Laiacona, & Luzzatti, 1980). No sex differences were observed in the severity of either fluent (global and Wernicke's) or nonfluent (Broca's) aphasias in a sample of 616 patients with vascular lesions. Miceli et al. (1981) observed no sex difference in the incidence of mild, moderate, and severe aphasia. Except for the reports of Edwards et al. (1976) and Sasanuma (1980), these studies are in agreement that there are no sex differences in severity of aphasia.

Are there sex differences in aphasia type? Six studies have explored this issue (DeRenzi, Faglioni, & Ferrari, 1980; Harasymiw & Halper, 1981; Eslinger & Damasio, 1981; Kertesz & Sheppard, 1981; Miceli et al., 1981; Brown & Gruber, 1983). The combined results of these studies are presented in Table 5. A χ^2 analysis of sex by aphasia type was performed for the combined data shown in Table 5. The sexes were found to significantly differ in aphasia type ($\chi^2 = 10.63$, $P<0.01$). Males were more frequently globals, and females nonfluent and fluent. Though the differences were quite small, they suggest relatively more "pure" aphasias (i.e., fluent and nonfluent) in females.

On the whole, the aphasia data do not suggest a striking sex difference, in accord with data presented in Table 3. Among the left-lesioned patients, the VIQs of the sexes were approximately equal. These data do not rule out, however, the possibility that there are differences within the left hemispheres of males and females.

Kimura (1983; Kimura & Harshman, 1984) has provided striking data on differences in the representation of language functions within the left hemisphere. She divided her sample of stroke patients into those with lesions confined to pre- ($N=40$) and postrolandic ($N=41$) regions. Males were equally likely to be aphasic after anterior or posterior lesions (40% and 41%, respectively). Females were much more likely to become aphasic after anterior than posterior lesions (62% versus 11%). Kimura suggested that females may have more anterior representation both language production and comprehension systems. To support her notion that posterior left hemisphere damage is less likely to lead to aphasia in females, Kimura reported that only 2 of 9 (22%) females with Wernicke's lesions were aphasic, compared with 13 of 18 (72%) for males. Further, of patients with restricted parietal lobe lesions, 0 of 6 females and 7 of 10 males were aphasic.

TABLE 5. Sex Differences in Aphasia Type[a]

	Fluent				Global				Nonfluent			
	Males		Females		Males		Females		Males		Females	
Study	N	%	N	%	N	%	N	%	N	%	N	%
DeRenzi et al. (1980)	60	56	66	72	18	17	7	8	30	28	19	21
Harasymiw et al. (1980)	41	24	27	19	40	24	20	14	87	52	93	66
Eslinger & Damasio (1981)	18	56	9	39	11	34	6	26	3	9	8	35
Kertesz & Sheppard (1981)	13	24	10	25	23	42	15	38	19	34	15	38
Miceli et al. (1981)	70	56	38	62	17	14	6	10	38	30	17	28
Brown & Gruber (1983)	33	15	9	10	76	35	26	30	106	49	52	60
Total	235	33	159	36	185	26	80	18	283	40	204	46

[a] $\chi^2 = 10.63$, $P<0.01$.

These data suggest an intrahemispheric sex difference in brain organization such that posterior regions subserve language functions to a greater extent in males than females. Apraxia and oral fluency (generating words beginning with a specific letter of the alphabet) data showed the same pattern: anterior and posterior lesions impaired males, whereas the effects of anterior lesions were relatively more severe in females.

Recent observations regarding global aphasia and its anatomic correlates are consistent with Kimura's data (Vignolo, Boccardi, & Caverni, 1986). In a large series of global aphasics, most were noted to have lesions involving both Wernicke's and Broca's area. Occasionally, some patients with limited posterior lesions were found to be globally aphasic. All eight of the global aphasics with limited anterior lesions were females, and all three of the global aphasics with posterior lesions were male.

These observations begin an important task for the neuropsychology of individual differences, to evaluate possible differences within, as opposed to between, hemispheres. Given the limited data available, intrahemispheric differences must at this point be considered speculative. It is of interest to note, however, that the electrical stimulation method of functional neuroanatomy has also suggest intrahemispheric sex differences (Mateer, Polen, & Ojemann, 1982). Among patients undergoing surgery for intractable epilepsy, the loci of electrical stimulation producing naming errors differed in men and women. Naming errors were evoked from 63% of the total sites sampled in males but from only 38% of the total sites in females. The sites evoking naming errors were more widespread in males, suggesting a more diffuse representation. These data are consistent with those presented in Table 5 concerning a sex difference in aphasia type. If females have more focal representation of language skills, one might expect them to be less likely to show a global aphasia. A given lesion may be slightly more likely to produce a selective impairment, as the representation of discrete cognitive functions may be less likely to overlap.

Gur et al. (1982) concluded from regional cerebral blood flow analyses that females show higher ratios of gray to white matter in the cerebral cortex than do males. Greater focal organization in females might require fewer long myelineated (white) fibers, as local processing might be relatively more prevalent than widely distributed processing. Given the difficulty of making observations about possible within-hemisphere differences in humans because of difficulty in obtaining large numbers of patients with small unilateral hemispheric lesions, investigation of this issue may require multicenter collaborative efforts.

Sex differences in human brains probably exist, but the nature of these differences remain unclear. Even less certain is whether the identified differences are relevant for sex differences in cognition. Clarification of this issue awaits both additional data, especially on possible intrahemispheric differences, and greater understanding of the psychobiology of cognition, for example, the nature and importance of interactions among the various dimensions along which brains may differ.

AGE DIFFERENCES

Most neuropsychological investigations of aging have focused on differences between children and adults. Studies of age differences in the sequelae of cerebral lesions within the adult range offer one way to begin unraveling the neurobiological mechanisms underlying age-related differences in behavior. In addition to the types of differences between brains outlined in the previous section, young and old adult brains also differ at the neuronal level. Several investigations have revealed extensive neuronal loss with aging (Brody, 1955, 1973; Henderson, Tomlinson, & Gibson, 1980), though a recent study with a more sophisticated method for neocortical cell counts observed only minimal neuronal loss through the adult age span (Terry, DeTeresa, & Hansen, 1987). This investigation also noted, however, that neuronal shrinkage is a prominent effect of aging and tends to be maximal in frontal and temporal lobes.

These considerations suggest that age-related differences in lesion sequelae, as compared to sex differences, are more apt to reflect structural than representational, organizational, or activational differences. The most prominent hypothesis concerning age differences with focal lesions is thus the older individuals will be more impaired after a given lesion than will younger adults. The degenerative changes accompanying normal aging may serve to increase the behavioral consequences of lesions and also to lessen recovery of function.

Some investigations support the hypothesis that older individuals suffer more severe consequences from brain lesions. Corkin (1979) reported that bilateral cingulotomy produced greater impairment on a complex figure-copying task in older patients than in younger patients. Performance on several other cognitive and sensory tests were not, however, more impaired in the older group. Benton (1977) measured visual and auditory reaction time in older (46–60) and younger (16–44) patients with cerebral damage matched for etiology of lesions as well as in young and old control subjects. Results indicated a significant interaction of age with group (patients versus controls) such that older patients with a brain lesion were especially impaired.

Cognitive abilities were examined in a second study (Hamsher & Benton, 1978). Again, older (mean age 54 years) and younger patients (mean age 33) were matched for side of lesion and etiology of lesion (30 lateralized neoplasms, 4 bilateral neoplasms, and 6 lateralized CVAs). Age groups were approximately matched for sex. Older and younger control groups were obtained and matched with the patient groups on education and sex. The cognitive tests included six subtests from the WAIS, five other verbal tests, two verbal memory tests, and a spatial test. An overall MANOVA revealed a significant group (patients versus controls)-by-age interaction. The older patients showed more severe deficits relative to controls. The specific tests showing significant interactions were all verbal, the WAIS subtests of Comprehension and Similarities, and paired-associates learning task. Benton's data suggest that lesions produce relatively greater deficits in older patients and, since the differences in lesion sequelae were verbal, raise the question of greater age differences for left hemisphere functioning. Unfortunately, Benton did not separate his patients into those with left versus right lesions specifically to evaluated this hypothesis. Also, no data were presented regarding lesion size and locus, leaving open the possibility that age differences in lesions rather than age differences in brains accounted for group differences.

The only investigation comparing the effects of left and right lesions separately, as well as the only investigation of age effects to quantify lesion paramenters, was that of Yeo et al. (1984). In this study of 78 unilateral-lesion patients (38 left, 40 right), the median age (47) was used to dichotomize the sample into young adult (mean age 34.3) and old adult (mean age 61.9) groups. Each subject was administered a comprehensive neuropsychological assessment from which scores were collapsed to form seven composite neuropsychological variables: verbal ability, nonverbal ability, perceptual motor skills, right-side motor skills, left-side motor skills, right-side sensory skills, and left-side sensory skills. Expressions of lesion size and locus did not differ across young and old groups.

The effects of a lesion were noted to differ across age samples on only a single variable, right motor skill. Given the number of comparisons made and the significance level of this observation ($P = 0.05$), the possibility of a type I error cannot be ruled out. The age groups compared in this analysis differed in terms of etiology, with the older group, predictably, showing relatively more strokes and fewer neoplastic lesions. When young and old subjects were matched for etiology (necessitating the elimination of some subjects), the only significant age effect was on the left sensory variable, with older lesioned subjects performing especially poorly. The results of this investigation do not suggest striking age-related differences in the sequelae of focal lesions. A variety of factors that might have masked greater deficits in the older group (i.e., lesion differences, etiology differences, education differences) were determined not to be relevant in this data set.

Aphasia studies have also examined the variable of age. There appears to be a relationship between age and aphasia type. Table 6 combines the results of several studies examining this

TABLE 6. Age Differences in Aphasia Types

Study	Fluent		Global		Nonfluent	
	N	Age	N	Age	N	Age
Obler *et al.* (1980)	29	62.8	27	56.2	61	52.6
DeRenzi *et al.* (1980)	126	62.6	25	59.2	49	56.8
Harasymiw & Halper (1981)	68	64.6	60	62.3	180	55.7
Eslinger & Damasio (1981)	27	62.1	17	58.0	11	45.5
Kertesz & Sheppard (1981)	23	67.7	38	66.8	34	59.0
Miceli *et al.* (1981)	108	59.5	23	53.1	55	51.2
Brown & Gruber (1983)	42	56.5	102	50.6	158	45.3
Total	423	61.8	292	57.0	547	51.9

issue (Obler, Albert, Goodglass, & Benson, 1978; DeRenzi *et al.*, 1980; Harasymiw & Halper, 1981; Kertesz & Sheppard, 1981; Eslinger & Damasio, 1981; Miceli *et al.*, 1981; Brown & Gruber, 1983). Wernicke's aphasics (mean age 61.8) and global aphasics (mean age 57.0) are clearly older than Broca's aphasics (mean age 51.9).

Why are the Broca's aphasics younger? Eslinger and Damasio (1981) offer four possible explanations: (1) There might be some neuropathological variation associated with increasing age, such that the preferred locus of infarction would shift to posterior brain sites; consequently, younger patients would demonstrate a higher incidence of relatively anterior strokes, and older patients would demonstrate a higher incidence of posterior strokes. (2) There might be changes in cerebral blood flow associated with aging, predisposing different brain regions to strokes. (3) The higher incidence of Wernicke's aphasia in older patients might be related to the cumulative effects of mental decline associated with aging and insult to a language area of the brain, regardless of its anatomic locus. (4) There might be continuous, age-related changes in the neurophysiological mechanisms subserving language functions, such that regardless of lesion location certain aphasia types become more prevalent with age.

The first two hypotheses, regarding gross neuropathological variations associated with aging and cerebral blood flow changes, are not mutually exclusive (Eslinger & Damasio, 1981, pp. 349–350). Two observations argue against hypotheses 1 and 2. Miceli *et al.* (1981) reported the same relationship of age and aphasia type among tumor patients as among vascular patients. Further, Brown and Gruber (1983) noticed the same trend in traumatic aphasia. Age differences in aphasia type do not appear to be related to different sites of vascular lesions. Hypothesis 3 is also not supported by clinical data: Wernicke's aphasics may be demented but are clearly not in many cases. The changing picture of aphasia type with age thus seems to be more a function of age-related changes in the brain than in the lesions.

Jason Brown has hypothesized (Brown & Jaffe, 1975; Brown & Gruber, 1983) that age-related changes in the representation of language functions account for age-related differences in types of aphasia. Brown suggests that throughout the life span language centers become more "focally" represented, gradually increasing the left hemisphere control of language, and, within the left hemisphere, gradually increasing the focalization of language centers. Thus, within the left hemisphere, language production and comprehension regions would become more and more localized and limited to the classical speech areas of Broca and Wernicke. The younger the individual, the more widespread and overlapping may be the processing centers.

One difficulty with Brown's hypothesis is that it would predict that more "pure" aphasia would be seen in older patients, i.e., more pure Broca's and pure Wernicke's, rather than a predominance of one type, as is revealed by Table 6. As the cortical regions for production and comprehension become more focally organized, they would overlap less, increasing the possibility that one may be damaged without the other. If Brown's hypothesis were limited to age-related changes in the speech production system, it would better account for the pattern revealed in Table 6 as well as for the observation that almost all cases of aphasia in children include production difficulties (Hecaen, 1976).

Though clearly speculative, it is interesting to note that Terry et al. (1987) reported that the major difference among brains differing in age is a reduction of large neurons in older brains. This effect was greater in midfrontal and superior temporal gyrus regions and may reflect in part a reduction in pyramidal (motor) neurons. Similarly, Scheibel and Scheibel (1975) have described selective loss of horizontally oriented dendrites in pyramidal neurons, a process they suggest may be related to deterioration of psychomotor performance in aging. Would neuronal loss or shrinkage be equivalently distributed throughout a given functional system? Physiological research suggests that neuronal loss is in part a function of the amount of neuronal activation (Hamburger & Levi-Montalcini, 1949). Perhaps, then, those neurons at the periphery of a neural network (i.e., in the motor system, those pyramidal neurons outside the motor strip) would be less frequently activated and thus more apt to show age-related changes. Such a process could conceivably lead to a progressively more focal representation of a given functional system. These considerations suggest a hypothesis regarding the functional correlates of degree of focal representation. If "diffuse" or "more overlapping" representational systems are characteristic of right more than left hemispheres, of men more than women, and of younger more than older people, then this pattern of neuronal development might be associated with relatively greater visuospatial than verbal ability.

Recovery of function appears to be reduced in older individuals in closed-head injuries (Overgaard et al., 1973), aphasic disorders (Sands et al., 1969; Teuber, 1974, 1975; Kertesz & McCabe, 1977), and sensorimotor impairment (Teuber, 1974, 1975). The mechanisms presumed to underlie reduced recovery (e.g., differential lesion-induced synaptogenesis: Hoff, Scheff, & Cotman, 1982) are outside the scope of this review, but reduced recovery of function could be responsible for the greater neuropsychological deficits sometimes observed in the initial assessment of older individuals (Benton, 1977; Hamsher & Benton, 1978).

ANATOMIC ASYMMETRIES

Anatomic asymmetries of the cerebral hemispheres have been linked to patterns of cognitive specialization (Yeo et al., 1987; Rosenberger & Hier, 1980; Luchens et al., 1982). To understand the mechanisms of this effect, we need ask whether brains that differ in anatomic asymmetries differ in other ways. One prominent hypothesis has been of a linkage between anatomic and functional hemispheric asymmetries (Pieniadz, Naeser, Koff, & Levine, 1983; Geschwind & Galaburda, 1985). Given the associations of anomalous asymmetries with psychopathology (e.g., Tsai, Nasrallah, & Jacoby, 1983), learning disabilities (e.g., Schacter, Ransil, & Geschwind, 1987), and recovery of function (Pieniadz, Naeser, Koff, & Levine, 1983; Schenkman, Butler, & Naeser, 1983), understanding the nature and correlates of anatomic asymmetries is a broadly important issue in neuropsychology.

Do focal-lesion patients who differ in anatomic asymmetry (as determined from CT scans) also differ in neuropsychological sequelae? Global aphasics with and without the typical cerebral asymmetries did not differ on intitial postlesion language assessment (Pieniadz, Naeser, Koff, & Levine, 1983). Similarly, severity of initial postlesion hemiplegia was noted to be unrelated to anatomic asymmetries (Shenkman, Butler, & Naeser, 1983). Crossed aphasics, i.e., those right handers with right hemisphere lesions impairing language, do not differ from left-lesion aphasics in terms of anatomic asymmetries (Henderson et al., 1985). Further, in a large series of focal-lesion patients (those reported in Yeo et al., 1984), we have noted no relationship between anatomic asymmetries and performance on cognitive, motor, or sensory tests in either right- or left-hemisphere-damaged patients. In contrast, Kertesz and Ferro (1984) suggested that variations in anatomic asymmetry may be related to severity of apraxia following a left hemisphere lesion. From a sample of 177 stroke patients, they isolated those with small lesions producing apraxia ($N = 9$) and those with large lesions that did not produce apraxia acutely ($N = 9$). Among the latter, two showed atypical frontal asymmetries, and four showed atypical posterior asymmetries. Because the role of anatomic asymmetries in mediating the effects of a left hemisphere lesion was not analyzed statistically and was reported for only a small number of subjects, caution seems warranted in interpreting these results.

On the whole, anatomic asymmetries appear to be unrelated to level or pattern of neuropsychological deficit on initial assessment. Two studies have, however, reported that subjects with reduced asymmetry recover better from aphasia (Pieniadz, Naeser, Koff, & Levine, 1983) and hemiplegia (Shenkman, Butler, & Naeser, 1983).

The mechanism for better recovery among those with reduced asymmetry and the range of neuropsychological variables that may be characterized by better recovery in those with reduced asymmetry are unknown. Naeser and Borod (1986) have suggested that the hemispheric representation of some language functions may be linked to anatomic asymmetries. Better recovery of certain language skills thus reflects more contralateral representation of those skills. Since initial postlesion deficits are unrelated to asymmetry, however, one must suppose that the verbal (or motoric) skills of the contralateral hemisphere become manifest rather slowly, perhaps secondary to diminution over time of the effects of diaschisis. Obviously, much more data are needed before clear elucidation of the meaning of anatomic asymmetries will be possible. As we noted in our discussion of sex differences, linking behavioral differences with brain differences may well require an understanding of the interactions among dimensions of differences among brains. In this regard, a recent study (M. Naeser, G. McAnulty, M. Albert, F. Duffy, C. Palumbo, & J. Fang, personal communication, 1987) of complex differences in patterns of cerebral activation among normal individuals varying in anatomic asymmetry represents an exciting direction for future research.

SUMMARY

In this review we have focused on three dimensions of difference among people (sex, age, and asymmetry) that have been associated with systematic differences in cognitive skills. For each of these individual difference variables, we have sought evidence for differences among brains using the focal lesion method, attempting to sort through both methodological and interpretive difficulties. Understanding the nature of individual differences may eventually offer significant insights regarding principles of brain–behavior relationships, especially with systematic efforts to evaluate the multiple dimensions along which brains may differ.

Acknowledgments. The author wishes to thank Mary Hungate for her assistance in preparation of this manuscript.

REFERENCES

Basso, A., Capitani, E., Laiacona, M., & Luzzatti, C. (1980). Factors influencing type and severity of aphasia. *Cortex, 16,* 631–636.
Bear, D., Schiff, D., Sauer, J., Greenberg, M., & Freeman, R. (1986). Quantitative analysis of cerebral asymmetries: Fronto-occipital correlation, sexual dimorphism and association with handedness. *Archives of Neurology, 43,* 598–603.

Benton, A. L. (1977). Interactive effects of age and brain disease on reaction time. *Archives of Neurology,* 34, 369–370.

Berg, C., Hertzog, C., & Hunt, E. (1982). Age differences in the speed of mental rotation. *Developmental Psychology, 18,* 95–107.

Blanton, P., & Gouvier, W. (1986). Sex differences in hemispatial performance following right hemisphere cerebrovascular accidents. Paper presented at the annual convention of the International Neuropsychological Society, Denver.

Bornstein, R. (1984). Unilateral lesions and the Wechsler Adult Intelligence Scale-Revised: No sex differences. *Journal of Consulting and Clinical Psychology, 52,* 604–608.

Bornstein, R., & Matarazzo, J. (1982). Wechsler VIQ versus PIQ differences in cerebral dysfunction: A literature review with emphasis on sex differences. *Journal of Clinical Neuropsychology, 4,* 319–334.

Brody, H. (1955). Organization of the cerebral cortex III. A study of aging in the human cerebral cortex. *Journal of Comparative Neurology, 102,* 511–555.

Brody, H. (1973). Aging of the vertebrate brain. In M. Rockstein (Ed.), *Development and aging in the nervous system* (pp. 39–64). New York: Academic Press.

Brown, J., & Gruber, E. (1983). Age, sex, and aphasia type. *Journal of Nervous and Mental Disease, 171,* 431–434.

Brown, J., & Jaffe, J. (1975). Hypothesis on cerebral dominance. *Neuropsychologia, 13,* 107–110.

Bryden, M. (1979). Evidence for sex-related differences in cognitive organization. In M. Wittig & A. Petersen (Eds.), *Sex-related differences in cognitive functioning* (pp. 121–143). New York: Academic Press.

Caplan, P., MacPherson, G., & Tobin, P. (1985). Do sex related differences in spatial ability exist? *American Psychologist, 40,* 786–799.

Chui, H., & Damasio, R. (1980). Human cerebral asymmetries evaluated by computed tomography. *Journal of Neurology, Neurosurgery, and Psychiatry, 43,* 873–878.

Cohen, J. (1977). *Statistical power analysis for the behavioral sciences.* New York, Academic Press.

Corkin, S. (1979). Hidden-figures-test performance: Lasting effects of unilateral penetrating head injury and transient effects of bilateral cingulotomy. *Neuropsychologia, 17,* 585–605.

Craik, F. (1977). Age differences in human memory. In J. Birren & K. Schaie (Eds.), *Handbook of the psychology of aging* (pp. 384–420). New York: Van Nostrand Reinhold.

Denenberg, V. (1980). General systems theory, brain organization, and early experiencs. *American Journal of Physiology, 7,* 3–13.

DeRenzi, E., Faglioni, P., & Ferrari, P. (1980). The influence of sex and age of the incidence and type of aphasia. *Cortex, 16,* 631–636.

Edwards, S., Ellams, J., & Thompson, J. (1976). Language and intelligence in aphasia: Are they related? *British Journal of Disorders in Communication, 11,* 83–94.

Endo, H., Larsen, B., & Lassen, N. (1977). Regional cerebral blood flow alterations remote from the site of intracranial tumors. *Journal of Neurosurgery, 46,* 271–281.

Eslinger, P., & Damasio, A. (1981). Age and type of aphasia in patients with stroke. *Journal of Neurology, Neurosurgery, and Psychiatry, 44,* 377–382.

Feeney, D., Gonzalez, A., & Law, W. (1982). Amphetamine, haloperidol, and experience interact to affect rate of recovery after motor cortex injury. *Science, 217,* 855–857.

Friedland, J., & Kershner, J. (1986). Sex linked lateralized central processor for hierarchically-structured material? Evidence from Broca's aphasia. *Neuropsychologia, 24,* 411–415.

Garai, J., & Scheinfeld, A. (1968). Sex differences in mental and behavioral traits. *Genetic Psychology Monographs, 77,* 169–299.

Geschwind, N. (1974) Late changes in the nervous system. In D. Stein (Ed.), *Plasticity and recovery of function in the central nervous system* (pp 294–314). NY, NY: Academic Press.

Geschwind, N., & Galaburda, A. (1985). Cerebral lateralization, biological mechanisms, associations and pathology: I. A hypothesis and a program for research. *Archives of Neurology, 42,* 428–459.

Geschwind, N., & Levitsky, W. (1968). Human brain: Left–right asymmetries in temproal speech regions. *Science, 161,* 186–187.

Gloning, I., Gloning, K., Haub, G., & Quatember, R. (1969). Comparison of verbal behavior in right-handed and nonright-handed patients with anatomically verified lesion of one hemisphere. *Cortex, 5,* 43–52.

Goldman, P. (1974). Functional development of the pre-frontal cortex in early life and the problem of neuronal plasticity. *Experimental Neurology, 32,* 366–387.

Gould, S. (1981). *The mismeasure of man.* New York: W. W. Norton.

Gur, R. C., Gur, R. E., Obrist, W., Hungerbuhler, D., Younkin, D., Rosen, A., Skolnick, B., & Reirich, M. (1982). Sex and handedness differences in cerebral blood flow during rest and cognitive activity. *Science, 217,* 659–660.

Hamburger, V., & Levi-Montalcini, R. (1949). Proliferation, differentiation, and degeneration in the spinal ganglia of the chick embryo under normal and experimental conditions. *Experimental Zoology, 111,* 457–501.

Hamsher, K., & Benton, A (1978). Interactive effects of age and cerebral disease on cognitive performances. *Journal of Neurology, 217,* 195–200.

Harasymiw, S., & Halper, S. (1981). Sex, age, and aphasia type. *Brain and Language, 12,* 190–198.

Hardyck, C., & Petrinovich, M. (1977). Left-handedness. *Psychological Bulletin, 84,* 385–404.

Harshman, R., Hampson, E., & Berenbaum, S. (1983). Individual differences in cognitive abilities and brain organization part one: Sex and handedness differences in ability. *Canadian Journal of Psychology, 37,* 144–192.

Hecaen, H. (1976). Acquired aphasia in children and the ontogenesis of hemispheric specialization. *Brain and Languages, 3,* 114–134.

Hecaen, H., & Sauget, J. (1971). Cerebral dominance in left-handed subjects. *Cortex, 7,* 19–48.

Hecaen, H., DeAgostini, M., & Monzon-Montes, A. (1981). Cerebral organization in left-handers. *Brain and Language, 12,* 261–284.

Helm-Estabrooks, N., Yeo, R., Geschwind, N., Freedman, M., & Weinstein, C. (1986). Stuttering: Disappearance and reappearance with acquired brain lesions. *Neurology, 36,* 1109–1112.

Henderson, G., Tomlinson, B., & Gibson, P. (1980). Cell counts in human cerebral cortex in normal adults throughout life using an image analyzing computer. *Journal of Neurological Science, 46,* 113–116.

Henderson, V., Naeser, M. A., Weiner, J. M., & Pieniadz, J. (1984). CT criteria of hemispheric asymmetry failed to predict language laterality. *Neurology, 34,* 1086–1089.

Herring, S., & Reitan, R. (1986). Sex similarities in verbal and performance IQ deficits following unilateral cerebral lesions. *Journal of Consulting and Clinical Psychology, 54,* 537–541.

Hoff, S. F., Scheff, S. W., & Cotman, C. W. (1982). Lesion-induced synaptogenesis in the dentate gyrus of aged rates. II. Demonstration of an imparied degeneration of clearing response. *Journal of Comparative Neurology, 205,* 253–259.

Inglis, J., & Lawson, J. (1981). Sex differences in the effects of unilateral brain damage of intelligence. *Science, 212,* 693–695.

Inglis, J., & Lawson, J. (1982). A meta-analysis of sex differences in the effects of unilateral brain damage on intelligence test results. *Canadian Journal of Psychology, 36,* 670–683.

Inglis, J., Ruckman, M., Lawson, J., MacLean, A., & Monga, T. (1982). Sex differences in the cognitive effects of unilateral brain damage. *Cortex, 18,* 257–276.

Isaacson, R. L. (1975). The myth of recovery from early brain damage. In N. Ellis (Ed.), *Aberrant development in infancy* (pp. 1–25). NY, NY: Wiley, 1975.

Juraska, J. (1984). Sex differences in developmental plasticity in the visual cortex and hippocampal dentate gyrus. In G. DeVries, J. P. C. De Bruin, H. B. M. Uylings, & M. A. Corner (Eds.). *Progress in Brain Research,* Vol. 61, Amsterdam: Elsevier.

Kaste, M., & Waltimo, O. (1976). Prognosis of patients with cerebral artery occlusion. *Stroke, 7,* 482–484.

Kertesz, A., & Ferro, J. (1984). Lesion size and location ideomotor apraxia. *Brain, 107,* 921–937.

Kertesz, A., & Geschwind, N. (1971). Patterns of pyramidal decussation and their relationship to handednes. *Archives of Neurology, 24,* 326–332.

Kertesz, A., & McCabe, P. (1977). Recovery patterns and prognosis in aphasia. *Brain, 100,* 1–18.

Kertesz, A., & Sheppard, A. (1981). The epidemiology of aphasia and cognitive impairment in stroke. Age, sex and aphasia type and laterality differences. *Brain, 104,* 117–128.

Kirmra, D. (1983). Sex differences in cerebral organization for speech and praxic functions. *Canadian Journal of Psychology, 37,* 19–35.

Kimura, D., & Harshman, R. (1984). Sex differences in brain organization for verbal and non-verbal functions. In G. DeVries, J. P. C. DeBruin, H. B. M. Uylings, & M. A. Corner (Eds.), *Progress in Brain Research,* V. 61, Amsterdam: Elsevier.

Koff, E., Naeser, M., Pieniadz, J., Foundas, A., & Levine, H. (1986). CT scan hemisphasic asymmetries in right-and left-handed males and females. *Archives of Neurology, 43,* 487–491.

Lacoste-Utamsing, C. de, & Holloway, R. (1982). Sexual dimorphism in the human corpus callosum. *Science, 216,* 1431–1432.

Lansdell, H. (1961). The effect of neurosurgery on a test of proverbs. *American Psychologist, 16,* 448.

Lansdell, H. (1962). A sex difference in the effect of temporal lobe neurosurgery on design preferences. *Nature, 194,* 852–854.

Lansdell, H. (1968a). Effect of temporal lobe ablations on two lateralized deficits. *Physiology and Behavior, 3,* 271–273.

Lansdell, H. (1968b). The use of factor scores from the Wechsler–Bellevue Scale of Intelligence in assessing patients with temporal lobe removals. *Cortex, 4,* 257–268.

Lansdell, H. (1973). Effect of neurosurgery on the ability to identify popular word associations. *Journal of Abnormal Psychology, 81,* 255–258.

Lansdell, H., & Urbach, N. (1965). Sex differences on a personality measure related to size and side of temporal lobe ablations. In *Proceedings of the American Psychological Association* (pp. 113–114). Washington, DC: APA Publications.

Lewis, R. S., & Kamptner, N. L. (1987). Sex differences in spatial task performance of patients with and without unilateral cerebral lesions. *Brain and Cognition, 6,* 142–152.

Linn, R. (1986). *Age and health status in relation to rigidity and fluid intelligence.* Unpublished doctoral dissertation, University of New Mexico, Albuquerque.

Luchins, D., Weinberger, D., & Wyatt, J. (1982). Schizophrenia and cerebral asymmetry detected by computed tomography. *American Journal of Psychiatry, 139,* 753–757.

Maccoby, E., & Jacklin, C. (1974). *The psychology of sex differences.* Stanford: Stanford University Press.

Matarazzo, J. (1972). *Wechsler's measurement and appraisal of adult intelligence.* New York: Oxford University Press.

Mateer, C., Polen, S., & Ojemann, G. (1982). Sexual Variation in cortical localization of naming as determined by stimulation mapping. *The Brain and Behavioral Sciences, 5,* 310–311.

McGlone, J. (1977). Sex differences in the cerebral organization of verbal functions in patients with unilateral brain lesions. *Brain, 100,* 775–793.

McGlone, J. (1978). Sex differences in functional brain asymmetry. *Cortex, 14,* 122–128.

McGlone, J. (1980). Sex differences in brain asymmetry. *The Behavioral and Brain Sciences, 3,* 215–264.

McGlone, J. (1984). The neuropsychology of sex differences in human brain organization. In G. Goldstein & R. Tarter (Eds.), *Advances in clinical neuropsychology,* Vol. III. New York: Plenum Press.

McGlone, J., & Kertesz, A. (1973). Sex differences in cerebral processing of visuospatial tasks. *Cortex, 9,* 313–330.

Mettler, F. A. (1956). *Culture and the structural evolution of the neural system.* New York: The American Museum of Natural History.

Miceli, G., Caltagirone, C., Gainotti, G., Masullo, C., Silveri, M., & Villa, G. (1981). Influence of sex, age, literacy, and pathologic lesion on incidence, severity, and type of aphasia. *Acta Neurologica Scandinavica, 64,* 370–382.

Naeser, M., & Borod, J. (1986). Aphasia in left-handers: CT scan lesion site, lesion side, and hemispheric asymmetries. *Neurology, 36,* 471–489.

Obler, L., Albert, M., Goodglass, H., & Benson, F. (1978). Aphasia type and aging. *Brain and Language, 6,* 318–322.

Ojemann, G. (1983). Brain organization for language from the perspective of electrical stimulation mapping. *The Brain and Behavioral Sciences, 6,* 189–230.

Oppenheim, J., Lee, B., Nass, R., & Gazzaniga, M. (1987). No sex-related differences in human corpus callosum based on magnetic resonance imaging. *Annals of Neurology, 21,* 604–606.

Overgaard, J., Christensen, S., Hvid-Hansen, O., Haase, J., Land, A., Pederson, K., & Tweed, W. (1973). Prognosis after head injury based on early clinical examination. *Lancet, 2,* 631–635.

Pieniadz, J. M., Naeser, M. A., Koff, E., & Levine, H. L. (1983). CT scan cerebral hemispheric asymmetry measurements in stroke cases with global aphasias. Atypical asymmetries associated with improved recovery. *Cortex, 19,* 371–379.

Pizzamiglio, L., Mammucari, A., & Razzaro, C. (1985). Evidence from recovery from aphasia for sex differences in brain organization. *Brain and Language, 25,* 213–223.

Rosenberger, G., & Hier, D. (1980). Cerebral asymmetry and verbal intellectual deficits. *Annals of Neurology, 3,* 216–220.

Sands, E., Sarno, M., & Schankweiler, D. (1969). Long-term assessment of language function in aphasia due to stroke. *Archives of Physical Medicine and Rehabilitation, 50,* 202–222.

Sasanuma, S. (1980). Do Japanese show sex differences in brain asymmetry? Supplementary findings. *The Behavioral and Brain Sciences, 3,* 247–248.

Schacter, S. C., Ransil, B. J., & Geschwind, N. (1987). Associations of handedness with eye color and learning disabilities. *Neuropsychologia, 25,* 269–276.

Scheibel, M. (1984). A dendritic correlate of human speech. In N. Geschwind & A. Galaburda (Eds.), *Cerebral dominance* (pp. 115–128). Cambridge, MA: Harvard University Press.

Scheibel, M., & Scheibel, A. (1975). Structural changes in the aging brain. In H. Brady, D. Harmon, & J. Ordy (Eds.), *Aging,* Vol. I (pp. 43–72). New York: Raven Press.

Schenkman, M., Butler, R. B., Naeser, M. A., & Kleefield, J. (1983). Cerebral asymmetry in CT and functional recovery from hemiplegia. *Neurology, 33,* 473–477.

Semmes, J. (1968). Hemispheric specialization. A possible clue to mechanism. *Neuropsychologia, 6,* 11–26.

Sinderman, F., Bechinger, D., & Dichgans, J. (1970). Occlusion of the internal carotid artery compared with those of the middle cerebral artery. *Brain, 93,* 199–210.

Snow, W., & Sheese, S. (1985). Lateralized brain damage, intelligence and memory: A failure to find sex differences. *Journal of Consulting and Clinical Psychology, 53,* 940–941.

Stern, J., & Baldinger, A. (1983). Hemispheric differences in preferred modes of information processing and the aging process. *International Journal of Neuroscience, 18,* 97–106.

Sundet, K. (1986) Sex differences in cognitive impairment following unilateral brain damage. *Journal of Clinical and Experimental Neuropsychology, 8,* 51–61.

Teuber, H. L. (1974). Recovery of function after brain injury. In *Outcome of severe damage to the nervous system,* Ciba Symposium 34 (pp. 197–211). New York: Elsevier.

Teuber, H. L. (1975). Effects of focal brain injury on human behavior. In D. Tower (Ed.), *The nervous system,* Vol. 2: *The clinical neurosciences* (pp. 344–367). New York: Raven Press.

Terry, R., DeTeresa, R., & Hansen, L. (1987). Neocortical cell counts in normal human adult aging. *Annals of Neurology, 21,* 530–539.

Tsai, L., Nasrallah, H., & Jacoby, C. (1983). Hemispheric asymmetries on computed tomographic scans in schizophrenia and mania. *Archives of General Psychiatry, 40,* 1286–1289.

Turkheimer, E., Yeo, R., & Bigler, E. (1983). Digital planimetry in APLSF. *Behavior Research Methods, 15,* 471–473.

Vignolo, L. (1964). Evolution of aphasia and language rehabilitation: A retrospective exploratory study. *Cortex, 1,* 344–367.

Vignolo, L., Boccardi, E., & Caverni, L. (1986). Unexpected CT scan findings in global aphasia. *Cortex, 22,* 55–69.

Wada, J., Clarke, R., & Hamm, A. (1975). Cerebral asymmetry in humans. *Archives of Neurology, 32,* 239–246.

Warrington, E., James, M., & Maciejewski, C. (1986). The WAIS as a lateralizing instrument: A study of 656 patients with unilateral cerebral lesions. *Neuropsychologia, 24,* 223–239.

Yeo, R., & Cohen, D. (1983). Familial sinistrality and sex differences in cognitive ability. *Cortex, 19,* 125–130.

Yeo, R., Melka, B., & Haaland, K. Y. (1988). *Intra-hemispheric differences in motor control: A re-evaluation of the Semmes hypothesis.* Paper presented at the 1988 meeting of the International Neuropsychological Society, New Orleans, LA.

Yeo, R., Turkheimer, E., & Bigler, E. (1983). Computer analysis of lesion volume: Reliability, utility, and neuropsychological applications. *Clinical Neuropsychology, 45,* 311.

Yeo, R., Turkheimer, E., & Bigler, E. (1984). The influence of sex and age on unilateral cerebral lesion sequelae. *International Journal of Neuroscience, 24,* 299–301.

Yeo, R., Turkheimer, E., Raz, N., & Bigler, E. (1987). Volumetric asymmetries of the human brain: Intellectual correlates. *Brain and Cognition, 6,* 15–23.

Zuckerman, M. (1983). A biological theory of sensation seeking. In M. Zuckerman (Ed.), *Biological bases of sensation seeking, impulsivity, and anxiety* (pp. 306–331). Hillsdale, NJ: Lawrence Erlbaum Associates.

12

Structural Anomalies and Neuropsychological Function

ERIN D. BIGLER, J. R. LOWE, and RONALD A. YEO

INTRODUCTION

The previous chapters have dealt with relating current topics in neuropsychology with brain imaging, toward the goal of improved understanding of the relationship between anatomy and function. Much of this research has dealt with "average" or "group" findings with patients who have been selected because they shared a common diagnosis. Thus, the focus has been on systematic variations that occur when a "normal" brain is injured or damaged. In Chapter 11 on individual differences, Yeo discusses normal differences in brain organization. In the present chapter we explore a variety of issues related to abnormal variations of the central nervous system, especially developmental or congenital abnormalities, and how such abnormalities may be reflected in neuropsychological functioning. Some of this variability is dependent on plasticity issues of the developing brain and its response to maldevelopment or early brain injury. Understanding these issues not only is clinically important but has implications for understanding the limits of neural plasticity and recovery of function.

CONGENITAL DISORDERS

Brain anomalies associated with dysgenic brain development, traumatic intrauterine influences, or perinatal complications may produce syndromes that are quite unlike their adult counterparts (Herskowitz & Rosman, 1982). Brain damage that occurs to the developing nervous system frequently has a different effect than that which occurs in the mature brain. Clinically, one may see the entire range of neuropsychological outcome, from predicted deficits of a specific brain lesion (based on structure–function analyses of adult brain injury) to the development of adaptive or normal functions despite an obvious lesion site. In these cases the nature and locus of structural brain anomaly or abnormality, does not always predict the deficits that may or may not be present. For example, Figure 1 depicts bilateral occipital porencephaly in a 13-year-old female who had a large occipital encephalocele surgically removed at 4 days of age. This patient has

ERIN D. BIGLER ● Austin Neurological Clinic and Department of Psychology, University of Texas at Austin, Austin, Texas 78705.   J. R. LOWE ● Department of Pediatrics, University of New Mexico, Albuquerque, New Mexico 87131.   RONALD A. YEO ● Department of Psychology, University of New Mexico, Albuquerque, New Mexico 87131.

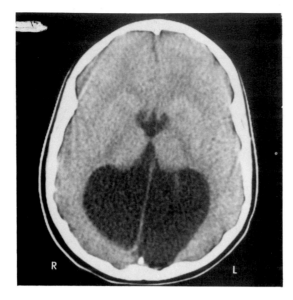

FIGURE 1. Horizontal CT scan of a 13-year-old female who at 4 days of age had a large occipital encephalocele removed, which included occipital tissue. The CT demonstrates bilateral porencephaly with marked dilation of the posterior horns bilaterally. There also appears to be agenesis of the corpus callosum. Despite the prominent visual cortical defect, the patient was capable of reading at above grade level and had a verbal intellectual quotient in the bright normal range (VIQ 117). Because of the visual dysgnosia she could not do the performance tests on the WISC-R but, as indicated, had no difficulty with reading. This is a case that demonstrates remarkable plasticity in the brain of a child with marked structural abnormalities.

demonstrated remarkable levels of intact functioning, presumably reflecting remarkable plasticity. Despite a prominent visual agnosia for objects, features of cortical blindness, CT findings that indicate an absence of the visual cortex, agenesis of the corpus callosum, and the previously described porencephaly, this patient has a WISC-R verbal IQ score of 117 and can read at the 10th- to 12th-grade level, depending on the specific test utilized to assess reading ability.

This is not to say that all errors of development result in such favorable outcome; clearly, many do not. The majority of children with such structural defects will have some sequelae, frequently prominent. For example, the case presented in Figure 2 is similar to that in Figure 1 in that there is bilateral porencephaly and agenesis of the corpus callosum, but this patient has a mild to moderate-level mental retardation (WAIS-R VIQ score 56, PIQ score 59, FS IQ score 54) and is alexic.

Typically, if a brain insult occurs *in utero* or early in life and is unilateral, there is better recovery of function than if the structural damage is bilateral or multifocal (Bigler & Naugle, 1985; Stringer & Fennell, 1987). For example, in Figure 3, this child has complete absence of the left frontal lobe along with most of the left temporal and parietal regions. Despite this enormous defect, her right-side hemiplegia is incomplete, and she has functional use of the right body side, can walk and run, and has functional finger movement as well as tactile perception. Language functions are basically intact, and through the second grade, when she was last tested (see Bigler & Naugle, 1985), she was ahead of grade level in reading and spelling (see Table 1). In such cases, there have been obvious compensatory changes that have occurred within the intact hemisphere and possibly in subcortical structures. There does appear to be a "crowding" effect in these children (see Stringer & Fennell, 1987) though, and they typically show some performance deficit in other areas. For example, in this child, even though the right hemisphere was fully intact, she had some delays in visual–motor development, particularly in the realm of constructional praxis, functions typically thought to reside primarily in the right hemisphere (Bigler, 1988).

The observation that unilateral lesions in the developing brain are associated with better outcomes than bilateral or diffuse involvement may not consistently apply in cases of hydrocephalus. Figure 4 provides an example of this. As can be seen in the CT depicted in this figure, there

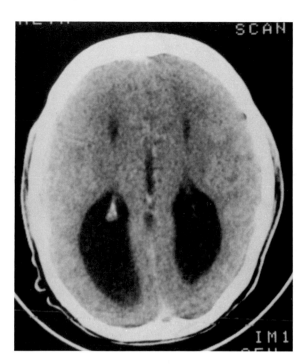

FIGURE 2. Horizontal CT scan in a patient with congenital agenesis of the corpus callosum and marked dilation of the posterior horns. Unlike the patient in Figure 1 who had some similar structural abnormalities, this patient was mentally retarded and alexic.

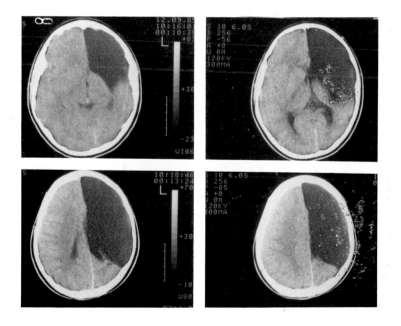

FIGURE 3. Horizontal CT scan demonstrating an enormous structural defect in the left hemisphere (note that the picture is taken in the radiological plane in which left is on the pictorial right). This scan demonstrates complete absence of the left frontal lobe along with most of the left parietal and temporal regions. Despite such marked defects, language function developed normally, including reading, spelling, and expressive language abilities (see Table 1).

TABLE 1. Intellectual and Academic Performance in a Child with a Massive Left
Hemisphere Porencephalic Cyst

	WPPSI, 4 years, 4 months	WPPSI, 5 years, 2 months	WISC-R, 6 years, 2 months	WISC-R, 7 years, 9 months
VIQ	102	100	105	98
PIQ	100	89	98	98
FSIQ	101	94	101	98
WRAT				
Reading grade level	N.1	1.0	1.3	4.4
Spelling grade level	N.8	K.0	1.4	4.1
Arithmetic grade level	P.8	K.7	K.4	2.7

is extreme thinning of the cortical mantle with enormous hydrocephalus. There is some sparing
in the basal frontal and temporal region, and subcortical structures appear intact. Despite the
generalized abnormalities present throughout all of the cerebral cortex, functionally the patient
was quite intact.

This patient was born to migrant parents, who sheltered her throughout her childhood, and
she never received any formal education or special training. She had lived only in rural farm

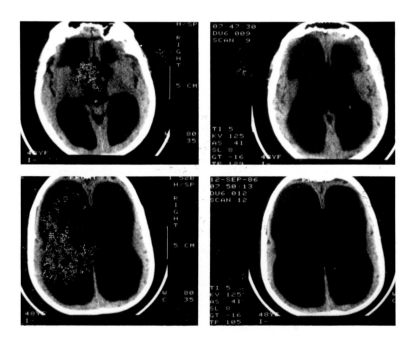

FIGURE 4. Severe congenital hydrocephalus in a 48-year-old Hispanic female with spina bifida. Note the
marked enlargement of the ventricular system in the various horizontal CT scans. Despite such prominent
hydrocephalus, the patient had intact conversational speech, was fluent in both Spanish and English, and had
no difficulty in recognizing objects on the Reitan–Indiana Aphasia Screening Test.

FIGURE 5. Reitan–Aphasia screening results of copying a cross, square, and triangle in the patient with severe hydrocephalus presented in Figure 4. Note that despite pronounced thinning of the cerebral cortex, even in the occipital region, this patient could visually recognize all of the shapes, name them, and correctly copy them.

areas. Despite these limitations she became fluent in English as well as her native Spanish. On the WAIS-R (given in English), she obtained a verbal IQ score of 66, performance IQ score of 67, and full-scale IQ score of 64. Although these scores are in the mild range of mental retardation, they likely are a somewhat lower estimate of her true functional level, given the lack of formal educational training and the fact that English was her second language. As an example of the level of her cognitive abilities, when asked to name four Presidents of the United States since 1950, she quickly responded with Reagan, Carter, Nixon, and Johnson." When asked on the WAIS-R comprehension subtest what "Strike while the iron is hot" meant, she quickly responded with "Do it while you can." Despite the low psychometric score obtained on intellectual tests, which undoubtedly represent an underestimate because of cultural and education factors, she was able to perform adaptively. What is striking, however, is that she was able to perform at this level with such a grossly abnormal brain. Further, despite having essentially no visual cortex, she could process visual information sufficiently to copy geometric forms (see Figure 5).

Why some of these hydrocephalic patients can function as well as they do remains an enigma. Whether the surviving tissue in the remaining thin cortical mantle has any functional abilities is uncertain (Kovnar, Coxe, & Volpe, 1984; Ogden, 1986). The role of subcortical areas in cognition is still unknown, but one must hypothesize that subcortical areas play an important role in the functional abilities of hydrocephalic individuals, since cortical tissue is grossly impaired.

When dealing with cases of mental retardation, one should not assume from the previous discussion that a "normal" scan necessarily carries with it a more optimistic outcome. The scan in Figure 6 is that of a 22-year-old female with severe mental retardation (IQ score less than 25) but who had a normal CT scan. This patient has had a prominent, generalized seizure disorder of an idiopathic nature throughout childhood and her adult life. There was no known cause for the mental retardation. Obviously the brain is dysfunctional in this patient, but from a morphological standpoint, the gross brain anatomy is "normal."

Occasionally, cases are encountered that seemingly contradict basic principles of brain–be-

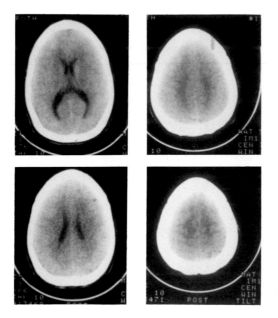

FIGURE 6. Normal CT scan in the horizontal plane in a patient with severe (IQ < 25) mental retardation. There are no gross structural abnormalities in the brain of this individual to account for the level of mental retardation. The posterior aspect of the ventricular system is somewhat generous but within the boundaries of what may be considered normal. This case demonstrates the point that a normal scan does not necessarily equate with normal function.

havior relationships. This was the situation for the patient presented in Figure 7. This patient presented with symptoms of depression and headache. Because of the latter symptom, he received a routine CT scan, which surprisingly demonstrated the various abnormalities (arachnoid cyst, ventriculomegaly, cortical atrophy, anomalous bitemporal lobe development) presented in Figure 7. Because of these abnormalities, detailed neurological studies, including EEG and spinal tap, were carried out, although his history was completely negative for any type of neurological syndrome or event, and likewise his physical examination was normal. All of these studies were negative. Similarly, neuropsychological studies were all well within normal limits. In fact, the patient was functioning in the superior range of intellectual and cognitive ability. His WAIS-R scores were as follows: VIQ 130, PIQ 124, FSIQ 132. Wechsler Memory Scale memory quotient was 143. His Halstead–Reitan Neuropsychological Test Battery impairment index was 0.0, which means no tests were performed in the impaired range. This patient showed no signs of any type of language or memory deficit. The CT abnormalities were not thought to be of a degenerative nature, as the patient had had a previous WAIS done 11 years prior to the CT and current neuropsychological examination, and, in comparison, there were no signs of deterioration or decline in level of intellectual functioning (prior WAIS results indicated the following: VIQ 137, PIQ 119, FSIQ 131). In this case, even though there are obvious structural abnormalities that are bilateral and diffuse, they did not correlate with any neurobehavioral measurement. The origin of these abnormalities was thought to be related to maldevelopment *in utero*.

Such developmental abnormalities are unlikely to be the effect of an acute and abrupt-onset lesion. More likely, they may reflect abnormal neuronal developmental over an extended period of time. Thus, the relative sparing of function in such cases may represent a prenatal example of the serial lesion effect, which is associated with relatively greater recovery of function than a single, abrupt-onset lesion of equivalent size (see Finger & Stein, 1982).

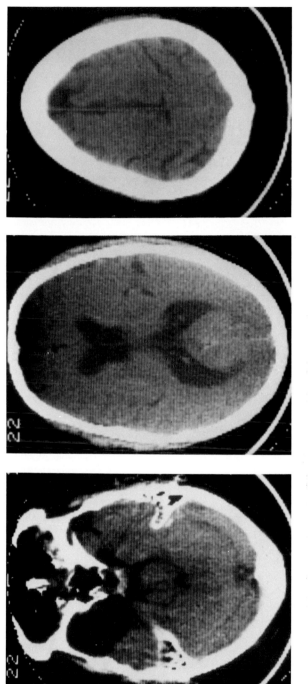

FIGURE 7. Horizontal CT scans in a patient with several pronounced structural abnormalities. The scan on the left depicts a large arachnoid cyst in the left temporal lobe with a smaller one in the right. The middle scan demonstrates enlarged ventricles and prominence of the interhemispheric and Sylvian fissures, a sign of brain atrophy, which is depicted in the scan on the right. Despite these structural abnormalities, this patient displayed normal cognitive functioning.

CONGENITAL CEREBROVASCULAR ABNORMALITIES AND SUBSEQUENT STROKE

There may be early brain adaptation in some cases of unilateral vascular abnormalities, such as an arteriovenous malformation (AVM). The presence of such an abnormality, particularly if it is large and in some way distorts or displaces the surrounding brain tissue, may lead to unusual patterns of functional organization. Accordingly, when a subsequent vascular lesion occurs within one of these anomalous areas, the full neurobehavioral syndrome that normally would be expected may not develop. Take, for example, the case presented in Figure 8 and Table 2, and contrast that with the case presented in Figure 9 and Table 3.

The patient in Figure 8 had an occult right middle cerebral artery AVM that subsequently ruptured at age 17. Throughout childhood, she was described as an impulsive child who was emotionally labile, and brief psychiatric hospitalization was necessary at the age of 15. Fortunately, premorbid intellectual data were available for this patient at age 16 (WISC-R: VIQ 105, PIQ 115, FSIQ 110). As can be seen in Figure 8, there are a large hemorrhagic mass, edema, and midline shift as a result of the ruptured AVM. A craniotomy was necessary to evacuate the hemorrhagic mass, and the CT shown on the right was obtained 5 months post-stroke. Despite a very large right hemisphere lesion, the only chronic deficits noted were a left hemiplegia, finger agnosia, and astereognosis. She never developed visual neglect, anosognosia, or other signs of prominent right hemisphere dysfunction at any time post-stroke. There was never any evidence of the constructional apraxia so frequently observed in patients with such brain lesions (see Figures 10 and 11 for comparison). The only other neuropsychological abnormality noted was on the Benton VRT, on which she had difficulty with accurate recall of some stimuli presented to the left hemispace. Her visual memory performance on the Wechsler Memory Scale and Rey–Osterrieth Complex Figure were well within normal limits (see Figure 11).

It is important to note that there was little neuropsychological change from 4 months post-stroke to reevaluation at 18 months post-stroke. Accordingly, this suggests that the typical right hemisphere cognitive skills were represented elsewhere in this patient, which would account for her limited deficits on examination. Her excellent outcome, given the size of the original lesion, thus reflects unusual representation of function. Compared with the premorbid WISC-R results, there was little change in intellectual function shortly after the CVA, and by 18 months, intellectual levels had fully returned to the premorbid measurement level. Such observations would be consistent with a premorbidly dysfunctional right frontotemporoparietal region that, when more completely damaged following the rupture of the AVM, did not result in a prominent neurobehavioral syndrome. Such cases suggest that the intended or "normal" functions of that region had developed elsewhere; hence, subsequent damage there had no appreciable effect on the cognitive skills that are commonly affected.

In contrast, adult patients with no such developmental history may develop very prominent and persistent neurobehavioral syndromes following similar lesions despite intensive rehabilitation efforts. In Figure 9 and Table 3, the CT and neuropsychological results are presented from a 35-year-old man who suffered a spontaneous hemorrhage as a result of a relatively small aneurysm at the level of the trifurcation of the right middle cerebral artery. Although the brain lesion area in this patient is larger than in the case presented in Figure 8 and Table 2, there is definite overlap and similarity of areas affected. This patient developed a similar left hemiplegia and hemisensory deficit as the patient in Figure 8 but also had features of the constructional apraxia and pronounced visual–spatial and visual–memory deficits as well as left-side neglect, which are the expected right-hemisphere-syndrome deficits. As can be seen in Table 3, these deficits did not improve with time or respond to rehabilitation efforts.

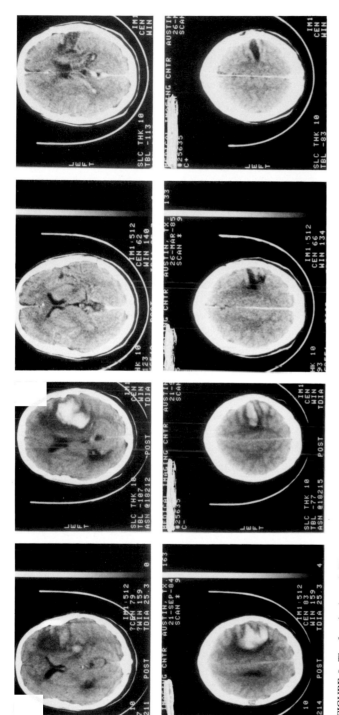

FIGURE 8. The four horizontal CT scans presented on the left were taken on the day of admission, approximately 4 hr after the onset of a severe headache followed by left hemiplegia and then stupor. These scans depict a large hemorrhagic mass with right-to-left midline shift and marked surrounding edema. The hemorrhagic mass was situated in the posterior frontal, anterior parietal, and anterodorsal temporal region. Typically such distortion effects may induce structural abnormalities in addition to infarction effects of the hemorrhage. The four scans on the right correspond to those on the left but were taken 5 months post-stroke. Note that there has been good resolution, but there is an area of infarction in the posterior frontal, anterodorsal temporal, and anterior parietal region. Despite the presence of right hemisphere pathology, this patient did not develop a right hemisphere neurobehavioral syndrome other than the mild left hemiparesis and dysstereognosis. See Table 2 for neuropsychological test results.

TABLE 2. Intellectual and Neurological Findings in the Stroke Patient Described in Figure 8

	4 months post-stroke	18 months post-stroke
Wechsler Adult Intelligence Scale		
VIQ	100	114
PIQ	102	110
FSIQ	102	114
Wechsler Memory Scale		
MQ	114	116
Logical memory	14	13
Visual memory	14	12
Associate learning	17	17
Halstead–Reitan Neuropsychological Test Battery		
Strength of grip		
L	7	8
R	31	36
Finger oscillation		
L	0	0
R	46	49
Sensory perceptual		
L	Astereognosis, finger agnosia	Astereognosis, finger agnosia
R	Normal	Normal
Tactual Performance Test		
L	NA	NA
R Trial 1	4.8	5.1
Trial 2	4.2	3.9
Trial 3	4.4	4.2
Memory	7.0	8.0
Localization	2.0	3.0
Category test		
errors	23.0	24.0
Seashore rhythm		
Raw score	24.0	23.0
Scale score	8.0	9.0
Speech sounds perception		
(errors)	3.0	1.0
Trail-making test		
A seconds (errors)	36.0 (0)	26.0 (0)
B seconds (errors)	69.0 (2)	43.0 (0)
Other neuropsychological measures		
Benton Visual Retention Test		
Correct	8.0	9.0
Errors	4.0 (1 RVF, 2 LVF)	3.0 (1 RVF, 2 LVF)
Raven Colored Progressive Matrices	34/36	34/36

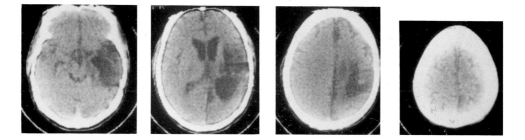

FIGURE 9. Horizontal CT scan in a patient with a history of spontaneous rupture of a middle cerebral artery aneurysm with corresponding infarction in frontal, temporal, and parietal regions. The CT scan was taken approximately 6 months post-stroke. Table 3 presents the neuropsychological data on this patient.

TABLE 3. Intellectual and Neurological Findings in the Stroke Patient Described in Figure 9

	3 months post-stroke	9 months post-stroke
Wechsler Adults Intelligence Scale		
VIQ	91	93
PIQ	51	60
FSIQ	72	76
Wechsler Memory Scale		
MQ	57	67
Logical memory	2	7
Visual memory	0	2
Associate learning	3	9
Halstead–Reitan Neuropsychological Test Battery		
Strength of grip		
L	0	0
R	32	34
Finger oscillation		
L	0	0
R	22	31
Sensory perceptual	Astereognosis, finger agnosia	Finger agnosia, astereognosis
Tactual performance test		
L	NA	NA
R Trial 1	5'DC	5'DC
Trial 2	5'DC	5'DC
Trial 3	NA	NA
Memory	2	5
Localization	0	0
Seashore rhythm		
Raw score	17	19
Scaled score	10	10
Speech sounds perception		
(errors)	8	7
Trail-making test		
A seconds (errors)	100 + DC	58
B seconds (error)	100 + DC	100 + DC

(continued)

TABLE 3. (Continued)

	3 months post-stroke	9 months post-stroke
Other neuropsychological measures		
Raven Colored Progressive Matrices	DC	12/36
Benton Visual Retention Tests	DC	DC

DC—discontinued.

PERIVENTRICULAR INTRAVENTRICULAR HEMORRHAGE

Much of the preceding discussion deals with errors in neural development or intrauterine-related neurological abnormalities. In the perinatal period of development, the preterm infant is at particular risk for stroke, which represents a somewhat different situation than an embryological error or an *in-utero* developmental deficit. These infants are "normal" up to the point of delivery but, because of immaturity, may suffer a variety of brain insults. Accordingly, periventricular intraventricular hemorrhage (PIVH) is the most common central nervous system lesion found at autopsy in preterm infants (Towbin, 1969). Cerebral intraventricular hemorrhage in these infants refers to subependymal germinal matrix hemorrhage, intraventricular hemorrhage (IVH), or IVH with parenchymal hemorrhage (PIVH; Papile, Burstein, Burstein, & Koffler, 1978).

Papile *et al.* (1978) were the first to use CT scan to determine the incidence, extent, and evolution of PIVH in 46 consecutively admitted low-birth-weight infants (less than 1500 g). In 1980, approximately 1% of liveborn infants weighed less than 1500 g and were considered very low birth weight (VLBW) (Kitchen *et al.*, 1980). The overall incidence of PIVH in the 46 infants studied was 43%, with a range of severity in PIVH identified. A procedure for grading the severity and type of PIVH was developed as follows:

Grade I: subependymal hemorrhage. This was an isolated germinal matric hemorrhage. The extent of the lesion was diminished on the follow-up CT at 1 week and not evident on CT scan at 3 weeks (Figure 12).

Grade II: periventricular intraventricular hemorrhage without ventricular dilation. The size of the ventricles still was found to be normal, though there was blood found within the ventricle. Subependymal hemorrhage (SEH) was also found. The CVH was present on CT scan 1 week later but not detected at 3 weeks (Figure 13).

Grade III: periventricular intraventricular hemorrhage with ventricular dilation. The ventricles were enlarged with the hemorrhage, with ventricular dilation being greater in nonsurviving infants. Progressive dilation of the ventricles was shown in serial CT scan. By 3 weeks, the hemorrhage was no longer evident on CT scanning. Infants required treatment for progressive hydrocephalus (Figure 14).

Grade IV: periventricular intraventricular hemorrhage with parenchymal hemorrhage. These were considered the most serious of all CVH, and the parenchymal hemorrhage was found on subsequent CT scans (Figure 15).

The subependymal germinal matrix is a vascular structure that is most pronounced in the fetus of 6 to 8 months gestation (Papile *et al.*, 1978). It is the source of spongioblasts, which help form the cerebral cortex, basal ganglia, and other forebrain structures (Towbin, 1968). The pathogenesis of PIVH in the preterm infant is still unknown, though there are numerous theories, including infarction and thromboses of deep cerebral vessels (Volpe *et al.*, 1983; Towbin, 1969), increased venous pressure (Cole *et al.*, 1974), and increased subependymal capillary pressure

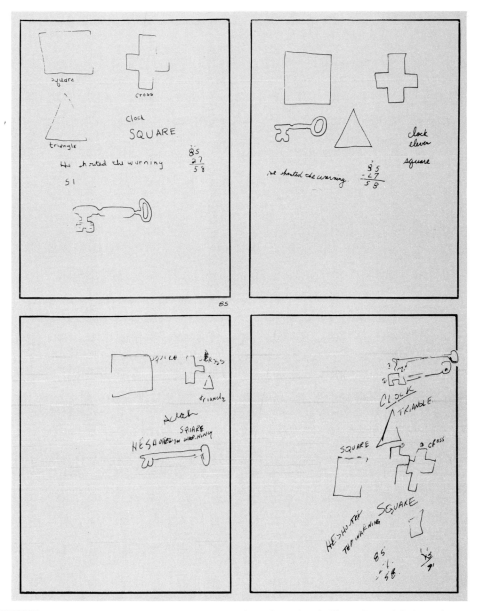

FIGURE 10. Reitan–Indiana Aphasia Screening Test from the patient in Figure 8 (top left 4 months post-stroke, top right 18 months post-stroke) and from the patient in Figure 9 (bottom left 3 months post-stroke, 9 months post-stroke bottom right). Note that with the patient in Figure 8 there is no substantial change between the initial evaluation and the 6-month follow-up. Note also the absence of constructional apraxia and left-side neglect. The results with the patient in Figure 9 demonstrate the common cognitive deficits in patients with right hemisphere damage and the chronicity of those deficits.

(Hambleton and Wigglesworth, 1976). The interhemispheric fissure appears prominent on CT in 83% of these infants, indicating some degree of associated cerebral atrophy (Papile *et al.*, 1978).

Papile and Skipper (1978) developed a multivariate model to show which infants may be at

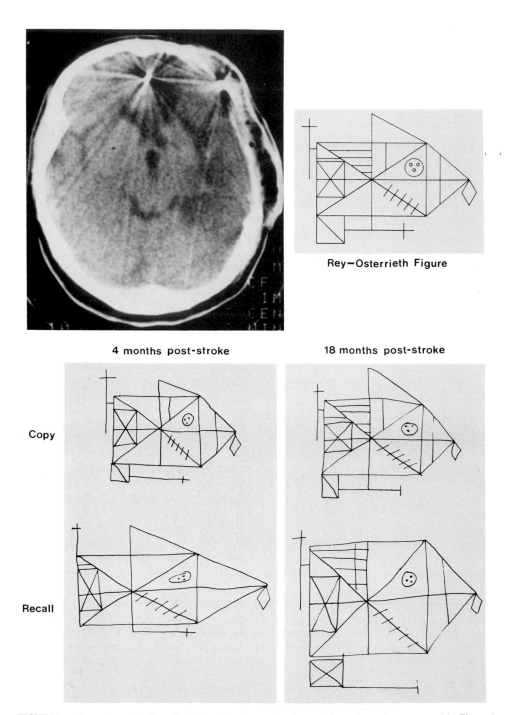

Rey–Osterrieth Figure

4 months post-stroke   **18 months post-stroke**

Copy

Recall

FIGURE 11. Rey–Osterrieth Complex Figure Design results (bottom) from the patient presented in Figure 8 and Table 2. Note that even with the earliest testing, no constructional apraxia is evident, and there is good visual memory retention There is little difference between the 14-month follow-up testing (right) and the original testing (left). In contrast, the Rey–Osterrieth figures presented on the facing page are those from a patient in whom there is prominent constructional apraxia and no retention of visual information after a delay procedure. These deficits persist even a ycar after the right-hemisphere brain injury. The CT scan (top left, this page) depicts the extent of the right-hemisphere involvement. The Rey–Osterrieth Complex Figure stimulus is presented at top right, this page.

3 months post-stroke   12 months post-stroke

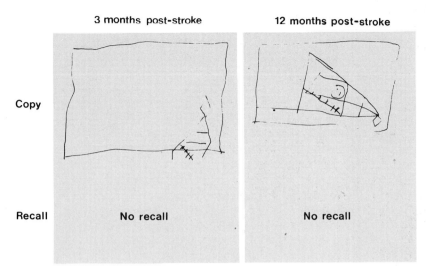

Copy

Recall   No recall   No recall

FIGURE 11. (*continued*)

risk for CVH, based on medical background data on the 46 infants. The infants who were at greatest risk for the development of PIVH were those with birth weight less than 1200 g, Apgar score less than 4 at 1 min, the product of multiple gestation (a twin, triplet, etc.), and a history of being transferred from another hospital. There appeared to be a greater incidence of PIVH in male than female births.

Data regarding the possible relationship of PIVH and later developmental or neuromotor handicaps suggested that outcome is correlated with the severity of the PIVH (Papile, Munsick-Bruno, & Schaefer, 1983; Williamson, Desmond, Wilson, Andrew, & Garcia-Prats, 1982; Catto-Smith, Yu, Bajuk, Orgill, & Astbury, 1985). In a study of 198 1- and 2-year-olds, Papile *et al.* (1983) found that 18% of the VLBW infants had a major handicap in early childhood. A major handicap was defined as (1) an abnormal Mental Developmental Index (MDI) or Performance Developmental Index (PDI) score on either of the Bayley Scales of Infant Development (score below 68), (2) an abnormal neuromotor examination, (3) blindness or sensorineural hearing loss, or (4) seizure disorder. A child considered to be multihandicapped had an abnormal MDI and PDI scale and abnormal neuromotor findings (Papile *et al.*, 1983).

The outcome for VLBW infants with grade I and II PIVH was identical to infants with VLBW and no PIVH. Major handicaps were identified in 11% of the infants with no PIVH, 9% of the infants with grade I CVH, and 11% of the infants with grade II PIVH. Infants with grade 3 and 4 PIVH had much poorer outcome. Fifty-eight percent of the infants with grade III and IV PIVH had major handicaps. Multihandicaps were present in 45% of the group with grades III and IV PIVH.

Twenty-two (52%) of the infants with grade III or IV PIVH developed hydrocephalus. Ten of the infants had grade III PIVH, and 12 had grade IV PIVH. Four (18%) of these infants with grade III or IV PIVH and hydrocephalus died within the first year of life; 3 had bronchopulmonary dysplasia, and 1 had meningitis. Of the remaining 18 infants, 17 were evaluated. Ten of the infants had a major handicap, 1 was considered normal, and 6 had subtle problems and were considered questionable. When those infants who had grade III or IV PIVH with and without hydrocephalus were compared, similar developmental outcomes were found in regard to major or minor handicaps. More of the infants who developed hydrocephalus had multihandicaps (10 in-

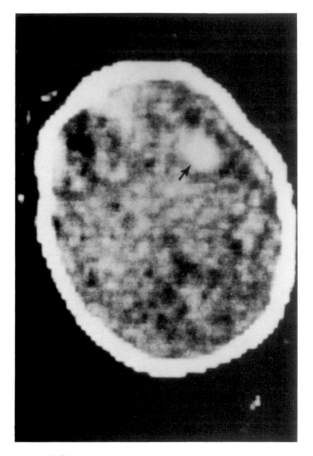

FIGURE 12. Grade I: subependymal hemorrhage.

fants) compared to the infants with grade III and IV PIVH who did not develop hydrocephalus (4 infants).

Williamson *et al.* (1983) evaluated the outcome of 20 LBW infants who had documented PIVH. The findings, similar to those of Papile *et al.* (1983), were that the infants with grade I and grade II PIVH had 29% abnormal outcome in contrast to 79% of the infants with grade III or IV PIVH. Poorest outcome (multihandicaps) was related to extreme prematurity (gestational age less than 27 weeks). Outcome for hydrocephalus was associated with grade of PIVH and with gestational age. Because of the small sample size, the trend could not be statistically tested.

In a follow-up study of extremely preterm infants born between 23 and 28 weeks of gestation, Catto-Smith *et al.* (1985) found 61% of the infants to have PIVH based on cerebral ultrasonography. Again, outcome was significantly worse in the group with interventricular or intracerebral hemorrhage, with 75% of the 12 infants having major handicaps. Only 3 (16%) of the 19 infants with normal scans or germinal layer hemorrhage had major disability. Extreme pessimism regarding developmental outcome of preterm infants with intracerebral hemorrhage was warranted according to the authors, though longer follow-up of larger numbers of infants is needed.

Serial neurodevelopmental testing at 6, 12, and 18 months of age (age was corrected for prematurity) was completed on a cohort of very-low-birth-weight preterm infants by Scott, Ment,

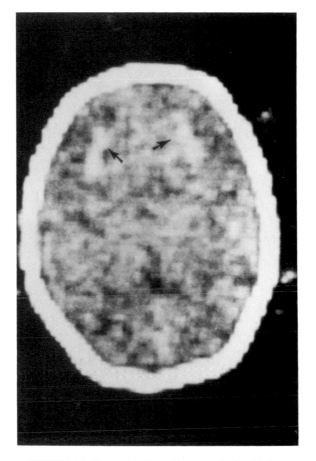

FIGURE 13. Grade II: PIVH without ventricular dilation.

Ehrenkranz, and Warshow (1984). Of the 102 scanned survivors of less than 1250 g birth weight, 45% were found to have PIVH by CT and echo studies. Follow-up developmental evaluations of 88 VLBW infants (86%) indicated that a significant downward trend over time on the Bayley Mental Index scores was found in those VLBW infants with PIVH. In this study the survivors had mainly grade I or II hemorrhages (42 of 46 infants), suggesting long-term neurodevelopmental abnormalities in neonates with germinal matrix hemorrhage.

Short-term follow-up studies of VLBW survivors with PIVH indicated that infants with moderate to severe PIVH have a higher incidence of neuromotor and developmental handicaps at 1 and 2 years of age than do infants with mild PIVH or no hemorrhage. The long-term developmental outcome of VLBW infants who incurred a PIVH is less well defined.

A study by Lowe (1985) compared the neuropsychological outcome of 5- and 6-year-olds with VLBW and mild PIVH (grades I and II) with VLBW survivors with no hemorrhage and a control group of children born full term with no medical complications. The 38 children with birth weights <1500 were a subgroup of children from the Papile et al. (1983) 1- and 2-year follow-up study. The 22 children in the control group were selected from local schools and matched with the study group for ethnicity, age, grade, and socioeconomic status (SES). The diagnostic battery administered to each child consisted of five measures: (1) cognition (the McCarthy Scale

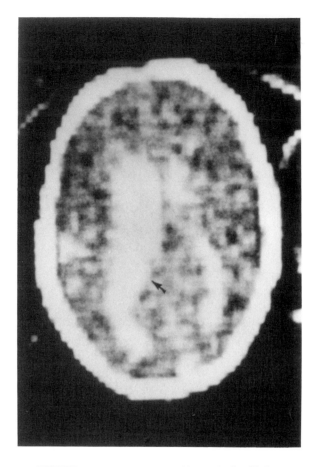

FIGURE 14. Grade III: PIVH with ventricular dilation.

of Children's Ability), (2) reading readiness (the Test of Early Reading Ability), (3) visual motor ability (Developmental Test of Visual Motor Integration), (4) finger agnosia, and (5) hyperactivity (the Conner's Parent–Teacher Symptom Questionnaire).

Multivariate analysis demonstrated a significant difference between the control group and children born VLBW with and without PIVH for the five dependent variables combined, accounting for 46% of the variance. On univariate analysis, there was a significant difference between the control group and the VLBW children on tests of finger agnosia, reading readiness, and cognition. When the children born VLBW with no PIVH were compared to those born with mild PIVH, there again was a significant difference using a multivariate contrast, accounting for 21% of the variance. However, univariate analysis indicated that the two VLBW groups did not differ significantly on any single measure. In contrast to the short-term outcome data (Papile et al., 1983), this follow-up study of the same sample at ages 5 and 6 (Lowe, 1985) indicated that very low birth weight and mild PIVH had a significant impact on neurodevelopmental outcome. The low-birth-weight children in general were found to be "at risk" for later learning problems.

In a follow-up study of 12 children born VLBW who developed posthemorrhage hydrocephalus following grade II PIVH, Krishnamoorthy, Kuehnle, Todres, and DeLong (1982) noted generally favorable outcomes on neurological and psychometry evaluation for half of their subjects.

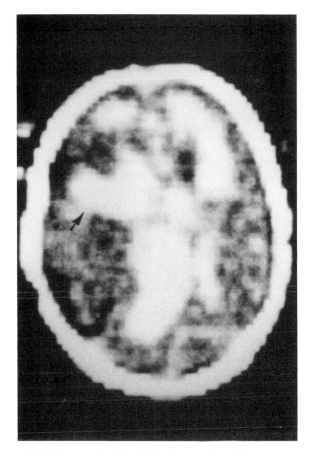

FIGURE 15. Grade IV: PIVH with parenchymal hemorrhage.

Excessive distractibility and short attention span were noted in 6 children (50%), with 3 of these children also having perceptual motor, fine motor, and gross motor difficulties; they also demonstrated lower IQ scores, short attention span, speech and language impairment, and perceptual and visual motor problems. Consistent with Lowe's (1985) report, these children appear to be at risk for attention deficit disorder, academic problems, and possible social difficulties.

The neurodevelopmental outcome of VLBW infants with grade III and grade IV PIVH who were followed longitudinally was assessed at 5 to 9 years of age (Lowe, Papile, & Munsick-Bruno, 1987). Of the original 29 infants who were evaluated at 1 and 2 years by Papile *et al.* (1983), 22 (76%) were evaluated at school age. In contrast to the high rate of handicapping conditions found at infancy, cognitive testing at school age indicated that over half of the VLBW children with grade III PIVH (56%) had average IQ scores, 33% were in the borderline range, and 1 child (11%) was severely retarded. Of those children born VLBW with grade IV PIVH, 38% scored in the average range on cognitive testing, 54% scored in the borderline range, and 8% (1 child) were mildly retarded. Neuromotor alterations were found in 33% of the VLBW children with grade III PIVH and 69% of the VLBW children with grade IV PIVH. Eight of the VLBW children (62%) with grade IV PIVH continued to have a diagnosis of cerebral palsy in contrast to only 1 of the VLBW children (11%) with grade III PIVH. Cognitive test data for all

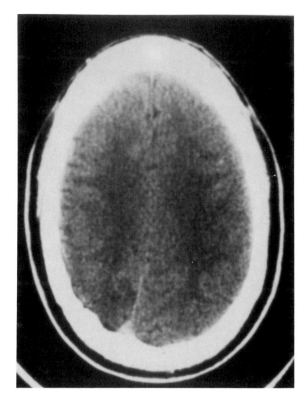

FIGURE 16. Horizontal CT scan demonstrating pronounced hemispheric asymmetry and density decrease in the left occipital region in a young adult male who had a longstanding learning disorder with prominent dyslexia, The normal asymmetry of the cerebral hemispheres is to have the left posterior aspect larger than its right counterpart and the opposite true for the frontal regions (see Koff, Naeser, Pieniadz, Foundas, & Levine, 1986; Yeo, Turkheimer, Raz, & Bigler, 1987). In some patients these cerebral asymmetries may correspond with functional learning deficits.

VLBW children with grade III or IV PIVH are shown in Figures 5 and 6 after 1 year, after 2 years, and at school age. Though patterns of testing indicated that children in both groups had test scores similar to learning-disabled children, the majority of the children (73%) were in regular school programs with minimal therapy services. Further follow-up through grade school and into adult years is necessary to understand better the long-term consequences of PIVH.

Another issue related to long-term outcome of infants with brain damage is the question of "maintenance" of recovery of function. Schallert (1983) has noted that elderly animals who received brain damage earlier in life, from which they had completely recovered, developed symptoms consistent with the lesion with advanced age. Whether such losses will be seen in elderly humans who sustained damage earlier in life remains to be determined. This may be a particular problem for the low-birth-weight infant who suffered PIVH.

LEARNING DISORDER

There is mounting evidence that a variety of structural CNS anomalies may exist in the learning-disabled (LD) individual and that these anomalies represent the anatomic basis for disturbed learning (Galaburda, Sherman, Rosen, Arboitiz, & Geschwind, 1985; Geschwind, 1984; Geschwind & Galaburda, 1985a,b,c). These studies have demonstrated deviations from the normal symmetry of the cerebrum (i.e., larger left planum temporale) along with neuronal ectopias and architectonic dysplasias in the perisylvian region. Although these abnormalities can be demonstrated at the histological level, CT-imaging studies have not consistently demonstrated similar findings, such as reversed asymmetries or other irregularities, in the LD children as a group

(Harlam, Dalby, & Johns, 1981; Harcherik *et al.*, 1985; Hier, LeMay, & Rosenberger, 1978; Denckla, LeMay, & Chapman, 1984; Ramsey *et al.*, 1986; Shaywitz *et al.*, 1983). Some of the lack of findings of these previous studies may again be related to the group data analysis. For example, on an individual, case-by-case basis, the occasional patient is seen in whom this appears to fit. For example, the young man presented in Figure 16 demonstrates a case of definite cerebral asymmetry with mild density changes in the posterior left cerebral hemisphere. This patient had a prominent dyslexia.

SUMMARY

Anomalous brain development or early brain injury may induce a functional plasticity such that expected neurobehavioral syndromes either do not develop or are incomplete. In such cases it is important not to overinterpret CT/MRI findings. Accordingly, structural abnormalities in such patients may not relate systematically with established guidelines of neuropsychological functioning. In such situations it is important to understand that structure and function may not be equated or predicted from one another. Thus, one should exercise caution in making some predictions about outcome in infants and children based solely on structural abnormalities of the brain. Likewise, the same can be stated about adult individuals who have neurological damage superimposed on an earlier congenital abnormality. Preterm infants who suffer serious intraventricular hemorrhaging also may show better recovery than would be predicted given the extent of original injury. These variations from what would be expected underscore the need for a better understanding of structure–function relationships in such unique patient populations.

REFERENCES

Bigler, E. D. (1988). *Diagnostic clinical neuropsychology* (2nd ed.). Austin, TX: University of Texas Press.

Bigler, E. D., & Naugle, R. I. (1985). Case studies in cerebral plasticity. *Clinical Neuropsychology, 7,* 12–23.

Catto-Smith, A. G., Yu, V. Y., Bajuk, B., Orgill, A. A. & Astbury, D., (1985). Effect of neonatal-periventricular hemorrhage on neurodevelopmental outcome. *Archives of Diseases of Childhood, 60,* 8–11.

Cole, V. A., Curbin, G. M., & Olaffson, A., (1974). Pathogenesis of intraventricular hemorrhage in newborn infants. *Archives of Diseases of Childhood, 49,* 722.

Denckla, M. B., LeMay, M., & Chapman, C. A. (1984). Few CT scan abnormalities found even in neurologically impaired learning disabled children. *Journal of Learning Disabilities, 18,* 132–135.

Finger, S., & Stein, D. G. (1982). *Brain damage and recovery.* New York: Academic Press.

Galaburda, A. M., Sherman, G. F., Rosen, G. D., Arboitiz, F., & Geschwind, N. (1985). Developmental dyslexia: Four consecutive patients with cortical anomalies. *Annual Neurology, 18,* 222–233.

Geschwind, N. (1984). The brain of a learning-disabled individual. *Annals of Dyslexia, 34,* 319–327.

Geschwind, N., & Galaburda, A. M. (1985a). Cerebral lateralization: Biological mechanisms, associations, and pathology: I. A hypothesis and a program for research. *Archives of Neurology, 42,* 427–500.

Geschwind, N., & Galaburda, A. M. (1985b). Cerebral lateralization: Biological mechanisms, associations, and pathology: II. A hypothesis and a program for research. *Archives of Neurology, 42,* 521–552.

Geschwind, N., & Galaburda, A. M. (1985c). Cerebral lateralization: Biological mechanisms, associations, and pathology: III. A hypothesis and a program for research. *Archives of Neurology, 42,* 634–654.

Hambleton, G., & Wigglesworth, J. S. (1976). Origin of intraventricular hemorrhage in the newborn. *Archives of Diseases of Childhood, 51,* 651–659.

Harcherik, D. F., Cohen, D. J., Ort, S., Paul, R., Shaywitz, B. A., Volkmar, F. R., Rothman, S. L. G., & Leckman, J. F. (1985). Computed tomographic brain scanning in four neuropsychiatric disorders of childhood. *American Journal of Psychiatry, 142,* 731–734.

Harlam, R., Dalby, H., & Johns, R. (1981). Cerebral asymmetry in developmental dyslexia. *Archives of Neurology, 38,* 679–682.

Herskowitz, J., & Rosman, N. P. (1982). *Pediatrics, neurology and psychiatry—common ground: Behavioral, cognitive affective and physical disorders in childhood and adolescence.* New York: Macmillan.

Hier, D., LeMay, M., & Rosenberger, P. (1978). Developmental dyslexia: Evidence for a subgroup with a reversal of cerebral asymmetry. *Archives of Neurology, 35,* 90–92.

Kitchen, W., Ryan, M. M., Rickards, A., McDougall, A. B., Billson, F. A. Keir, E. H., & Naylor, F. D. (1980). Longitudinal study of very low birthweight infants. IV: An overview of performance at eight years of age. *Developmental Medicine and Child Neurology, 22,* 172–188.

Koff, E., Naeser, M. A., Piedniadz, J. M., Foundas, A. L., & Levine, H. L. (1986). Computed tomographic scan hemispheric asymmetries in right- and left-handed male and female subjects. *Archives of Neurology, 43,* 487–491.

Kovnar, E. H., Coxe, W. S., & Volpe, J. J. (1984). Normal neurologic development and marked restitution of cerebral mantle after postnatal treatment of intrauterine hydrocephalus. *Neurology, 34,* 840–842.

Krishnamoorthy, K. S., Kuehnle, K. J., Todres, I. D., & DeLong, G. R. (1982). Neuro-developmental outcome of survivors with post-hemorrhagic hydrocephalus following grade II intraventricular hemorrhage (IVH). In *The Second Special Ross Laboratories Conference on Perinatal Intracranial Hemorrhage.* Columbus, OH: Ross Laboratories.

Lowe, J. R. (1985). *Early detection of reading disorders in 5 and 6 year olds born very low birth weight with and without cerebral intraventricular hemorrhage.* Unpublished doctoral dissertation, University of New Mexico, Albuquerque.

Lowe, J. R., Papile, L., & Munsick-Bruno, G. (1987). Follow-up at school age of very low birthweight infants with grades 3 and 4 intraventricular hemorrhage. Paper presented at the 12th Annual Conference on Neonatal/Perinatal Medicine, Banff, Canada.

Ogden, J. A. (1986). Neuropsychological and psychological sequelae of shunt surgery in young adults with hydrocephalus. *Journal of Clinical and Experimental Neuropsychology, 8,* 657–679.

Papile, L., & Skipper, B. (1978). Cerebral intraventricular hemorrhage: A multivariate model. *Pediatric Research, 12,* 544.

Papile, L., Burstein, J., Burstein, R., & Koffler, H. (1978). Incidence and evolution of subependymal and intraventricular hemorrhage: A study of infants with birth weights less than 1,500 gm. *The Journal of Pediatrics, 92(4),* 529–534.

Papile, L., Munsick-Bruno, G., & Schaefer, A. (1983). Relationship of cerebral intraventricular hemorrhage and early childhood neurologic handicaps. *The Journal of Pediatrics, 103(2),* 273–277.

Ramsey, J., Dorwart, R., Vermess, M., Denckla, M., Kruesi, M., & Rapoport, J. (1986). Magnetic resonance imaging of brain anatomy in severe developmental dyslexia. *Archives of Neurology, 43,* 1045–1046.

Schallert, T. (1983). Sensorimotor impairment and recovery of function in brain-damaged rats. Reappearance of symptoms during old age. *Behavioral Neuroscience 97,* 159–164.

Scott, D. T., Ment, L. R., Ehrenkranz, R. A., & Warshaw, J. B. (1984). Evidence for late developmental deficit in very low birthweight infants surviving intraventricular hemorrhage. *Child's Brain, 11,* 261–269.

Shaywitz, B. A., Shaywitz, S. E., Byrne, T., Cohen, D. J., & Rothman, S. (1983). Attention deficit disorder: Quantitative analysis of CT. *Neurology, 33,* 1500–1503.

Stringer, A. Y., & Fennell, E. B. (1987). Hemispheric compensation in a child with left cerebral hypoplasia. *The Clinical Neuropsychologist, 1,* 124–138.

Towbin, A. (1968). Cerbral intraventricular hemorrhage and subependymal matrix infarction in the fetus and premature newborn. *American Journal of Pathology, 52,* 121–133.

Towbin, A. (1969). Cerebral hypoxic damage in fetus and newborn. *Archives of Neurology, 20,* 35–43.

Volpe, J. J., Herscovitch, P., Perlman, J., & Raichle, M. E. (1983). Positron emission tomography in newborn: Extensive impairment of regional cerebral blood flow with intraventricular hemorrhage and hemorrhagic intracerebral involvment. *Pediatrics, 72,* 589–601.

Williamson, W. D., Desmond, M. M., Wilson, G. S., Andrew, L. P., & Garcia-Prats, J. A. (1982). Low birthweight infants surviving neonatal IVH: Outcome in the preschool years. *Journal of Perinatal Medicine, 10,* 34–41.

Williamson, W. D., Desmond, M. M., Wilson, G. S., Murphy, M. A., Rozelle, J., & Garcia-Prats, J. A. (1983). Survival of low birthweight infants with intraventricular hemorrhage. *American Journal of Disease of children, 137,* 1181–1184.

Yeo, R., Turkheimer, E., Raz, N., & Bigler, E. D. (1987). Volumetric asymmetries of the brain: Intellectual correlates. *Brain and Cognition, 6,* 15–23.

13

Neuropsychological Functioning and Brain Imaging

Concluding Remarks and Synthesis

ERIN D. BIGLER, RONALD A. YEO, and ERIC TURKHEIMER

As reviewed in this text, tremendous progress has been made over the past 15 years in the development of brain-imaging techniques and our ability to link imaging data with important aspects of behavior. These advances have had a great impact on both theoretical and clinical issues in the neurosciences and have greatly refined neurological diagnostics, as has been discussed by Rutledge (Chapter 2). Likewise, the greater specificity and precision of current brain-imaging techniques have provided a more complete paradigm for the study of the neurological patient in which the effects of focal, lateralized, and/or generalized neurological damage/dysfunction can be compared with neuropsychological function. Much past research in cognitive neuroscience had been hampered by the inability to study simultaneously anatomy and pathology of the brain and function. As the current generation of research unfolds, we have to rethink our understanding of the brain and its relationship to cognition and neuropsychological function. For example, as pointed out by Knopman, Selnes, and Rubens (Chapter 5), the language system of the brain may not be as localized as was once thought, and Haaland and Yeo (Chapter 8) make similar statements about motor control in their attempt to elucidate the complex factors that are involved in motor function.

A central theme of this volume is the difficulty in drawing valid structure–function relationships, a difficulty that, as Turkheimer (Chapter 3) points out, reflects conceptual as well as technological issues. Because of the unique development of each human brain along with the diversity of functional neural systems and pathways, a clinically distinct syndrome may result from several lesion sites, rendering specification of component processes and their neuroanatomic loci difficult indeed. Furthermore, there are major technical problems in determining the exact locus and extent of brain damage. For example, the physiological changes associated with structural brain abnormalities identified by CT or MRI may far exceed the boundaries of the structural deficit (see Figure 1). Pawlik and Heiss (Chapter 4) elegantly demonstrate this point in their chapter and suggest that the rigid framework of precise localization theory simply is not tenable in many brain disorders. One clear example of this is their PET work in normal subjects performing the Wis-

ERIN D. BIGLER ● Austin Neurological Clinic and Department of Psychology, University of Texas at Austin, Austin, Texas 78705. RONALD A. YEO ● Department of Psychology, University of New Mexico, Albuquerque, New Mexico 87131. ERIC TURKHEIMER ● Department of Psychology, University of Virginia, Charlottesville, Virginia 22903.

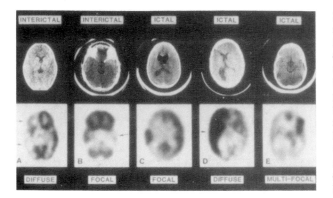

FIGURE 1. X-ray computed tomography (CT) and fluorodeoxyglucose images from patients with focal and diffuse forms of epilepsy studied ictally or interictally. The first patient (A) has a normal x-ray CT and hypometabolism in the left frontal, temporal, and occipital cortices (arrows). The second patient (B), who had partial complex seizures, has a normal x-ray CT and diffuse hypometabolism in the right temporal cortex (arrow). The third patient (C) had right focal motor seizures accompanied by a hypermetabolic focus on the ictal study; the x-ray CT is normal, but air in the ventricles from a previous pneumoencephalogram (which was also normal) is evident. The fourth patient (D) has left hemiatrophy (arrow) and diffuse hypermetabolism on the ictal study, and the fifth patient (E) has a normal x-ray CT and several hypermetabolic foci on the ictal study (from Engel, Brown, & Kuhl, 1982; reproduced by permission). These metabolic studies demonstrate the potential lack of relationship between what appears to be a structurally "normal" brain and distinctly abnormal physiological functioning. Such studies underscore the need to integrate structural, metabolic/physiological, and cognitive function in the neuropsychology of brain disorders.

consin Card-Sorting Test, which has a long tradition in clinical neuropsychology as a test of frontal lobe function (Lezak, 1983). However, PET studies demonstrate whole-brain activation in a rather uniform fashion, not just the frontal lobes. Such brain-imaging studies indicate the need for reevaluation of traditional neuropsychological assessment techniques that aim to localize dysfunction and their implications for brain impairment.

Brain imaging and neuropsychological studies have had a profound impact on our understanding of the neurological mechanisms in emotional control and, by implication, our understanding of psychopathology. Because of the complexity of human emotional expression, it has been anticipated that the unraveling of neuromechanisms in regulating emotion likewise will be complex. Accordingly, Cullum (Chapter 10) demonstrated the difficulty of models positing a simple lesion-localization relationship in emotional control. Rather, emotional dysregulation in neurological disease or disorder occurs in a multifaceted fashion associated with numerous brain lesion sites. Neuropsychiatric disorders present a similar picture. As pointed out by Raz (Chapter 9), there are a number of anatomic brain abnormalities present in the major psychoses. It is interesting to take a historical perspective on this, as it was not too long ago that clinicians and researchers considered the neurobiological component relatively insignificant.

Neuroimaging research has helped to overcome an excessive concern regarding the distinction between "organic" patients (who were thought truly to have "structural brain damage") and "functional" patients (who were thought to have only an "emotional" illness or disorder that had no neurobiological basis). Goldstein (1986) reviews the history of this particular distinction between "organic" and the so-called "functional" schizophrenic patients. In the late 1960s and early 1970s, when clinical neuropsychology began more completely to address this problem of distinguishing between patients with "real brain damage" and those with just "schizophrenic illness," the studies indicated that the groups really could not be separated by neuropsychological testing. We now realize, in retrospect, that the reason the "schizophrenic group" could not be separated from the "organic group" was that the schizophrenic group in fact had underlying neurological deficits, as reviewed by Raz (Chapter 9). This issue could not have been clarified without the brain-imaging technology of today (Kelsoe, Cadet, Pickar, & Weinberger, 1988). It is anticipated that future research in this area may facilitate the development of diagnostic markers and the identification of biologically valid subtypes. This especially may be true as imaging techniques come on line that allow *in vivo* measurement of neurotransmitter activity.

The diagnosis of various dementias has been greatly aided by the interface between neuropsychological assessment and brain imaging. The behavioral correlates of volumetric analyses of the human brain, as discussed by Naugle and Bigler (Chapter 7), is in its infancy. Preliminary research certainly suggests that there may be clinical and empirical applications of various volumetric procedures in the assessment and diagnosis of Alzheimer's disease and related dementias. There is a need, however, for further longitudinal studies of the degenerative process as well as determination of the neuropathological significance of cortical atrophy and ventricular dilation. As this basic neurobiological work unravels the relationship between neuronal-level degenerative changes and what is observed as cortical atrophy and ventricular dilation, it will likely pave the way for a better understanding of the degenerative processes as they relate to cognition.

For example, with improvements in MRI resolution, specific nuclei can now be identified more precisely. This has been impossible with CT scanning techniques, even with the most recent generation of technology. Recently, Kim *et al.* (1988) have demonstrated that the size of the hippocampus is more discriminating than volumetric measures of atrophy or ventricular dilation in assessing dementia. Luxenberg, Haxby, Creasey, Sundaram, & Rapoport (1987) have demonstrated that it is not the size of ventricular enlargement that best differentiates Alzheimer's patients from controls but the rate of volumetric ventricular dilation over time. Such research will undoubtedly lead to a better understanding of the clinical significance of volumetric changes identified by CT and MRI.

The relationship between individual differences and anomalous brain development presents a challenge in making generalized statements about brain function based on brain imaging. As pointed out in the chapters by Yeo (Chapter 11) and Bigler, Lowe, and Yeo (Chapter 12), there is a problem in referring to the "average" brain, as certain cognitive operations may be greatly affected by sex differences, age differences, and the prior presence of early brain insult or developmental error. These individual differences to a certain extent detract from the predictability of any type of lesion-localization theory of brain function. A number of cases were discussed in which particular lesions were clearly identified by CT or MRI technique but the lesion had little, if any, correspondence to what would have been predicted clinically. Thus, accounting for individual differences will certainly play a major role in future studies relating lesion location to function. Dimensions of differences among normal brains offer much more to neuropsychology than "nuisance" variables serving to limit generalizability. As Yeo pointed out (Chapter 11), such differences may help reveal principles of higher cortical function in terms of elucidating the advantages and disadvantages of a given design.

As a better understanding of the relationship between structure and function of the nervous system evolves, there should be at least two major effects on the clinical neurosciences. The first is improved diagnostics. We have already witnessed, in the past decade, marked refinement in the detection of space-occupying lesions, stroke, and certain degenerative disorders, to name a few (see Figure 2). Brain-imaging tests have become the diagnostic procedure of choice for determining the presence or absence of many of these disorders. The second effect, a consequence of the first, is a direct connection between diagnostic capabilities and improved treatment, particularly with respect to rehabilitation efforts in neurologically compromised patients. For example, based on current technology, the patient with Alzheimer's disease is being diagnosed at an earlier stage than ever before because of the improvements in brain imaging and neuropsychological assessment. If there is an improvement in the treatment of Alzheimer's disease, early detection is certain to be a critical variable. The biochemical and physiological abnormalities that precede the development of cognitive symptoms and neuropathological changes in brains of Alzheimer's patients occur before there are demonstrable symptoms. If we are going to be able to effect treatment of Alzheimer's disease, we will have to detect the degenerative changes at the earliest possible

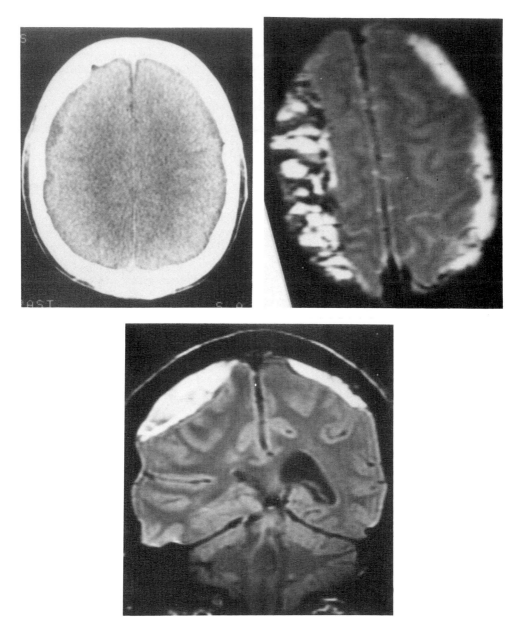

FIGURE 2. Improved detection of a subdural hematoma by MRI over CT (see also Figure 9 in Chapter 1). The CT on the upper left was interpreted as being within normal limits. Obviously, in retrospect, after comparison with the MRI scans, the shadows on the lateral surface of the CT were actually bilateral subdural hematomas. The MRI scans (upper right horizontal plane; bottom coronal view) clearly depict the extent of the bilateral hematomas as well as the structural deformations caused by the position and size of the hematomas. Such improvements will likely lead to a refinement in understanding brain–behavior relationships.

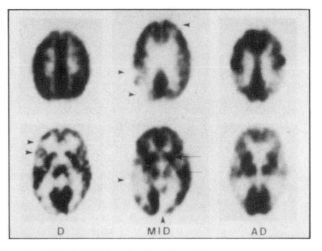

FIGURE 3. Fluorodeoxyglucose brain images of elderly patients with depression (D), multiple-infarct dementia (MID), and Alzheimer's disease (AD). Depressed patient has left frontal hypometabolism but is otherwise normal. Patient with MID has numerous hypometabolic foci (cortical, arrowheads; subcortical, arrows). Patient with AD has characteristic pattern of parietal and temporal hypometabolism (from Kuhl, 1984; reproduced by permission). Such metabolic studies in conjunction with computerized EEG and standardized CT/MRI imaging should permit more refined diagnostic capabilities for the dementias and a more precise understanding between neuropathological changes and cognitive impairment.

stage because of the lack of regeneration within the central nervous system. There is the potential that pharmacological treatment might become available that could either stave off or actually halt progression of neuronal degeneration in Alzheimer's disease (Scheibel & Wechsler, 1986). There is also some speculation that brain grafting of acetylcholine-rich tissues may have some potential role in treating Alzheimer's disease, similar to the initial positive treatment effects of dopamine-rich tissues grafted into brain tissue in the Parkinson's disease patient (Bjorklund et al., 1987; Lewin, 1987). But for these to be effective, the diagnosis will have to be made at the earliest possible stage. Based on the work summarized in this book, and the work of others (see McGeer et al., 1986), we are making progress towards earlier detection of the various dementias (see Figure 3).

Research on the diagnosis and treatment of traumatic disorders raises similar issues. As discussed by Ruff, Cullum, and Luerssen (Chapter 6), brain imaging has revolutionized the early treatment and diagnosis of intracranial pathology in the traumatically brain-injured individual. This has had a direct impact on improved recovery of function and reduced morbidity. However, the residual cognitive deficits in traumatic brain injury have a marked effect on quality of life (Bigler, 1987a,b). In this regard, tremendous interest in cognitive rehabilitation has emerged over the past half decade (Prigatano, 1987). It may be that brain imaging and neuropsychological testing will play the critical role in evaluating the efficacy of cognitive rehabilitation for a given patient and in the assessment of appropriate treatment modalities for cognitive rehabilitation training. Brain-imaging and neuropsychological research may provide further insights into the recovery process of brain injury and the adaptability of the brain to retraining. Likewise, the whole issue of recovery of function following a brain injury has been fraught with many problems until the recent generation of brain imaging has permitted a greater standardization with respect to identifying pathology and documenting structural integrity/abnormality.

Neuropsychological testing used to be done in isolation from a knowledge of underlying neuropathology. Prior to the advent of CT scanning (see Chapter 1), it was rare that the brain could be studied at a postmortem examination close to the time when the neuropsychological studies were conducted. For the most part, neuropsychological studies prior to 1975 had to infer the location and extent of cerebral damage from behavioral signs and their purported relationship to neurological examination, EEG, or pneumoencephalographic findings. For these reasons, neuropsychology developed a tradition of localizing dysfunction independent of specific knowledge of precise areas of actual structural damage. Much has changed. We no longer have to wait for the postmortem examination to determine the structural integrity of the brain, as the various neuroimaging techniques that have been discussed in this text are widely available. It is now rare for any patient who has significant neurological abnormality not to have had one of the neuro-imaging procedures. Neuroimaging diagnostics have become a standard mode of evaluation of the neurological patient, and one might predict that such procedures will be even more frequently utilized in the future. In fact, the neurological patient of the future may well receive assessment incorporating brain imaging, neuropsychological evaluation, and direct metabolic and physiolog-ical functioning (see Gibbins *et al.*, 1987), and the data from each of these will be interfaced together to produce a composite assessment that interrelates all of this information.

Much past research in neuropsychology has been based on standardization of neuropsycho-metric procedures (see Yeudall, Reddon, Gill, & Stefanyk, 1987). This has been necessary be-cause most neuropsychological diagnostic schemes are based on deviation from ''normal.'' Thus, some level of precision has been achieved in neuropsychology by applying statistical principles to the decision of whether a score is aberrant or not. The same process needs to take place in brain imaging. As discussed by Turkheimer (Chapter 3), a variety of quantification techniques are available, allowing researchers and clinicians to go far beyond simple rating scales. As these techniques become more and more automated, we should be able to quantify directly a variety of parameters of the brain, including the amount of white versus gray matter, ventricular volume, hemispheric volume, subcortical nuclei, normal asymmetries, etc. These anatomic measurements may be critical parameters with respect to manifestations of certain neurological disorder states. Accordingly, the statistical emphasis that has played such an important role in neuropsychology needs to be applied to brain imaging as well.

One goal of this text is to give some direction to future research and practice in neuropsy-chology and behavioral neurology. It is apparent that the future of neuropsychological assessment will take a very different path than in the past. Much current research in neuropsychology is based on measures that were developed prior to 1965. As brain-imaging techniques became avail-able, it was important first to investigate and quantify the significance of such brain abnormalities as visualized by brain-imaging methods with standardized neuropsychological measures. How-ever, the current neuropsychological armamentarium is somewhat archaic and insufficiently re-lated to current theories of cognitive function (Posner, 1986; Stillings *et al.*, 1987). Further, neuropsychology has just begun to utilize the advances in computer technology. There has been a wave of enthusiasm for computerized assessment, but to date, most have simply put ''paper and pencil'' tests on the computer without making them adaptive, interactive, or an expert system (see Morrison, Schaeffer, & Russell, 1987; Russel *et al.*, 1987). The assessment battery of the future likely will be, in large part, computer-based and will probably include on-line metabolic (i.e., PET or MRI) and/or physiological (computerized EEG) measurement. Such an evaluation could document anatomic integrity via specific measures of major brain structures and nuclei and integrate this with neuropsychological function and associated metabolic and physiological activity that correspond with certain cognitive states. This is the future. Such clinical advances will be paralleled by advances in the basic neurosciences, allowing simultaneous assessment of differing facets of brain activity and the development of a more integrated science of the brain.

REFERENCES

Bigler, E. D. (1987a). Acquired cerebral trauma: An introduction to the special series. *Journal of Learning Disabilities, 20,*, 454–457.

Bigler, E. D. (1987b). Clinical significance of cerebral atrophy in traumatic brain injury. *Archives of Clinical Neuropsychology, 2,* 293–304.

Bjorklund, A., Linduall, O., Isacson, O., Brundin, P., Wictorin, K., Strecker, R. E., Clarke, D. J., & Dunnett, S. B. (1987). *Trends in Neurosciences, 10*, 509–516.

Engel, J., Jr., Brown, W. J., & Kuhl, D. E. (1982). Pathological findings underlying focal temporal lobe hypometabolism in partial epilepsy. *Annals of Neurology, 12*, 518–528.

Gibbins, A. S., Morgan, N. H., Bressler, S. L., Cutillo, B. A., White, R. M., Illes, J., Greer, D. S., Doyle, J. C., and Zeitlin, G. M. (1987). Human neuroelectric patterns predict performance accuracy. *Science, 235*, 580–585.

Goldstein, G. (1986). The neuropsychology of schizophrenia. In N. I. Grant & K. M. Adams (Eds.), *Neuropsychological and neuropsychiatric disorders* (pp. 147–171). New York: Oxford University Press.

Kelsoe, J. R., Cadet, J. L., Pickar, D., & Weinberger, D. R. (1988). Quantitative neuroanatomy in schizophrenia. *Archives of General Psychiatry, 45*, 533–541.

Kim, Y., Zito, J., Huber, D., Kane, J., Reife, R., Schaul, N., Whitney, J., Grande, A., & Alvir, J. (1988). Can MRI tell us more about dementia than CT? *Journal of Clinical and Experimental Neuropsychology, 10*, 61.

Kuhl, D. E. (1984). Imaging local brain function with emission computed tomography. *Radiology, 150*, 625–631.

Lewin, R. (1987). Brain graphs benefit Parkinson's patients. *Science, 236*, 149.

Lezak, M. D. (1983). *Neuropsychological Assessment* (pp. 487–491). New York: Oxford University Press.

Luxenberg, J. S., Haxby, J. V., Creasey, H., Sundaram, M., & Rapoport, S. I. (1987). Rate of ventricular enlargement in dementia of the Alzheimer type correlates with rate of neuropsychological deterioration. *Neurology, 37*, 1135–1140.

Morrison, I. R., Schaefer, B. A., & Russell, D. L. (1987). An expert system environment tailored for neuropsychology. *Journal of Clinical and Experimental Neuropsychology, 9*, 37–38.

McGeer, P. L., Kamo, H., Harrop, R., McGeer, E. G., Martin, W. R. W., Pate, B. D., & Li, D. K. B. (1986). Comparison of PET, MRI and CT with pathology in a proven case of Alzheimer's disease. *Neurology, 36*, 1569–1574.

Posner, M. I. (1986). A framework for relating cognitive to neural systems. In W. C. McCallum, R. Zappoli, & F. Denoth (Eds.), *Cerebral psychophysiology: Studies in event related potentials* (pp. 297–318). New York: Elsevier.

Prigatano, G. P. (1987). Recovery and cognitive retraining after craniocerebral trauma. *Journal of Learning Disabilities, 20*, 603–613.

Russell, D. L., Schaefer, B. A., Morrison, I. R., Siegel, J., Morrison, R. N., Side, R. S., & Joschko, M. (1987). A progress report of the SAVANT system: An expert system environment tailored for neuropsychology. *Journal of Clinical and Experimental Neuropsychology, 9*, 273.

Scheibel, A. B., & Wechsler, A. F. (1986). *The biological substrates of Alzheimer's disease*. Orlando, FL: Academic Press.

Stillings, N. A., Feinstein, M. H., Garfield, J. L., Rissland, E. L., Rosenbaum, D. A., Weisler, S. E., & Baker-Ward, L. (1987). *Cognitive science: An introduction*. Cambridge, MA: MIT Press.

Yeudall, L. T., Reddon, J. R., Gill, D. M., & Stefanyk, W. O. (1987). Normative data for the Halstead–Reitan neuropsychological test stratified by age and sex. *Journal of Clinical Psychology, 43*, 346–367.

Index

Acoustic neuroma, 40
Acquired immune deficiency syndrome
  brain abscesses, 40
  lymphoma, 40
Affective disorders
  computed tomography, abnormalities in, 248
  drug response in, morphological brain abnormalities and, 257–258
  morphological brain abnormalities
    mental status and, 255
    neurotransmitters and, 255
    relationship between mental status and brain morphology, 255
Agnosia, visual object, 100
Alcoholism, brain abnormalities, 247
Alzheimer's disease, 110
  cerebral atrophy, 210
  cerebral energy metabolism, 110
  diagnosis, 341
  Down's syndrome, 111
  fluorodeoxyglucose brain image, 343
  frontal cortex, 110
  Global Deterioration Scale, 112
  hypothalamus, 205
  vs. multiinfarct dementia, 119
  parietal cortex, 110
  vs. Pick's disease, 114
  positron emission tomography, 114
  temporal cortex, 110
  thalamus, 205
Amnesia
  cerebral blood flow, 103
  herpes simplex encephalitis, 116
  lacunar retrograde, 107
  magnetic resonance imaging, 104
  oxygen utilization, cerebral, 103
  transient global, 103
Amygdala, 81
Anorexia nervosa, brain abnormalities, 246
Anxiety, positron emission tomography, 72
Aphasia
  Broca's, 85

Aphasia (cont.)
  computed tomography, 140–141
  depressive symptomatology and, 278
  etiology, 85
  male, following left hemisphere damage, 306
  nonfluent speech, defined, 155
  sentence comprehension, 146
  single-word comprehension, 144
  subcortical, 88
    symptoms, 91
  syndromes, 85
  temporal lobe lesions, 148
  test profiles in, differences of, 90
  Token Test, 229
  transcortical, mixed, dementia and, 86
Aphasic syndromes, 54, 85
Aqueduct of Sylvius, 25
Arachnoid cysts, 32
  temporal lobe, 323
Arterial–venous malformation
  computed tomography, 33
  magnetic resonance imaging, 33
  ruptured, 324
    computed tomography, 327
Attention
  cerebral glucose metabolic rates, 73
  disorders, Moyamoya disease, 124
  frontal lobes and, 122
  hematoma and, 176
  limbic system and, 122
  right hemisphere and, 237
Autism, infantile, 128

Balint's syndrome, 97
Basal ganglia, 29
  computed tomography, 18
  formation, 328
  lymphoma, 40
  positron emission tomography, 68
Bayley Scales of Infant Development, 330
Benton Visual Retention Test, 326, 328
Benton's Judgement of Line-Orientation Test, 74

347

Bilateral porencephaly, 318
Binswanger's disease, 119
Block Design, 238
Boston Diagnostic Aphasia Examination, 85, 141
    confrontation naming, 151–152
    naming subtest, 111
    sentence repetition, 152–155
Brachium conjunctivium, 25
Brain
    activational differences, defined, 298
    cortical atrophy: see Cortical atrophy
    electrophysiological mapping, 65
    embryological development, 13–17
    energy storage capacity, 66
    epilepsy and organization of, 301
    formation, 13
    glucose consumption
        measurement of, 67
        See also Glucose metabolic rate
    oxygen consumption, measurement of, 67
    structural abnormalities, cognitive function, 323
    ventricular enlargement: see Ventricular enlarge-
        ment
Brain abscesses
    acquired immune deficiency syndrome, 40
    blood–brain barrier disruption, 41
    computed tomography, 40
    magnetic resonance imaging, 40
Brain density, 52
    aging and, 53
Brain injury: see Traumatic brain injury
Brain lesion
    aphasia
        sentence comprehension and, 146
        single-word comprehension, 144
    description, complete, 57
    epilepsy, 301
        personality attributes and, 272
    left hemisphere
        Minnesota Multiphasic Personality Inventory,
            284
        vs. right hemisphere, 279–280, 281–282, 283
    location, 54–56
    Minnesota Multiphasic Personality Inventory pro-
        files, 278
    models, 57
    naturally occurring, 58
    nonfluent speech, 155
    pre- and postcentral gyrus, nonfluent speech, 157
    predictive accuracy of regional, in aphasia, 144,
        154, 155
    right hemisphere, Minnesota Multiphasic Person-
        ality Inventory, 284
    sex differences in sequelae, 301
    size, 56–57
        expression of, 225

Brain lesion (cont.)
    measurement, 271
    size and location, motor deficits and, 238
    Wechsler Adult Intelligence Scale, 302
Brain size
    cognitive function and, 296
    sex differences in, 304
    variations in, 296
Brain tumors, 37
Brain volume, measurement, 195–198
Brainstem
    glioma, magnetic resonance imaging, 38
    glucose metabolic rate, 69
Broca's aphasia, 85
    age and, 70, 310
    cerebral glucose metabolic rate, 87
    computed tomography, 87
    glucose metabolic rate, 69
    lesions, 56
    metabolic activity, test performance and, 82
    positron emission tomography, 87
    sex differences in, 307

Caudate, 29
Caudate nucleus, glucose metabolic rate, 69
Cerebellum, 23
    computed tomography, 18
    formation from metencephalon, 16
    glucose metabolic rate, 69
    parts of, 24–25
Cerebral blood flow
    amnesia, 103
    measurement, 65
    Moyamoya disease, 124
Cerebral contusion
    computed tomography, 36
    magnetic resonance imaging, 36
Cerebral cortex
    anatomic asymmetries of hemispheres, cognitive
        specialization, 311
    computed tomography, 19
    formation, 16, 328
    glucose metabolic rate: see Glucose metabolic
        rate
    hemispheres, interactive nature, 287
    as landmark of head, computed tomography, 48
    left hemisphere
        focalization of language centers, 310
        positive emotions, 272
    left hemisphere damage
        aphasia in males, 306
        MMPI evidence of depression, 281, 282
        sex differences in sequelae, aphasia, 307
    right hemisphere
        attention and, 237
        emotional behavior, 286

Cerebral cortex (*cont.*)
  negative emotions, 272
  ratio of white to gray matter, 297
  right hemisphere damage
    indifference reactions, 283
  sleep and activity of, 70
Cerebral syphilis, 119
Cerebritis, 42
Cerebrovascular accident
  depressive symptoms, 274–275
    lesion parameters and, 275
    site of lesion and, 286
  Minnesota Multiphasic Personality Inventory, 283
  preterm infant, 328
  right hemisphere damage vs. left hemisphere
    damage, 283
  Wechsler Adult Intelligence Scale, 303
  Wechsler Memory Scale, 303
Chiari malformation, 32
Childhood-onset pervasive development disorder,
    129
Choriocarcinoma, 37
Choroid plexus, 26
Cognitive function
  age and, 296
  brain structural abnormalities, 323
  cerebral glucose utilization, 74
  cortical atrophy, 200–204
Computed tomography
  affective disorders, 248
  aphasia, 140–141
    mixed transcortical, 86
  arterial–venous malformation, 33
    ruptured, 327
  basal ganglia, 18
  brain abscesses, 40
  brain density, 52
  Broca's aphasia, 87
  cerebellum, 18
  cerebral contusion, 36
  cerebral cortex, 19
  childhood-onset pervasive developmental disor-
    der, 129
  corpus callosum, agenesis, 319
  dementia, 193
  description, 17
  dyslexia, 336
  epidural hematoma, 36
  first generation vs. current generation, 7
  hematoma
    intrasylvian, 35
    traumatic brain injury, 163
  hemorrhage detection, 162
  hydrocephalus, congenital, 320
  interpretation, 30
  Kluver–Bucy-like syndrome, 117

Computed tomography (*cont.*)
  vs. magnetic resonance imaging
    cortical atrophy, 167
    hemorrhage, 162–163
    hydrocephalus, 165
    subdural hematoma, 342
  mental retardation, 321, 322
  Moyamoya disease, 124
  multiinfarct dementia, 119
  vs. positron emission tomography
    amnesia, 103
    aphasia, 140
    multiinfarct dementia, 119
    progressive supranuclear palsy, 114–115
  in schizophrenia, 6, 246; *see also* Schizophrenia,
    morphological brain abnormalities
  single photon emission, 65
  temporal lobe lesions, 148
  traumatic brain injury, 162
Conner's Parent–Teacher Symptom Questionnaire,
    334
Corpus callosum, 28
  agenesis, 31, 318
  sex differences in size, 304
Cortical atrophy, 167; *see also* Ventricular atrophy
  anorexia nervosa, 246
  atrophy volume index, 203
  behavioral changes from, 53
  cognitive function, 200–204
  Cushing's syndrome, 246
  defined, 53
  dementia of the Alzheimer's type, 189
  measurement, 47–48, 53–54
  migraine headaches, 246
  Parkinson's disease, 246
  protein malnutrition, 246
  senile dementia, 246
  traumatic brain injury, 167, 173, 175–176
Craniopharyngioma, 37
  calcification in, 40
Creutzfeldt–Jacob disease, 119
Cushing's syndrome, brain abnormalities, 246

Dandy–Walker syndrome, 32
Dementia, 109–122
  alcohol-induced, 192
  Alzheimer's type
    clinical presentation, 185
    cortical atrophy, 189
    degenerative changes, 185
    diagnosis, 188
    differential diagnosis, 192
    etiology, 189
    language impairment, 199
    magnetic resonance imaging, 194
    pericerebral atrophy, 202

Dementia (*cont.*)
   prediction of progression, 212
   prevalence, 188
   relationship between cerebral atrophy and cognitive impairment, 207
   vs. senile dementia, 204
   Token Test, 199
   visuoperceptual deficits, 198
  cerebral vasculitis, 119
  computed tomography, 193
  hydrocephalus, 121
  infectious etiologies, 116
  multiinfarct: *see* Multiinfarct dementia
  senile, 246
  traumatic, 122, 123
Dementia paralytica, 119
Depression, fluorodeoxyglucose brain images, 343
Developmental Test of Visual Motor Integration, 334
Diencephalon
  description, 26–27
  formation, 13
Down's syndrome, 111
Dyslexia, 336, 337

Echoencephalography, 2
Edema, 163
Emphysema, subdural, 42
Epilepsy
  brain lesions, personality attributes and, 272
  brain organization, 301
  temporal lobe, 79
Epithalamus, 26
Extrapyramidal motor system, 25

Foramen of Lushka, 25
Foramen of Magendie, 25
Foramina of Monro, 26
Forebrain, formation, 13
Frontal cortex, 28
  Alzheimer's disease, 110
  attention and, 122
  glucose metabolic rate, 69
    age and, 70
  Huntington's disease, 116
  limb apraxia, 233
  neuronal shrinkage, age in, 308
  positron emission tomography, 68

Gilles de la Tourette syndrome, 127
Glioblastoma multiforme, 38
Glioma, 37
  brainstem, 38
  grade II, 39
Global Deterioration Scale, 112
Globus pallidus, 29

Glucose metabolic rate, 69
  age and, 70
  auditory stimulation, 76
  Broca's aphasia, 87
  cognitive function, 74
  frontal cortex, 70
  Huntington's disease, 116
  infantile autism, 128
  Kluver–Bucy-like syndrome, 117–118
  memory, 74
  mental retardation, 128
  Moyamoya disease, 124
  parietal cortex, 70
  Parkinson's disease, 116
  temporal cortex, 70
  Word Fluency Test, 78
Graves Design Judgement Test, 301

Halstead–Reitan Category Test, 255
Halstead–Reitan Neuropsychological Test Battery, 1, 326, 327
Hematoma, 42
  attention and, 176
  detection, magnetic resonance imaging vs. computed tomography in, 163
  epidural, 36
  intraparenchymal, 34
  intrasylvian, 35
  memory, 176
Hemianopias, 101
Herpes simplex, 42
Herpes simplex encephalitis, 116
Hindbrain
  formation, 13
  major components, 23
Hippocampus, 69
Huntington's disease
  frontal cortex, 116
  parietal cortex, 116
  temporal cortex, 116
  thalamus, 116
  whole-brain glucose consumption, 116
Hydrocephalus, 121
  communicating, 32
  congenital, computer tomography in, 320
  obstructive, 32
  periventricular intraventricular hemorrhage, 332
Hygroma, 192
Hypokinesia, 237
Hypophysis, 27
Hypothalamus, 26
  in Alzheimer's disease, 205
  parts, 27

Infantile autism, 128
Infundibulum, 27

Insular cortex, 27
    glucose metabolic rate, age and, 70

Kluver–Bucy-like syndrome, 116, 118
Korsakoff's syndrome
    etiology, 106
    positron emission tomography, 106

Language deficits, motor deficits and, 229
Learning, glucose consumption, temporal cortex, 81
Learning disability, central nervous system anoma-
        lies and, 336
Lesions: see Brain lesions
Leukoencephalopathy, progressive multifocal, 45
Limb apraxia, 219
    cerebral hemispheres involved in, 233, 236
    defined, 220
    frontal lobes, 233
    language tasks and, 232
    motor skills and, 221, 232
    parietal lobes, 233
    physical examination, 220–221
    subcortical lesions, 223
    testing, hand posture tasks, 236
    types of, 222
Limbic system, 122
Lipoma, 32
Lissauer's cerebral sclerosis, 119
Lupus erythematosus, systemic, ventriculomegaly
        in, 246
Lymphoma
    acquired immune deficiency syndrome, 40
    immunity disorders, 40

Magnetic resonance imaging, 8
    advantages, 22
    amnesia, 104
    arterial–venous malformation, 33
    brain abscesses, 40
    cerebral contusion, 36
    childhood-onset pervasive developmental disor-
        der, 129
    vs. computed tomography, 65
        cortical atrophy, 167
        hemorrhage, 162–163
        hydrocephalus, 165
        subdural hematoma, 342
    dementia, Alzheimer's type, 194
    dementia paralytica, 119
    description, 17, 20
    epidural hematoma, 36
    glioblastoma multiforme, 38
    globus pallidus, 29
    hematoma, 42
        intraparenchymal, 34
        traumatic brain injury, 163

Magnetic resonance imaging (cont.)
    hemorrhage detection, 162
    interpretation, 30
    Kluver–Bucy-like syndrome, 117
    major anatomic sites, brain, 10
    meningioma, 39
    meningitis, 42
    multiinfarct dementia, 119
    principle, 17, 20
    in schizophrenia, 245
    traumatic brain injury, 162
Mamillary bodies, 27
McCarthy Scale of Children's Ability, 333–334
Medulla, 23
    surface features, 24
Medulla oblongata, 23
Medulloblastoma, 37
Melanoma, 37
Memory
    cerebral glucose utilization, 74
    disorders, 103–109, see also Amnesia
    hematoma, 176
    measurement, 198
    sequencing task and, 232
    techniques, metabolic activity and, 82
    temporal cortex and, 105
Meningioma, 37
    frontal, 39
Meningitis, 42
    magnetic resonance imaging, 42
Mental retardation, 128
    computer tomography, 321, 322
Metachromatic leukodystrophy, 45
Metencephalon, formation, 13, 16
Midbrain
    description, 25
    formation, 13
Middle cerebral artery infarction, 85
    apraxia, 228
Migraine headaches, brain abnormalities, 246
Miller Analogies Test, 74
Mini-Mental State Examination, 255
Minnesota Multiphasic Personality Inventory, 277,
        278, 283, 284
Mooney's Closure Faces Test, 301
Motor control, anatomic basis for hemisphere differ-
        ences in, 241
Motor deficits
    correlation with language deficits, 229
    size and location of brain lesion, 238
Moyamoya disease, 124, 125
    cerebral blood flow, 124
    cerebral glucose utilization, 124
Multiinfarct, 192
    vs. Alzheimer's disease, 119
    clinical picture, 120

Multiinfarct (*cont.*)
  computed tomography, 119
  differential diagnosis, 119
  fluorodeoxyglucose images, 343
  magnetic resonance imaging, 119
  positron emission tomography, 119
Multiple sclerosis
  differential diagnosis, 45
  periventricular focal plaques, 42
Myelencephalon, formation, 13

Nephroma, 37
Nervous system, genesis from ectoderm, 14
Neuroma, acoustic, 40
Neurophysiological tests
  Bayley Scales of Infant Development, 330–331
  Benton Visual Retention Test, 326, 328
  Benton's Judgement of Line-Orientation Test, 74
  Block Design, 238
  Boston Diagnostic Aphasia Examination, 111,
    141, 151–152, 152–155
  Conner's Parent–Teacher Symptom Question-
    naire, 334
  Developmental Test of Visual Motor Integration,
    334
  Fitts Tapping Task, 239–241
  Global Deterioration Scale, 112
  Graves Design Judgement Test, 301
  Halstead–Reitan Category Test, 1, 255, 326, 327
  Miller Analogies Test, 74
  Mini-Mental State Examination, 255
  Minnesota Multiphasic Personality Inventory,
    277, 278, 283, 284
  Mooney's Closure Faces Test, 301
  Raven Colored Progressive Matrices, 326, 328
  Reitan–Indiana Aphasia Screening Test, 204,
    278, 320, 321, 329
  Rey–Osterrieth Complex Figure, 324, 330
  Scale for Assessment of Negative Symptoms, 248
  Schuell Aphasia Battery, 306
  Seashore Tonal Memory Test, 76
  Spielberger's state–trait anxiety inventory, 72
  Test of Early Reading Ability, 334
  Token Test, 141, 148, 199, 229
  Wechsler Adult Intelligence Scale, 198, 301,
    302, 303, 326, 327
  Wechsler Memory Scale, 114, 303, 324, 326,
    327
  Western Aphasia Battery, 307
  Wisconsin Card-Sorting Test, 74, 339–340
  Word Fluency Test, 78

Occipital encephalocele, 317, 318
Occipital lobe, 27
Occipital porencephaly, 317

Oligodendroglioma, 37
Optic chasm, 27

Panic disorder, 127–128
Parietal cortex, 28
  Alzheimer's disease, 110
  glucose metabolic rate, age and, 70
  Huntington's disease, 116
  left, single posture generation, 233
  limb apraxia, 233
  Parkinson's disease, 116
  positron emission tomography, 68
Parkinson's disease
  brain abnormalities, 246
  cerebral glucose utilization, 116
  parietal cortex, 116
Periventricular intraventricular hemorrhage, 328–
    336
  grade I, subependymal hemorrhage, 332
  grading, 328
  hydrocephalus, 331–332
  infants at risk for the development of, 330
  neurodevelopmental outcome, vs. controls, 334
  ventricular dilation, 334
Pick's disease
  vs. Alzheimer's disease, 114
  dementia-type syndromes, 192
  positron emission tomography, 114
  vs. progressive supranuclear palsy, 115
Pneumoencephalography, 2
  in schizophrenia, 245
Pons, 23
  description, 24
Porch Index of Communicative Ability, 85
Positron emission tomography
  advantages, 66, 129–130
  Alzheimer's disease, 110, 114
  amnesias, 103–109
  aphasia, mild conduction, 85
  apprehensive vs. relaxed resting state, 73
  auditory stimulation, 76
  Balint's syndrome, 97, 98
  basal ganglia, 68
  Benton's Judgement of Line-Orientation Test, 74
  cerebral blood flow imaging, 67
  Childhood-onset pervasive developmental disor-
    der, 129
  coincidence detection, 66
  vs. computed tomography
    amnesia, 103
    aphasia, 140
    multiinfarct dementia, 119
    progressive supranuclear palsy, 114–115
  in differential diagnosis
    Pick's vs. Alzheimer's disease, 114

Positron emission tomography (*cont.*)
    Pick's disease vs. progressive supranuclear
        palsy, 115–116
    frontal cortex, 68
    Gilles de la Tourette syndrome, 127
    glucose consumption, quantification of, 67
    glucose metabolic rate
        auditory stimulation, 76
        memory, 82
    image analysis, 68
    infantile autism, 128
    isotope production, 66
    Kluver–Bucy-like syndrome, 118
    Korsakoff's syndrome, 106, 107
    memorizing process, 81
    mental retardation, 128
    Miller Analogies Test, 74
    Moyamoya disease, 124
    multiinfarct dementia, 119
    multivariate nature of data, 130
    oxygen metabolic images, 75
    panic disorder, 127–128
    parietal cortex, 68
    Pick's disease, 114
    scanners, 66
    sleep and, 70
    Spielberger's state–trait anxiety inventory, 72
    temporal cortex, 68
    traumatic dementia, 123
    verbal auditory stimulation, 76
    visual cortex, metabolic responses in, 75
    visual object agnosia, 100
    visual stimulation, 75
    Wisconsin Card-Sorting Test, 74, 339–340
    Word Fluency Test, 78
Posterior cerebral artery infarction
    diabetes mellitus, 94
    visual system, 101
Progressive multifocal leukoencephalopathy, 45
Progressive supranuclear palsy
    clinical symptoms, 114
    Pick's disease, 115–116
    positron emission tomography, 115–116
Prosencephalon, formation, 13
Prosopagnosia, 103
Protein malnutrition, brain abnormalities, 246
Psychological Tests: *see* Neuropsychological Tests
Psychosis, relationship between mental status and
    brain morphology, 255
Putamen, 29

Quadrantanopias, 101

Raven Colored Progressive Matrices, 326, 328
Red nucleus, 25

Reitan–Indiana Aphasia Screening Test, 204, 278,
    329
    congenital hydrocephalus, 320, 321
Rey–Osterrieth Complex Figure, 324, 330
Rhomboencephalon, 13

Scale for Assessment of Negative Symptoms, 248
Schizophrenia
    brain abnormalities, 246
        genetic factors and, 251
    computed tomography, 6, 245
    drug response in morphological brain abnormali-
        ties and, 256–257
    magnetic resonance imaging in, 245
    morphological brain abnormalities
        mental status and, 256
        neurochemical dysregulation and, 258–260
    paranoid, 250
    perinatal complications and, 252
    pneumoencephalography in, 245
    relationship between mental status and brain mor-
        phology, 255
    two-syndrome concept, 247
    type I vs. type II, neurotransmitter dysregulation,
        258
    ventricular volume, 51, 54, 242
Schuell Aphasia Battery, 306
Seashore Tonal Memory Test, 76
Senile dementia, brain abnormalities, 246
Sequencing task, memory and, 232
Single photo emission computed tomography, 65
Sleep
    cerebral cortex activity, 70
    positron emission tomography, 70
Speech
    handedness and, 78
    temporal lobe epilepsy, 79
Spielberger's state–trait anxiety inventory, 72
Spongioblasts, 328
Steele–Richardson–Olszewski syndrome, 114
Subcortical arteriosclerotic encephalopathy, 119
Substantia nigra, 25
Syphilis, cerebral, 119

Tactile learning, 74
Telencephalon, 13, 16, 27–29
Temporal cortex, 28
    Alzheimer's disease, 110
    arachnoid cyst, 323
    epilepsy, 79
    glucose metabolic rate
        age and, 70
        memory and, 81
    Huntington's disease, 116
    infection, 116

Temporal cortex (*cont.*)
  lesions, computed tomography, 148
  in memory, 105
  neuronal shrinkage, age and, 308
  positron emission tomography, 68
Test of Early Reading Ability, 334
Thalamic pain syndrome, 126
Thalamus, 26
  in Alzheimer's disease, 205
  in aphasia, 89
  glucose metabolic rate, 69
  Huntington's disease, 116
Token Test, 141, 229
  dementia of Alzheimer's type, 199
  temporal lobe lesions and impairment of, 148
Traumatic brain injury
  brain function, localization of, 161
  computed tomography, 162
  cortical atrophy, 167, 173, 175–176
  coup–contrecoup effects, 174
  diffuse axonal injury, 171
  diffuse closed-head injury, 167
  edema, 163
  hematoma, 163
    evaluation of, 176
  hypoxic damage, 171
  ischemic vascular lesions, 169
  left hemisphere damage, Minnesota Multiphasic
      Personality Inventory, 284
  magnetic resonance imaging, 162, 177–180
  Minnesota Multiphasic Personality Inventory, 283
  regions most susceptible to, 176
  right hemisphere damage, Minnesota Multiphasic
      Personality Inventory, 283
  in utero, 318

Traumatic brain injury (*cont.*)
  ventricular enlargement, 165
    assessment, 171–172

Ventricle–brain ratio, 50
Ventricular atrophy, 50
Ventricular enlargement, 165
  age and, 246
  anorexia nervosa, 246
  Cushing's syndrome, 246
  migraine headaches, 246
  Parkinson's disease, 246
  protein malnutrition, 246
  senile dementia, 246
  systemic lupus erythematosus, 246
Ventricular volume
  measurement, 48–49
  schizophrenia, 51

Wechsler Adult Intelligence Scale, 198, 301, 326,
      327
  cerebrovascular accident, 303
Wechsler Memory Scale, 198, 324, 326, 327
  Alzheimer's disease, 114
  cerebrovascular accident, 303
Wernicke's aphasia, 103
  metabolic pattern, 85
  sex differences in, 307
Wernicke's area
  glucose metabolic rate, 69
  regional side-to-side differences, 74
Western Aphasia Battery, 307
White matter disease, 44–45
Wisconsin Card-Sorting Test, 74, 339–340
Word Association Test, 301
Word Fluency Test, 78